PENGUIN BOOKS

MOBY-DUCK

Donovan Hohn is a journalist whose work has appeared in *Harper's Magazine*, *The New York Times Magazine*, *Outside*, and the *Best Creative Nonfiction*. The recipient of a Whiting Writers' Award and a National Endowment for the Arts Creative Writing Fellowship, he is a former high school English teacher and currently the features editor at *GQ*. He lives in New York with his wife and children.

Overwhelming Acclaim for *Moby-Duck* and Donovan Hohn

A National Indie, NPR, and *Wall Street Journal* bestseller

""Hohn seems to have it all: deep intelligence, a strikingly original voice, humility, and a hunger to suss out everything a yellow duck may literally or metaphorically touch. . . . Hohn guides us expertly through the swirling currents of his own psyche, upon which so much of this book's considerable charm depends. . . . *Moby-Duck* succeeds as harebrained adventure, as a cautionary environmental tale, as a deconstruction of consumer demand, and as a meditation on wilderness and imagination. Hohn moves easily between the micro and the macro, weaving personal histories into science and industry as he roams." —Elizabeth Royte, *The New York Times Book Review*

"A metaphysical quest, an encyclopedic rummage through the mysteries of the ocean and the history of plastics, a meditation on the meaning of toys and childhood and fatherhood, and a close-up look at what we are doing to our planet. Like its literary eponym, *Moby-Duck* plunges into the depths of what it is to be land-walking mortals in a world surrounded by water." —Bill Marvel, *The Dallas Morning News*

"*Moby-Duck* is highly readable and, importantly, alive with a sense of intellectual curiosity. Beyond just reporting the facts, Hohn engages with them philosophically. It's a comprehensive account of everything connected to the spill of those toys. Indeed, what Melville did for whaling, Hohn has done for plastic bath toys lost at sea." —*The Boston Globe*

"Hohn navigates the complicated fields of oceanography, environmentalism, globalization, and maritime shipping with surprising humor and ease, raising pressing questions about these topics without giving any clear answers to them—because there aren't any. Hohn cleverly uses the deceptively whimsical premise of chasing a little plastic duck to provoke a massively complicated and thought-provoking conversation. Who knew spilled bath toys could be so important?" —*Chicago Sun-Times*

"A finely spun chronicle . . . a gladdening, artful journey of discovery." —*Kirkus Reviews* (starred review)

"Whimsical curiosity begets a quixotic odyssey and troubling revelations. . . . Charming . . . and packed with seafaring lore and astute reporting, this enthralling narrative is the *Moby-Dick* of drifting ducks." —*Publishers Weekly* (starred review)

"Like Bill Bryson on hard science, or John McPhee with attitude, journalist Hohn travels from beaches to factories to the northern seas in pursuit of a treasure that mystifies as much as it provokes. . . . Rubber ducks as harmless, ubiquitous symbols of childhood? Not anymore, not by a long shot. This dazzles from start to finish." —*Booklist* (starred review)

"A wonderfully willful and picaresque adventure, *Moby-Duck* takes us on a roller coaster transoceanic trip through a modern world threatened by its own addiction to plastic. This is a wry and witty tale of heroes and villains and bath toys. And if Donovan Hohn charts the unconscionable at least he does it in highly readable and supremely entertaining style." —Philip Hoare, author of *The Whale*

"*Moby-Duck* is a mind-blowing book, a rare page-turner that makes you think, feel, and laugh out loud, often all at once. Here is an adventure story, an important environmental book, a big piece of reportage, and a cabinet of wonders in which Hohn, the essayist and argonaut is at his finest, hashing, synthesizing, and reflecting the most pressing concerns of our world today. But most of all, this book is an exquisite delight to read." —Michael Paterniti, author of *Driving Mr. Albert*

"What to do with a book like *Moby-Duck*? Its structure seems to be quite traditional but is in fact audaciously strange. Its narrator claims to be meek and clumsy but is in fact funny and brave. Its central quest appears to be silly but is in fact a matter of great human and environmental seriousness. So here is what you do with a book like *Moby-Duck*: Read it twice. Then tell everyone you know that you have discovered a writer of immense skill and originality."
　　　　　　　　　　　　　　—Tom Bissell, author of *Chasing the Sea*

"*Moby-Duck* is an impossibly, distressingly wonderful read by one of our very best contemporary essayists."
　　　　　　　　　—Rivka Galchen, author of *Atmospheric Disturbances*

"Captain Ahab chased the Great White Whale; Donovan Hohn hunts the Tiny Yellow Duck. In tracking the mysterious fate of more than 28,000 plastic bath toys that tumbled into the Pacific, Hohn takes us on a journey almost as epic as *Moby-Dick*, a revelatory adventure over the high seas and into the murky backwaters of our throwaway consumer culture. As a storyteller, he's every bit as perceptive and entertaining—and a hell of a lot funnier—than Melville's Ishmael. Call me impressed."
　　　　　　　　　—Miles Harvey, author of *The Island of Lost Maps*

"Some years ago a cargo container of rubber bath toys fell overboard and dispersed its contents far and wide, and something about this book is as turbulent and as abundant as that incident, as though a whole cargo container of exquisite sentences were washing up in the reader's mind, a whole Pacific current of engaging ideas and encounters was sweeping that reader away. It's a marvelous journey, in which rubber ducks only bob on the surface of an exploration of the ocean, the container ships that traverse it, and the characters who still plunge into its still-perilous waters."
　　　　　　　　　—Rebecca Solnit, author of *A Paradise Built in Hell*

DONOVAN HOHN

PENGUIN BOOKS

MOBY-DUCK

The True Story of 28,800 Bath Toys Lost at Sea

and of the Beachcombers, Oceanographers,

Environmentalists, and Fools,

Including the Author,

Who Went in Search of Them

PENGUIN BOOKS

Published by the Penguin Group

Penguin Group (USA) Inc., 375 Hudson Street, New York, New York 10014, U.S.A.
Penguin Group (Canada), 90 Eglinton Avenue East, Suite 700, Toronto, Ontario, Canada M4P 2Y3
(a division of Pearson Penguin Canada Inc.)
Penguin Books Ltd, 80 Strand, London WC2R 0RL, England
Penguin Ireland, 25 St. Stephen's Green, Dublin 2, Ireland (a division of Penguin Books Ltd)
Penguin Books Australia Ltd, 250 Camberwell Road, Camberwell, Victoria 3124, Australia
(a division of Pearson Australia Group Pty Ltd)
Penguin Books India Pvt Ltd, 11 Community Centre, Panchsheel Park,
New Delhi – 110 017, India
Penguin Group (NZ), 67 Apollo Drive, Rosedale, Auckland 0632, New Zealand
(a division of Pearson New Zealand Ltd)
Penguin Books (South Africa) (Pty) Ltd, 24 Sturdee Avenue, Rosebank,
Johannesburg 2196, South Africa

Penguin Books Ltd, Registered Offices: 80 Strand, London WC2R 0RL, England

First published in the United States of America by Viking Penguin,
a member of Penguin Group (USA) Inc. 2011
Published in Penguin Books 2012

3 5 7 9 10 8 6 4

Copyright © Donovan Hohn, 2011
All rights reserved

Portions of this book first appeared in *Harper's*, *The New York Times Magazine*, and *Outside*.

Map illustrations by David Cain

THE LIBRARY OF CONGRESS HAS CATALOGED THE HARDCOVER EDITION AS FOLLOWS:
Hohn, Donovan.
Moby-duck : an accidental odyssey : the true story of 28,800 bath toys lost at sea and of the
beachcombers, oceanographers, environmentalists, and fools, including the author, who went
in search of them / Donovan Hohn.
p. cm.
ISBN 978-0-670-02219-9 (hc.)
ISBN 978-0-14-312050-6 (pbk.)
1. Ocean currents. 2. Marine debris. 3. Plastic toys. 4. Hohn, Donovan—Anecdotes. I. Title.
GC231.2.H65 2010
551.46'2—dc2   2010033608

Printed in the United States of America
Designed by Nancy Resnick

For Beth,
and for my father,
and for my sons.

Facing west from California's shores,
Inquiring, tireless, seeking what is yet unfound,
I, a child, very old, over waves, towards the house of maternity,
the land of migrations, look afar . . .

—Walt Whitman

There are more consequences to a shipwreck than the underwriters
notice.

—Henry D. Thoreau

CONTENTS

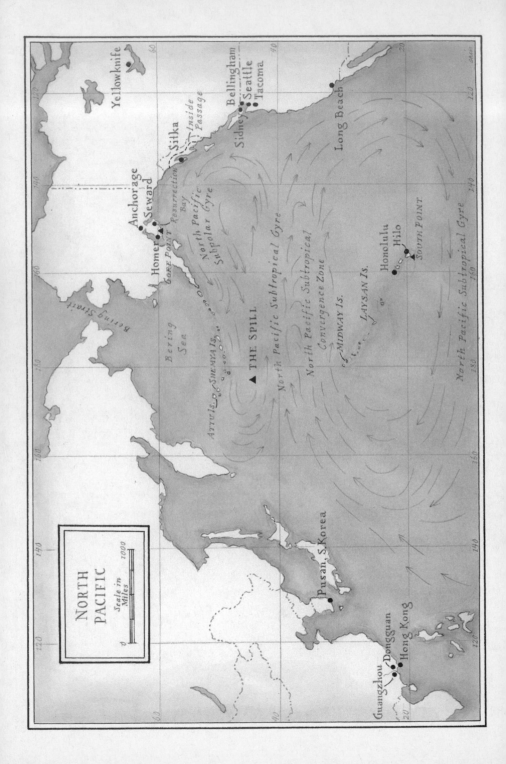

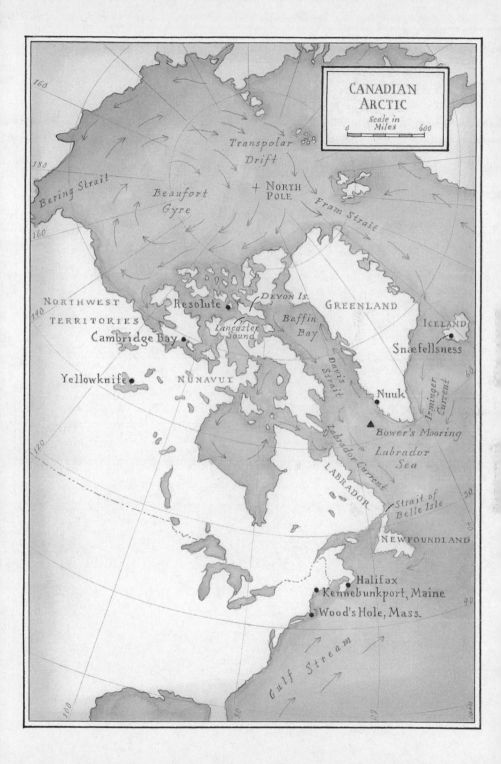

PROLOGUE

At the outset, I felt no need to acquaint myself with the six degrees of freedom. I'd never heard of the Great North Pacific Garbage Patch. I liked my job and loved my wife and was inclined to agree with Emerson that travel is a fool's paradise. I just wanted to learn what had really happened, where the toys had drifted and why. I loved the part about containers falling off a ship, the part about the oceanographers tracking the castaways with the help of far-flung beachcombers. I especially loved the part about the rubber duckies crossing the Arctic, going cheerfully where explorers had gone boldly and disastrously before.

At the outset, I had no intention of doing what I eventually did: quit my job, kiss my wife farewell, and ramble about the Northern Hemisphere aboard all manner of watercraft. I certainly never expected to join the crew of a fifty-one-foot catamaran captained by a charismatic environmentalist, the Ahab of plastic hunters, who had the charming habit of exterminating the fruit flies clouding around his stash of organic fruit by hoovering them out of the air with a vacuum cleaner.

Certainly I never expected to transit the Northwest Passage aboard a Canadian icebreaker in the company of scientists investigating the Arctic's changing climate and polar bears lunching on seals. Or to cross the Graveyard of the Pacific on a container ship at the height of the winter storm season. Or to ride a high-speed ferry through the smoggy, industrial backwaters of China's Pearl River Delta, where, inside the Po Sing plastic factory, I would witness yellow pellets of polyethylene resin transmogrify into icons of childhood.

I'd never given the plight of the Laysan albatross a moment's thought. Having never taken organic chemistry, I didn't know and therefore didn't care that pelagic plastic has the peculiar propensity to adsorb hydrophobic, lipophilic, polysyllabic toxins such as dichlorodiphenyltrichloroethane (a.k.a. DDT) and polychlorinated biphenyls (a.k.a. PCBs). Nor did I know or care that such toxins are surprisingly abundant at the ocean's surface, or that they bioaccumulate as they move up the food chain. Honestly, I didn't know what "pelagic" or "adsorb" meant, and if asked to use "lipophilic" and "hydrophobic" in a sentence I'd have applied them to someone with a weight problem and a debilitating fear of drowning.

If asked to define the "six degrees of freedom," I would have assumed they had something to do with existential philosophy or constitutional law. Now, years later, I know: the six degrees of freedom—delicious phrase!—are what naval architects call the six different motions floating vessels make. Now, not only can I name and define them, I've experienced them firsthand. One night, sleep-deprived and nearly broken, in thirty-five-knot winds and twelve-foot seas, I would overindulge all six—rolling, pitching, yawing, heaving, swaying, and surging like a drunken libertine—and, after buckling myself into an emergency harness and helping to lower the mainsail, I would sway and surge and pitch as if drunkenly into the head, where, heaving, I would liberate my dinner into a bucket.

At the outset, I figured I'd interview a few oceanographers, talk to a few beachcombers, read up on ocean currents and Arctic geography, and then write an account of the incredible journey of the bath toys lost at sea, an account more detailed and whimsical than the tantalizingly brief summaries that had previously appeared in news stories. And all this I would do, I hoped, without leaving my desk, so that I could be sure to be present at the birth of my first child.

But questions, I've learned since, can be like ocean currents. Wade in a little too far and they can carry you away. Follow one line of inquiry and it will lead you to another, and another. Spot a yellow duck dropped atop the seaweed at the tide line, ask yourself where it came from, and the next thing you know you're way out at sea, no land in sight, dog-paddling around in mysteries four miles deep. You're wondering when and why yellow ducks became icons of childhood. You want to know what it's like inside the toy factories of Guangdong. You're marveling at the scale of humanity's impact on this terraqueous globe and at the oceanic magnitude of your own ignorance. You're giving the plight of the Laysan albatross many moments of thought.

The next thing you know, it's the middle of the night and you're on

the outer decks of a post Panamax freighter due south of the Aleutian
island where, in 1741, shipwrecked, Vitus Bering perished from scurvy
and hunger. The winds are gale force. The water is deep and black, and
so is the sky. It's snowing. The decks are slick. Your ears ache, your fingers are numb. Solitary, nocturnal circumambulations of the outer decks
by supernumerary passengers are strictly forbidden, for good reason.
Fall overboard and no one would miss you. You'd inhale the ocean and
go down, alone. Nevertheless, there you are, not a goner yet, gazing up
at the shipping containers stacked six-high overhead, and from them
cataracts of snowmelt and rain are spattering on your head. There you
are, listening to the stacked containers strain against their lashings,
creaking and groaning and cataracting with every roll, and with every
roll you are wondering what in the name of Neptune it would take to
make stacks of steel—or for that matter aluminum—containers fall. Or
you're learning how to tie a bowline knot and say thank you in both Inuk-
titut and Cantonese.

Or you're spending three days and nights in a shabby hotel room in
Pusan, South Korea, waiting for your ship to come in, and you're wondering what you could possibly have been thinking when you embarked
on this harebrained journey, this wild duckie chase, and you're drinking
Scotch, and looking sentimentally at photos of your wife and son on
your laptop, your wife and son who, on the other side of the planet, on
the far side of the international date line, are doing and feeling and
drinking God knows what. Probably not Scotch. And you're remembering the scene near the end of *Moby-Dick* when Starbuck, family man,
first officer of the *Pequod*, tries in vain to convince mad Ahab to abandon his doomed hunt. "Away with me!" Starbuck pleads, "let us fly these
deadly waters! let us home!"

And you're dreaming nostalgically of your former life of chalk-
boards and Emily Dickinson and parent-teacher conferences, and
wishing you could go back to it, wishing you'd never contacted the
heavyset Dr. E., or learned of the Great Pacific Garbage Patch, or met
the Ahab of plastic hunters, or the heartsick conservationist or the
foulmouthed beachcomber or the blind oceanographer, any of them.
You're wishing you'd never given Big Poppa the chance to write about
Luck Duck, because if you hadn't you'd never have heard the fable of
the rubber ducks lost at sea. You'd still be teaching *Moby-Dick* to

American teenagers. But that's the thing about strong currents: there's no swimming against them.

The next thing you know years have passed, and you're still adrift, still waiting to see where the questions take you. At least that's what happens if you're a nearsighted, school-teaching, would-be archaeologist of the ordinary, with an indulgent, long-suffering wife and a juvenile imagination, and you receive in the mail a manila envelope, and inside this envelope you find a dozen back issues of a cheaply produced newsletter, and in one of those newsletters you discover a wonderful map—if, in other words, you're me.

GOING OVERBOARD

[T]he great flood-gates of the wonder-world swung open.

—Herman Melville, *Moby-Dick*

A RIDDLE ON THE SAND

We know where the spill occurred: 44.7°N, 178.1°E, south of the Aleutians, near the international date line, in the stormy latitudes renowned in the age of sail as the Graveyard of the Pacific, just north of what oceanographers, who are, on the whole, less poetic than mariners of the age of sail, call the subarctic front. We know the date—January 10, 1992—but not the hour.

For years the identity of the ship was a well-kept secret, but by consulting old shipping schedules published in the *Journal of Commerce* and preserved on scratched spools of microfiche in a library basement, I, by process of elimination, solved this particular riddle: the ship was the Evergreen *Ever Laurel,* owned by a Greek company called Technomar Shipping and operated by the Taiwanese Evergreen Marine Corporation, whose fir-green containers, with the company's curiously sylvan name emblazoned across them in white block letters, can be seen around harbors all over the world. No spools of microfiche have preserved the identities of the officers and crew, however, let alone their memories of what happened that stormy day or night, and if the logbook from the voyage still exists, it has been secreted away to some corporate archive, consigned, for all intents and purposes, to oblivion.

We know that the ship departed Hong Kong on January 6, that it arrived in the Port of Tacoma on January 16, a day behind schedule, and that the likely cause for this delay was rough weather. How rough exactly remains unclear. Although it did so on other days, on January 10, the *Ever Laurel* did not fax a weather report to the National Weather Service in Washington, D.C., but the following morning a ship in its vicinity did, describing hurricane-force winds and waves thirty-six feet high. If the *Ever Laurel* had encountered similarly tempestuous conditions, we can imagine, if only vaguely, what might have transpired: despite its grandeur, rocked by waves as tall as brownstones, the colossal vessel—a floating warehouse weighing 28,904 deadweight tons and powered by a diesel engine the size of a barn—would have rolled and pitched and yawed about like a toy in a Jacuzzi.

At some point, on a steep roll, two columns of containers stacked six high above deck snapped loose from their steel lashings and tumbled overboard. We can safely assume that the subsequent splash was terrific, like the splash a train would make were you to drive it off a seaside cliff. We know that each of the twelve containers measured eight feet wide and either twenty or forty feet long, and that at least one of them— perhaps when it careened into another container, perhaps when it struck the ship's rails—burst or buckled open as it fell.

We know that as the water gushed in and the container sank, dozens of cardboard boxes would have come bobbing to the surface; that one by one, they too would have come apart, discharging thousands of little packages onto the sea; that every package comprised a plastic shell and a cardboard back; that every shell housed four hollow plastic animals—a red beaver, a blue turtle, a green frog, and a yellow duck—each about three inches long; and that printed on the cardboard in colorful letters in a bubbly, childlike font were the following words: THE FIRST YEARS. FLOATEES. THEY FLOAT IN TUB OR POOL. PLAY & DISCOVER. MADE IN CHINA. DISHWASHER SAFE.

From a low-flying plane on a clear day, the packages would have looked like confetti, a great drift of colorful squares, exploding in slow motion across the waves. Within twenty-four hours, the water would have dissolved the glue. The action of the waves would have separated the plastic shell from the cardboard back. There, in seas almost four miles deep, more than five hundred miles south of Attu Island at the western tip of the Aleutian tail, more than a thousand miles east of Hokkaido, the northern extreme of Japan, and more than two thousand miles west of the insular Alaskan city of Sitka, 28,800 plastic animals produced in Chinese factories for the bathtubs of America—7,200 red beavers, 7,200 green frogs, 7,200 blue turtles, and 7,200 yellow ducks— hatched from their plastic shells and drifted free.

Eleven years later, ten thousand circuitous miles to the east, a beach-comber named Bethe Hagens and her boyfriend, Waynn Welton, spotted something small and bright perched atop the seaweed at the southwest end of Gooch's Beach near the entrance to Kennebunk Harbor in Maine. Its body was approximately the size and shape of a bar of soap, its head the size of a Ping-Pong ball. Welton bent down and picked it up. A brand name, The First Years, was embossed on its belly. The plastic was "white, incredibly weathered, and very worn," Hagens would later recall. Welton remembered it differently. The duck had been, he insisted, still yellow. "Parts of it had started to fade," he says. "But not a great deal. Whatever they'd used for the dye of the plastic had held up pretty well."

Much depends on this disagreement. If white, the duck Hagens and Welton saw could well have been one of the 7,200 let loose on the North Pacific. If yellow, it was nothing but a figment, a phantom, a will-o'-the-wisp. To complicate matters, The First Years had by then discontinued the Floatees, replacing them with a single yellow Floaty Ducky that also bore the company's logo. Yellow or white, the thing did look as though it had crossed the ocean; on that Hagens and Welton agreed. It was fun to imagine: a lone duck, drifting across the Atlantic, like something out of a fairy tale or a children's book—fun but also preposterous. "There were still kids playing on the beach," Welton remembered. "I thought, okay, some kid lost his toy and would come back for it." Sensibly, he and Hagens left the toy where they found it and walked on.

METAMORPHOSIS

The classified ads in the July 14, 1993, edition of the *Daily Sitka Sentinel* do not make for exciting reading, though they do convey something of what summertime in Alaska's maritime provinces is like. That week, the Tenakee Tavern, "in Tenakee," was accepting applications "for cheerful bartenders." The Baranof Berry Patch was buying berries— "huckleberries, blueberries, strawberries, raspberries." The National Marine Fisheries Service hereby gave notice that the winners of the 1992 Sablefish Tag Recovery Drawing, an annual event held to encourage the reporting of tagged sablefish, would be selected at 1 P.M. on July 19 at the Auke Bay Laboratory. "Tired of shaving, tweezing, waxing?!" asked Jolene Gerard, R.N., R.E., enticing the hirsute citizens of the Alaska

Panhandle (a region known to the people who live there as "Southeast") with the promise of "Permanent hair removel [*sic*]." Then, under the catchall heading of "Announcements," between "Business Services" and "Boats for Sale," an unusual listing appeared.

> ANYONE WHO has found plastic toy animals on beaches in Southeast please call the Sentinel at 747-3219.

The author of the ad was Eben Punderson, then a high school English teacher who moonlighted as a journalist, now a lawyer in rural Vermont. On Thanksgiving Day 1992, a party of beachcombers strolling along Chichagof Island had discovered several dozen hollow plastic animals amid the usual wrack of bottle caps, fishing tackle, and driftwood deposited at the tide line by a recent storm. After ten months at sea, the ducks had whitened, and the beavers had yellowed, but the frogs were still green as ever, and the turtles, still blue.

Now that summer had returned, the beachcombers were out in force, and on the windward side of Chichagof, as on other islands in the vicinity of Sitka, they found toys, hundreds of them—frogs half-buried under pebbles, beavers poised atop driftwood, turtles tangled in derelict fishing nets, ducks blown past the tide line into the purple fireweed. Beachcombing in the Alaskan wilderness had suddenly come to resemble an Easter egg hunt. A party game for children. Four animals, each one a different color, delivered as if supernaturally by the waves: collect them all!

Laurie Lee of South Baranof Island filled an unused skiff with the hoard of toys she scavenged. Signe Wilson filled a hot tub. Betsy Knudson had so many to spare she started giving them to her dog. It appeared that even the wild animals of Sitka Sound were collecting them: one toy had been plucked from a river otter's nest. On a single beachcombing excursion with friends, Mary Stensvold, a botanist with the Tongass National Forest who normally spent her days hunting rare specimens of liverwort, gathered forty of the animals. Word of the invasion spread. Dozens of correspondents answered the *Sentinel*'s ad. Toys had been found as far north as Kayak Island, as far south as Coronation Island, a range extending hundreds of miles. Where had they come from?

Eben Punderson was pretty sure he knew. Three years earlier, in May of 1990, an eastbound freighter, the *Hansa Carrier*, had collided with a storm five hundred miles south of the Alaskan Peninsula. Several containers had gone overboard, including a shipment of eighty thousand Nikes. Five months later, sneakers began washing up along Vancouver Island. The story had received national attention after a pair of oceanographers in Seattle—James Ingraham of the National Oceanographic and Atmospheric Administration (NOAA) and Curtis Ebbesmeyer, a scientist with a private consulting firm that assessed the environmental risks and impacts of engineering projects (sewage outflows, oil rigs)—turned the sneaker spill into an accidental oceanographic experiment. By feeding coordinates collected from beachcombers into NOAA's Ocean Surface Current Simulator, or OSCURS, a computer modeling system built from a century's worth of U.S. Navy weather data, Ebbesmeyer and Ingraham had reconstructed the drift routes of some two hundred shoes. In the process, the basement of Ebbesmeyer's bungalow had become the central intelligence agency of what would eventually grow into a global network of coastal informants. If anyone knew anything about the plague of plastic animals, it would be Ebbesmeyer, but when the *Sentinel*'s moonlighting reporter contacted him in the summer of 1993, it was the first the oceanographer had heard of the toys.

Punderson still had another lead. The ducks—and for some reason only the ducks—had been embossed with the logo of their manufacturer, The First Years. A local toy store was unable to find the logo in its merchandise catalogs, but the director of the Sheldon Jackson College library traced the brand back to its parent company, Kiddie Products, based in Avon, Massachusetts. Punderson spoke to the company's marketing manager, who somewhat reluctantly confirmed the reporter's speculations. Yes, indeed, a shipment of Floatees had been lost at sea. "Solved: Mystery of the Wandering Bathtub Toys," ran the lead headline in the *Sentinel*'s Weekend section a month after Punderson's ad first appeared. And that is where the story should have ended—as an entertaining anecdote in the back pages of a provincial newspaper. Mystery solved. Case closed. But then something else unexpected happened. The story kept going.

The story kept going in part because Ebbesmeyer and his beachcombers joined the hunt, in part because the toys themselves kept going. Years later, new specimens and new mysteries were still turning up. In the autumn of 1993, Floatees suddenly began sprinkling the shores of Shemya, a tiny Aleutian island that lies about 1,500 miles closer to Siberia than to Sitka, not far from the site of the spill. In 1995, beachcombers in Washington State found a blue turtle and a sun-bleached duck. Dean and Tyler Orbison, a father-son beachcombing team who annually scour uninhabited islands along the Alaskan coast, added more toys to their growing collection every summer—dozens in 1992, three in 1993, twenty-five in 1994, until, in 1995, they found none. The slump continued in 1996, and the Orbisons assumed they'd seen the last of the plastic animals. Then, in 1997, the toys suddenly returned in large numbers.

Thousands more were yet to be accounted for. Where had they gone? Into the Arctic? Around the globe? Were they still out there, traveling the currents of the North Pacific? Or did they lie buried under wrack and sand along Alaska's wild, sparsely populated shores? Or, succumbing to the elements—freezing temperatures, the endless battering of the waves, prolonged exposure to the sun—had they cracked, filled with water, gone under? All 28,800 toys had emerged from that sinking container into the same acre of water. Each member of the four species was all but identical to the others—each duck was just as light as the other ducks, each frog as thick as the other frogs, each beaver as aerodynamic as the next. And yet one turtle had ended up in Signe Wilson's hot tub, another in the jaws of Betsy Knudson's Labrador, another in an otter's nest, while a fourth had floated almost all the way to Russia, and a fifth traveled south of Puget Sound. Why? What tangled calculus of causes and effects could explain—or predict—such disparate fates?

There were still other reasons why the story of the toys kept going, reasons that had nothing to do with oceanography and everything to do with the human imagination, which can be as powerful and as inscrutable as the sea. In making sense of chaotic data, in following a slightly tangled thread of narrative to its source, Eben Punderson had set the plastic animals adrift all over again—not upon the waters of the North Pacific, but upon currents of information. The Associated Press picked up the *Daily Sitka Sentinel*'s story and far more swiftly than the ocean currents carried the castaway toys around the globe.

The Floatees made brief appearances in the *Guardian* and the *New*

York Times Magazine, and a considerably longer appearance in the *Smithsonian.* Like migrating salmon, they returned almost seasonally to the pages of *Scholastic News,* the magazine for kids, which has reported on the story seven times. They were spotted in the shallows of *People* and MSNBC, and in the tide pools of *All Things Considered.* They swirled through the sewers of the Internet and bobbed up in such exotic lagoons as a newsletter for the collectors of duck-themed stamps, an oceanography textbook for undergraduates, and a trade magazine for the builders of swimming pools.

These travels wrought strange changes. Dishwasher safe the toys may have been, but news-media safe they were not. By the time they drifted into my own imagination late one winter night several years ago, the plastic animals that had fallen into the Pacific in 1992 were scarcely recognizable. For one thing, the plastic had turned into rubber. For another thing, the beavers, frogs, and turtles had all turned into ducks. The day Eben Punderson published an unusual ad in the pages of the Sitka *Sentinel* a metamorphosis had begun, the metamorphosis of happenstance into narrative and narrative into fable—the Fable of the Rubber Ducks Lost at Sea.

Far across the ocean, in a toy factory made of red brick, a pinkly Caucasian woman in a brick-red dress and a racially ambiguous brown man in a sky-blue shirt work side by side at an assembly line. From a gray machine, yellow-billed and lacking irises in the whites of their eyes, rubber ducks emerge, one by one, onto a conveyor belt. *Chuckedy-chuckedy-chuck* goes the rubber duck machine. As the ducks roll past, the woman in the brick-red dress paints their bills brick red with a little brush. The man in the sky-blue shirt paints their irises sky blue. It is beautiful, this unnamed country across the sea. Green grass grows around the factory. A grass-green truck carries the ducks to a waiting ship named the *Bobbie.* Away the *Bobbie* chugs, carrying five cardboard boxes across a blue-green sea, a white streamer of smoke trailing behind it. Smiling overhead is an enormous sun the color of a rubber duck. Then a storm blows up. Waves leap. The *Bobbie* tosses about. The white-bearded captain cries and throws his hands out a porthole. Down goes a cardboard box. Ducks spill like candy from a piñata. Slowly, they drift apart. One frolics with a spotted dolphin. A second receives a come-hither look from a blueberry seal in a lime-green sea. A polar bear standing on an ice floe ogles a third. And so their journeys go, each duck encountering a different picturesque animal—a flamingo, a pelican, a sea turtle, an octopus, a gull, a whale. Finally, who should the tenth rubber duck meet but a brood of real ducks. "Quack!" says the mother duck. "Quack! Quack! Quack!" say the ducklings. "Squeak," says the rubber duck. So ends Eric Carle's *Ten Little Rubber Ducks*.

Carle's picture book was, perhaps inevitably, inspired by one of the many newspaper articles that appeared after Ebbesmeyer put the beachcombers of New England on alert. Having crossed the Arctic, having drifted south on the Labrador Current, along the wild coast of Newfoundland, past Nova Scotia and the Grand Banks, some of the Floatees would reach the Eastern Seaboard of the United States in the summer of 2003, the clairvoyant oceanographer predicted. A savvy publicist at The First Years, smelling a marketing opportunity, sent out a press release advertising a bounty: Kiddie Products would give a U.S. savings bond worth $100 to any beachcomber who found one of the castaway toys on an East Coast beach. All along the coasts of Massachusetts and Maine, people began to hunt. There was, as there usually is, a catch— two catches, actually: (1) to claim the reward, the lucky beachcombers would have to surrender the evidence, and (2) Ebbesmeyer would have to confirm a positive match. As intended, the press release provoked a flurry of coverage. Once again, news organizations large and small recounted the "rubber duck saga," as the *Montreal Gazette* dubbed it that summer.

A scrap of the article that Carle happened on, torn from an uncredited source, accompanies his author's note:

RUBBER DUCKS LOST AT SEA

In 1992, a shipment of 29,000 rubber bathtub toys including ducks, beavers, turtles and frogs, fell overboard from a container ship.

Some of these rubber toys have washed up on the shores of Alaska, while others have made their way through the Bering

Strait, past icebergs, around the northern coast of Greenland and into the Atlantic Ocean.

"I could not resist making a story out of this newspaper report," Carle's note explains. "I hope you like my story." Beautifully illustrated with Carle's signature mix of paint and paper tearings, *Ten Little Rubber Ducks* is hard not to like. Studies have shown that the primary colors, smiling faces, and cute animals in which Carle's book abounds—and of which the rubber duck may well be the consummate embodiment—have the almost narcotic power to induce feelings of happiness in the human brain. The metamorphosis that had begun in the pages of the *Daily Sitka Sentinel* was complete: in Carle, the fable had finally found its Aesop.

It's easy to see why Carle found the story irresistible. It was an incredible story, a *fabulous* story; the sort of head-shaking, who'd-a-thunk anecdote suited to an entry in *Ripley's Believe It or Not*, perhaps, or to cocktail party banter, or to a lighthearted closing segment on the evening news, or most of all to a picture book for children.

Visit the kids section of your local public library and you'll find dozens or possibly even hundreds of stories about inanimate objects that come magically to life or go on incredible journeys. Such stories are so common, in fact, that they constitute a genre—the "it-narrative," literary scholars have called it. Think of *Pinocchio*. Or *The Velveteen Rabbit*. Or *Winnie-the-Pooh*. Or the improbable eighteenth-century bestseller *The Adventures of a Pincushion*. The it-narrative that the legend of the castaway ducks most resembles surely must be Holling Clancy Holling's *Paddle-to-the-Sea*, the 1941 Caldecott winner in which a boy in the Canadian wilderness carves a wooden Indian man in a wooden canoe, carries his creation up a nearby mountain, and sets it atop a bank of snow. "The Sun Spirit will look down at the snow," the boy says. "The snow will melt and the water will run downhill to the river, on down to the Great Lakes, down again and on at last to the sea. You will go with the water and you will have adventures that I would like to have."

What distinguishes *Paddle-to-the-Sea* from most other it-narratives is its painstaking realism—realism so painstaking that the book feels like nonfiction. Carle, by contrast, has always preferred allegory to real-

ism. Think of *The Very Hungry Caterpillar*, his best-known book, the
protagonist of which, a gluttonous larva with eyes like lemon-lime lolli-
pops, is an entomological embodiment of childish appetites. He's born
on a Sunday, binges for a week, and then the following Sunday nibbles
contritely on a leaf, in reward for which penance, he pupates, abraca-
dabra, into a butterfly, an *angelic* butterfly. It's a Christian allegory with
which any American child can identify, an allegory about conspicuous
consumption: *The Prodigal Caterpillar*, Carle might have called that
book, or *The Caterpillar's Progress*.

In *Ten Little Rubber Ducks*, on the other hand, there are no choices,
no consequences. There is only chance. The human imagination is by
nature animistic. It can even bring a pincushion to life. But Carle's ten
identical rubber ducks remain inanimate—psychologically empty, de-
void of distinguishing characteristics, appetite, emotion, or charm. Carried along by ocean currents rather than by the lineaments of desire,
they drift passively about, facial expressions never changing.

BIG POPPA

A few months before Carle's book hit bookstore shelves, I likewise happened on the legend of the rubber ducks lost at sea, not in a newspaper
report but in an essay by one of my students. Late one night, after my
wife, Beth, had gone to sleep, when few windows remained lit in the
building across the street, I stayed up as usual to grade papers. Mostly I
taught the sorts of poems and novels and plays typically prescribed to
American teenagers as remedies for short attention spans and atrophied
vocabularies—*Hamlet*, for instance, or *Their Eyes Were Watching God*, or
Leaves of Grass. And mostly the papers I graded were the sorts of essays
English teachers typically ask American teenagers to write—five or six
paragraphs on the role of prophecy in *Macbeth* or the motif of walls in
"Bartleby, the Scrivener," that sort of thing. But every spring I also
taught a journalism course.

One of my favorite assignments I'd devised asked students to practice what James Agee called the archaeology of the ordinary. In *Let Us
Now Praise Famous Men*, of the overalls that Depression-era sharecrop-
pers wore, Agee writes, "I saw no two which did not hold some world of

exquisiteness of its own." Everywhere Agee went during his Alabaman travels, he found exquisite worlds, doing for the material lives of share-croppers what Thoreau did for Walden Pond, or Melville for whaling. If like Agee my students could learn to study a thing—any particular thing—"almost illimitably long," as Agee recommended, they too might begin to perceive "the cruel radiance of what is" rather than the narcotic shimmer of what isn't. Or so my hopeful thinking went.

One year one student chose to write about a venerable brand of shoe polish, discovering therein the lost world of the New York City shoe-shine boy, who in most instances wasn't a boy at all. Another chose a taxidermy crocodile. Another a charm bracelet her mother had given her. Another a baseball he'd caught in the stands of Yankee Stadium. This last student had studied his subject matter so illimitably he'd sawed the damn thing in half. And one student, a pudgy, myopic kid who'd given himself the nickname "Big Poppa," chose to write about the rubber duckie he carried around in his pocket for good luck. Luck Duck, he called it. It was his mojo, his talisman, his totem, his charm.

I myself was a struggling, part-time archaeologist of the ordinary. Like Professor Indiana Jones—or so I sometimes fancied—I lived a double life. Summers, after classes had ended, on a magazine assignment, I would hang up my olive-green corduroy blazer with the torn lining and the baggy pockets full of chalk nubbins, pull on my hiking boots, pick up my notebook and voice recorder, and head off in search of exquisite worlds. The worlds I tended to seek and find were those on the borderlands between the natural and the man-made, the civilized and the wild. I liked such borderlands because within them interesting questions and contradictions tended to flourish, like wildflowers on a vacant city lot. I also liked them because I have since childhood found natural history more enchanting than nature, whatever that was. I've never heard the howl of a wolf or felt a strong desire to answer its wild call, but I have often found myself entranced before the diorama at the American Museum of Natural History in which taxidermy wolves, though sus-pended from wires, appear to be racing over the Alaskan tundra under a black-light moon, leaving footprints in the plaster-of-paris snow.

More than most of the students at the private Quaker school in Manhattan where I taught, Big Poppa could have used a little luck. The only child of divorced parents, he was forever shuttling back and forth between his mom's apartment and his father's studio, leaving behind a trail of unfinished homework and misplaced books. His backpack was an experiment in chaos. When he was supposed to be studying, he instead stayed up all night playing fantasy sports games on the Internet. He loved playing real sports too, baseball especially, and he was by reputation a good infielder, and the team's best batter, but he was chronically late to class, and his attendance record was so poor that midway through the baseball season during the spring of his senior year, he had to be temporarily benched, per school policy. Most of his classmates were Yankee fans, whereas Big Poppa rooted avidly, hopelessly, for the underdog Mets, whose paraphernalia constituted a large portion of his wardrobe.

A class clown of the masochistic rather than the sadistic variety, he liked to amuse his peers at his own expense—for instance, by announcing one day that he wished henceforth to be known by the nickname Big Poppa (a self-deprecating reference to the babyish pudge that was for him a source of shame, but also to the rapper Notorious B.I.G., a.k.a. Biggie Smalls, a.k.a. Big Poppa); or, for instance, by rooting too avidly for the Mets; or by arriving late to class prematurely suited up in his baseball uniform (the cleats, the red stirrups, the white tights), fielding imaginary grounders and swinging at imaginary pitches as he crossed the room, before dropping himself into a chair, looking about at his smirking classmates, and inquiring, with an interrogative shrug, "What?" He'd once asked a French teacher for permission to go home sick because he'd eaten "a bad knish." It had become an inside joke. "What'sa matter?" his friends liked to ask him. "Bad knish?"

He could also be poignantly emotional. He felt strongly about bicycle helmets and upbraided teachers who biked to school without them. The degradation of the environment upset him, sometimes almost to tears. Once, when another of my students discarded her water bottle in the trash can, Big Poppa made a great fuss, rescuing it and depositing it in a recycling bin intended—the bottle's owner wryly observed—for paper products only.

He dreamed of becoming a sportswriter someday, and I, faculty adviser to the student newspaper, had encouraged him in this dream. I encouraged him because he could turn a phrase and because when it came to baseball, he knew his stuff, and because I sympathized with him (I, too, had been a pudgy, myopic, late-to-pubesce child of divorce), but also because Big Poppa was charming, and bright, and kind—just a little rudderless, a little juvenile, a little lost. Old enough to carry a rifle in Afghanistan or Iraq, he instead carried a rubber duckie for good luck. Earlier that winter, during a snowball fight in a classmate's backyard, his glasses had disappeared into a snowdrift. This wasn't the first time they'd gone missing. Afraid to ask his parents to buy a new pair, he'd decided to wait for the snow to melt and his glasses to resurface, which they eventually did. In the meantime, he'd spent three weeks stumbling through the halls, his face contorted into a squint. He lost Luck Duck for a while too, and the loss inspired what appeared to be genuine distress.

While researching his essay, Big Poppa had happened on a newspaper report, perhaps the very same one that Eric Carle had happened on. In a paragraph cataloging rubber duck trivia, he'd included a four-sentence synopsis—the container spill, the oceanographers in Seattle, the journey through the Arctic. The toys were supposed to have reached the coast of New England by the summer of 2003. It was now March 2005. Had they made it? Big Poppa didn't say. Neither did he mention anything about beavers, turtles, or frogs.

It was well after midnight by the time I finished marking his essay, and because I am prone to nocturnal flights of fancy, I sat there for a while at my desk thinking about those ducks. I tried to imagine their journey from beginning to end. I pictured the container falling—*splash!*—into the sea. I pictured the ducks afloat like yellow pixels on the vast, gray acreage of the waves, or skiing down the glassy slopes of fifty-foot swells, or coasting through the Arctic on floes of ice. I imagined standing on a beach somewhere in Newfoundland or Maine—places I had never visited or given much thought. I imagined looking out and seeing a thousand tiny nodding yellow faces, white triangles glinting in their cartoon eyes, insipid smiles molded into the orange rubber of their clownish bills. I imagined a bobbing armada so huge it stretched to the horizon, and possibly beyond. I imagined them washing ashore, littering the sand, a yellow tide of ducks.

Getting ready for bed that night, I noticed anew the rubber ducks roosting on our bathroom shelf. For years they'd perched there, between a jar of cotton balls and a bottle of facial cleanser, two yellow ducks of the classic variety and one red duck with horns. I couldn't remember when or why we'd acquired them. I picked one of the yellow ones up and gave it a squeeze. Air hissed from an abdominal hole. "Quack," I said, and returned it to its shelf.

At work the next day, I noticed as if for the first time the several yellow ducks of diminishing sizes processing single file across a colleague's desk—gifts from students, she explained when I asked. I began seeing them everywhere. In the course of a single afternoon, I came upon a great, neon rubber duck aglow in the window of an Old Navy store and a mother-duckling pair afloat in the margins of a brochure on vaccinations that our health insurance provider sent. My wife and I that spring were, as the euphemism has it, "expecting," and the baby outfitter where we'd registered for shower gifts—across the blue awning of which yet another duck swam, a pacifier abob like a fishing float beside it—appeared to be the epicenter of this avian plague. In addition to rubber ducks themselves, the store sold yellow towels and sweatshirts with orange bills for hoods, yellow rubber rain boots with beaks for toes, yellow pajamas with orange, webbed feet. There was the Diaper Duck, a duck-shaped dispenser of odor-proof trash bags, as well as numerous other implements—brushes, soap dishes, etc.—that incorporated the likeness and the yellowness of the duck.

Elsewhere, in drugstores and catalogs and the bathrooms of friends, I spotted exotic varieties in which strange, often ironical mutations had occurred—momma ducks with ducklings nested on their backs; black ducks, sparkly ducks, ducks with the face of Moses or Allen Iverson or Betty Boop; ducks sporting sunglasses or eyelashes or lipstick or black leather; ducks playing golf. Every powerful icon invites both idolatry and iconoclasm, and in the bestiary of American childhood, there is now no creature more iconic than the rubber duck. The more I thought about its golden, graven image, the more it seemed to me a kind of animistic god—but of what? Of happiness? Of nostalgia? Of innocence never lost?

THE MAP

"So," I asked the retired oceanographer when I reached him at his Seattle home, "did any of the toys make it through the Arctic?" I had by then read every article about the incredible journey I could find. As of October 2003, according to the news archives, not one of the 28,800 castaway toys had been discovered on the Atlantic Seaboard, not one savings bond had been handed out. Bounty-hunting beachcombers had found plenty of toy ducks, just not of the right species. After October 2003 the news archives fell silent.

Oh, yes, Curtis Ebbesmeyer assured me, yes, they'd made it. Right on schedule, in the summer of 2003, he'd received a highly credible eyewitness report from an anthropologist in Maine, which he'd published in his quarterly newsletter, *Beachcombers' Alert!* He promised to send me a copy. But before we hung up he dangled before my ears a tantalizing lure: if I really wanted to learn about things that float, then I should join him in Sitka that July. "You can't go beachcombing by phone," he said. "You have to get out there and look."

Since the summer of 2001, Sitka had played host to an annual Beachcombers' Fair, over which Ebbesmeyer—part guru, part impresario—presided. Beachcombers would bring him things they'd scavenged from the sand, and Ebbesmeyer, like some scientific psychic, would illuminate these discoveries as best he could. "Everything has a story," he likes to say. When a beachcomber presented Ebbesmeyer with flotsam of mysterious provenance, he'd investigate. At that year's fair in Sitka, a local fisherman named Larry Calvin would be ferrying a select group of beachcombers to the wild shores of Kruzof Island, where some of the toys had washed up. Ebbesmeyer, who would be leading the expedition, offered me a spot aboard Calvin's boat, the *Morning Mist.*

Alaska—snowcapped mountains, icebergs, breaching whales, wild beaches strewn with yellow ducks. How could I say no? There was only one problem. The Beachcombers' Fair ended July 24, and Beth's due date was August 1, which was cutting it pretty close. I told Ebbesmeyer I'd get back to him.

Soon thereafter an envelope with a Seattle postmark arrived. Inside,

printed on blue paper, were a half-dozen issues of Ebbesmeyer's newsletter, *Beachcombers' Alert!* Thumbing through this digest of the miscellaneous and arcane was a bit like beachcombing amid the wreckage of a
storm. Alongside stories about derelict vessels and messages in bottles,
the oceanographer had arrayed a photographic scrapbook of strange,
sea-battered oddities, natural and man-made—Japanese birch-bark
fishing floats, the heart-shaped seed of a baobab tree, land mines, tele-
visions, a torn wet suit, a 350-pound safe. Many of these artifacts had accumulated colonies of gooseneck barnacles. Some were so encrusted
they seemed to be made of the creatures: a derelict skiff of barnacles, a
hockey glove of barnacles.

At the end of an article titled "Where the Toys Are," Ebbesmeyer
had published the letter from that anthropologist in Maine. Bethe Hagens was her name. "You won't believe this," she'd written after hearing
about the castaway toys on NPR, "but two weeks ago, I found one of
your ducks." In fact, Ebbesmeyer *had* believed her, or wanted to. She
hadn't kept the evidence, so there was, she'd written, "no science, no
proof. But they're here!" Was there proof or wasn't there? Were they
here or weren't they? Accompanying the article was a world map indicating where and when the toys had been recovered by beachcombers.
Off the coast of Kennebunkport, Ebbesmeyer had printed a pair of
question marks the size of barrier reefs.

From a dusty bookshelf I fetched down our *Atlas of the World*, a neglected wedding gift, opened it to the Atlantic, and found Kennebunkport. Then I traced my finger out across the Gulf of Maine, around
Newfoundland and Labrador and—flipping to the map of the Arctic—
across Baffin Bay, westward past the pole, all the while pronouncing the
unfamiliar syllables (Point Hope, Spitsbergen Bank) as if the names of
these places could conjure up visions of their shores. What does the air
smell like in the Arctic? I wondered. Can you hear the creeping progress
of the ice?

"The loss of fantasy is the price we have paid for precision," I'd read
late one night in an outdated *Ocean Almanac*, "and today we have navi-
gation maps based on an accurate 1:1,000,000 scale of the entire world."
Surveying the colorful, oversize landscape of my atlas, a cartographic
wonder made—its dust jacket boasted—from high-resolution satellite

photographs and "sophisticated computer algorithms," I was unconvinced; fantasy did not strike me as extinct, or remotely endangered. The ocean was far less fathomable to my generation of Americans than it was when Melville explored that "watery wilderness" a century and a half ago. Most of us were better acquainted with cloud tops than with waves. What our migrant ancestors thought of as the winds, we thought of as turbulence, and fastened our seat belts when the orange light came on. Gale force, hurricane force—encountering such terms we comprehended only that the weather was really, really bad, and in our minds replayed the special-effects sequences of disaster films or news footage of palm trees blown inside out like cheap umbrellas. In growing more precise, humanity's knowledge had also grown more specialized, and more imaginary: unlike that of my unborn child, the seas of my consciousness teemed with images and symbols and half-remembered trivia as fabulous as those chimerical beasts cavorting at the edges of ancient charts. Not even satellite photographs and computer algorithms could burn away the mystifying fogs of ambient information and fantasy through which from birth I had sailed.

Not long ago on the Op-Ed page of the *New York Times*, the novelist Julia Glass worried that her fellow Americans, "impatient with flights of fancy," had lost the ability to be carried away by the "illusory adventure" of fiction, preferring the tabloid titillation of the "so-called truth." Perhaps, concluded Glass, "there is a growing consensus, however sad, that the wayward realm of make-believe belongs only to our children." I'd reached different conclusions. Hadn't we adults, like the imaginative preschoolers Glass admires, also been "encouraged"—by our government, by advertisers, by the fabulists of the cable news—"to mingle fact with fiction"?

"If men would steadily observe realities only, and not allow themselves to be deluded, life, to compare it to such things as we know, would be like a fairy tale." So wrote Thoreau, and for a number of years I'd been inclined to agree with him. I'd been inclined to agree, but despite my experiments in the archaeology of the ordinary, I'd also been more inclined to be deluded than to steadily observe realities only. Ask me where plastic came from and I'd have pictured Day-Glo fluids bubbling in vats, or doing loop-the-loops through glass tubes curly as Krazy

Straws. If you'd asked me how rubber ducks were made, I might well have pictured them emerging onto a conveyor belt—*chuckedy-chuckedy-chuck*—out of a gray machine.

Looking at the face of my unborn daughter or son adrift on a sonogram screen, I hadn't felt the sorts of emotions expecting parents are supposed to feel—joy, giddiness, pride, all that. Instead I'd felt a fatalistic conviction that either I or the world and probably both would let down that little big-headed alien wriggling around in those uterine grottoes. How safe and snug he or she looked in there. How peacefully oblivious, no doubts and vanities bubbling through his or her gray matter, no advertising jingles or licensed characters or boogeymen, no fantasies, not even dreams—at least none of the sort that would animate his or her postpartum inner life. It seemed cruel somehow, this conjuring act of incarnation, this impulse to summon out of one's DNA a person who'd had no choice in the matter. I'd had a choice, and I'd enthusiastically chosen to become a father. Now that the deed was done, I found my own paternity difficult to believe in. I could no more imagine being somebody's father than I could imagine performing the Eucharist or surgery.

Truth be told, it wasn't only my unborn child whom I was worried for. For months, a quote from one of Hawthorne's letters had been bothering me. It came to mind at unexpected moments—during faculty meetings, or as I trudged home beneath the fruitless pear trees and proprietary brownstones of Greenwich Village, or browsed among aisles of Bugaboos and Gymborees at BuyBuy Baby. It had drifted there, upon my inward seas, like a message in a bottle, a warning cast overboard by a shipwrecked seafarer years ago: "When a man has taken upon himself to beget children," Nathaniel Hawthorne wrote to Sophia Peabody, his fiancée, in 1841, "he has no longer any right to a life of his own."

At the hospital, Beth hadn't seemed to share my gloomy presentiments. Supine on the examination table beside me, gazing beatifically at the sonogram screen while a sullen West Indian nurse prodded her ballooning abdomen with a wand, Beth kept giving my hand little squeezes of motherly delight, squeezes that had the peculiar effect of making me gloomier still. Why? Guilt had something to do with it, no doubt. Self-loathing, perhaps. I think also that there exists a kind of chiaroscuro of the human heart whereby the light that another's joy gives off, instead of shining brightly upon us, casts us more deeply into shadow.

Riffling the pages of my atlas, I turned to the North Pacific, found the coordinates—44.7°N, 178.1°E—at which, on that January day or night in 1992, the toys became castaways, and marked the spot with a yellow shred of Post-it. How placid—how truly pacific—that vaguely triangular ocean seemed in the cartographer's abstract rendering. Its waters were so transparent, as though the basin had been drained and its mountainous floor painted various shades of swimming-pool blue. Way over there, to the east, afloat on its green speck of land like a bug on a leaf, was Sitka. And way over there, huge as a continent, was China, where, odds were, someone in some factory was at that very moment bringing new rubber ducks into the world. It was then, as I studied my map, trying in vain to imagine the journey of the toys, that there swam into my mind the most bewitching question I know of—*What if?*

What if I followed the trail of the toys wherever it led, from that factory in China, across the Pacific, into the Arctic? I wouldn't be able to do it in a single summer. It would require many months, maybe an entire year. I might have to take a leave of absence, or quit teaching altogether. I wasn't sure how or if I'd manage to get to all the places on my map, but perhaps that would be the point. The toys had gone adrift. I'd go adrift, too. The winds and currents would chart my course. Happenstance would be my travel agent. If nothing else, it would be an adventure, and adventures are hard to come by these days. And if I were lucky it might be a genuine voyage of discovery. Medieval Europeans divided the human lifetime into five ages, the first of which was known as the Age of Toys. It seemed to me that in twenty-first-century America, the Age of Toys never ends. Yes, stories fictional and otherwise can take us on illusory odysseys, but they can also take us on disillusory ones, and it was the latter sort of journey that I craved. It wasn't that I wanted, like Cook and Amundsen and Vancouver and Bering and all those other dead explorers, to turn terra incognita into terra cognita, the world into a map. Quite the opposite. I wanted to turn a map into a world.

THE FIRST CHASE

One day Mr. Mallard decided he'd like to take a trip to see what the rest of the river was like, further on. So off he set.
—Robert McCloskey, *Make Way for Ducklings*

THE HEAVYSET DR. E.

There are two ways to get to the insular city of Sitka—by air and by sea. In my dreams, I would have picked up the frayed end of that imaginary, ten-thousand-mile-long trail that led from Sitka to Kennebunkport and followed it backward, Theseus style, to its source—backward across the Gulf of Maine, backward through the Northwest Passage, that legendary waterway which the historian Pierre Berton has described as a "maze of drifting, misshapen bergs," a "crystalline world of azure and emerald, indigo and alabaster—dazzling to the eye, disturbing to the soul," a "glittering metropolis of moving ice." To Lieutenant William Edward Parry of the Royal Navy, who captained the *Alexander* into the maze in 1818, the slabs of ice looked like the pillars of Stonehenge.

By that summer, the summer of 2005, global warming had gone a long way toward turning Berton's maze of bergs into the open shipping channel of which Victorian imperialists dreamed. The following September, climatologists would announce that the annual summer melt had reduced the floating ice cap to its smallest size in a century of record keeping. Nevertheless, a transarctic journey, even aboard an icebreaker, was out of the question if I wanted to make it to and from Sitka by the first of August.

Instead I'd booked passage on the M/V *Malaspina*, part of the Alaskan Marine Highway, which is in fact not a highway at all but a state-operated fleet of ferries. Sailing from Bellingham, Washington, the *Malaspina* would reach Sitka five days before the Beachcombers' Fair began. If I flew home as soon as the fair ended, I would be in Manhattan a week before the baby arrived—assuming he or she did not arrive early, which, my wife's obstetrician warned us, was altogether possible. Although not at all happy about my plan, Beth had consented, on two conditions: one, that if she felt a contraction or her water broke, I would catch the next flight to New York, no matter the cost; and two, that I would call her by cell phone at least once a day.

Although I would soon be joining up with him in Sitka, I was eager to
meet the gumshoe oceanographer in whose footsteps I was following.
On my way to Bellingham I made time for a stopover in Seattle. A reporter for the *Oregonian* once referred to Curtis Ebbesmeyer's beachcombing correspondents as "disciples," and he is an unlikely prophet
figure of sorts, mailing out his epistles to the faithful, preaching the cir-
cuit of beachcombers' fairs, making oracular pronouncements in the
press, pronouncements like, "I think the ocean is writing to us. And I'm
trying to figure out what it all means." Or, "Why do we like to walk on
the beach? . . . All the cells inside our bodies realize they're close to
their mom." Or, my personal favorite: "The literature of things that float
from here to there is so scattered, it makes no sense until you compress
it. Then it begins to take on a glow, like radium." When you've read too
many issues of *Beachcombers' Alert!* back to back, as I had, a faint, sickly
glow did seem to radiate from the newsletter's pages. Even its exclama-
tory title conveys a hint of apocalyptic alarm, and there are troubling
ecological portents strewn among all the ruins and marvels.

But Ebbesmeyer is a conflicted prophet. He prefers the role of enter-
tainer, taking corny, avuncular delight in his fabulous stories of messages in bottles and derelict boats. He enjoys the nicknames people have
given him over the years, names that make him sound like a character
from a comic book—Dr. Curt, Dr. Duck, Dr. Froggie. Jacques Cousteau's movies made Ebbesmeyer first consider a career in oceanography,
and he shares with Cousteau a self-dramatizing flair. Every oceanographic study Cousteau conducted was an adventure. Ebbesmeyer, on
the other hand, prefers the genre of mystery. He is a great fan of Arthur
Conan Doyle, whose rationalistic, Victorian version of the gothic has
exerted an obvious influence on the oceanographer's imagination. When
bodies or body parts wash up on the shores of the Pacific Northwest, detectives often give Ebbesmeyer a call, and he seems to take macabre delight in investigating these mysteries, mysteries like the case of the
dismembered feet, or of the corpse packed into a suitcase.

"Shrill violin sounded from the second-story window at 221B Baker
Street," he begins one dispatch in *Beachcombers' Alert!* written in the
form of a lost Sherlock Holmes story, "The Case of the Baobab and the

Bottle." Ebbesmeyer appears in this fictional parody as the "heavyset" Dr. E., driftological specialist in messages in bottles (or MIBs, as he likes to call them), summoned by Holmes and Watson to Baker Street to provide expert testimony. But the oceanographer's real alter ego in the parody is Holmes himself. "He desperately required a puzzle to occupy his mind which saw universes in the most trivial fact," Ebbesmeyer says of the detective. The same could be said of him.

His own Baker Street is a quiet block in a quiet neighborhood near the main campus of the University of Washington, where back in the sixties he earned his Ph.D. Smaller than the adjacent houses, his bungalow, purchased from the fisherman who built it, exhibits a kind of cultivated slovenliness. Its peaked roof rests atop two squat brick columns. On the day I visited, a white rose trellis, bare and slightly askew, leaned against one of these columns as if left there by a distracted gardener. In the middle of the front lawn, inside a concrete planter box almost as big as the lawn itself, Ebbesmeyer's wife was growing a miscellaneous assortment of vegetables and flowers—lavender, agapanthus, squash—above which the purple pom-poms of onion blossoms swayed atop their stalks. Navy-blue awnings overshadowed the porch, and peering into the semidarkness I could see four matching forest-green Adirondack chairs, lined up, side by side, as if to behold the vista of the lawn. Next to the front steps, a small American flag protruded from a terra-cotta pot. (It had been only a week since Independence Day.) At the top of the steps, a cat dish sat on a ledge beneath a little bell. Ebbesmeyer himself greeted me at the door. "Come in, come in," he said.

His face was familiar to me from photographs I'd seen in the press and in the pages of *Beachcombers' Alert!* where he makes frequent cameo appearances, displaying a water-stained basketball, hoisting a plastic canister that was supposed to have delivered Taiwanese propaganda to the Chinese mainland, gazing down deifically at the four Floatees perched on his furry forearm. He has a white beard, a Cheshire grin, and close-set eyes that together make his face a bit triangular. In the portrait that decorates the masthead of *Beachcombers Alert!*—small as Washington's on a dollar bill—little pennants of light fly across the lenses of his glasses, which are large and square. Since Ebbesmeyer likes to wear Hawaiian shirts and a necklace of what appear to be roasted chestnuts but which are in fact sea beans, the waterborne seeds of tropical trees that ocean currents disseminate to distant shores, pictures of him often bring to mind cartoons of Santa Claus on vacation.

He brewed us each a cup of coffee and suggested we adjourn to the backyard, which he refers to as his "office." Passing through his basement, I saw many of the objects I'd read about in *Beachcombers' Alert!* Piled high on a bookshelf were dozens of Nikes. Some of them had survived the 1990 container spill—the first Ebbesmeyer ever investigated—in which 80,000 shoes had gone adrift. Others came from later accidents: 18,000 Nike sneakers fell overboard in 1999; 33,000 more in December of 2002. In January of 2000, some 26,000 Nike sandals— along with 10,000 children's shoes and 3,000 computer monitors, which float screen up and are popular with barnacles—plunged into the drink.

Nike's maritime fortunes are not unusually calamitous; thousands of containers spill from cargo ships every year, exactly how many no one knows, perhaps 2,000, perhaps as many as 10,000. No one knows because shipping lines and their lawyers, fearing bad publicity or liability, like to keep container spills hush-hush. But few commodities are both as seaworthy and as traceable as a pair of Air Jordans, each shoe of which conveniently comes with a numerical record of its provenance stitched to the underside of its tongue, and which—soles up, ankles down, laces aswirl—will drift for years. It helped, too, that Nike had thus far been cooperative. A lawyer gave Ebbesmeyer the serial numbers for all the shoes in the 1990 spill and then taught him how to "read the tongue."

Now, in his basement, Ebbesmeyer selected a high-top at random. "See the ID?" he asked. "'021012.' The '02' is the year. '10' is October. '12' is December. Nike ordered these from Indonesia in October of '02 for delivery in December."

Next he pulled down a black flip-flop, and then a matching one that he'd sliced in half. Inside the black rubber was a jagged yellow core resembling a lightning bolt—a perfect identifying characteristic. If Ebbesmeyer had discovered the coordinates of this particular spill, the sandals would have provided a windfall of data. Unfortunately, the shipping company had "stonewalled" him "like usual." He didn't really hold it against them, he said. For a container ship's crew—an "Oriental" crew especially—a spill incurred a "loss of face" that could easily lead to a loss of job, a price the oceanographer considered too steep, no matter how precious the data.

It took Ebbesmeyer a year of diplomacy and detective work to find out when and where the Floatees fell overboard. Initially, the shipping company stonewalled him like usual. Then one day he received a phone call. The container ship in question was at port in Tacoma. Ebbesmeyer was welcome to come aboard, on one condition: he had to swear to keep secret the names of both the ship and its owner.

For four hours, Ebbesmeyer sat in the ship's bridge interviewing the captain, a "very gracious" Chinese man who had a Ph.D. in meteorology and spoke fluent English. The day of the spill, the ship had encountered a severe winter storm and heavy seas, the captain said. The readings on the clinometer told the story best. When a ship is perfectly level in the water, its clinometer reads 0 degrees. If a ship were keeled on its side, the clinometer would read 90 degrees. When this particular spill occurred, the clinometer had registered a roll of 55 degrees to port, then a roll of 55 degrees to starboard. At that inclination, the stacks of containers, each one six containers tall, tall as a four-story townhouse, would have been more horizontal than vertical. Perhaps Dr. Ebbesmeyer would like to have a peek at the logbook, the captain discreetly suggested. He'd already opened it to January 10, 1992. There were the coordinates, the magic coordinates.

THE GREAT PACIFIC GARBAGE PATCH

OSCURS could now reconstruct, or "hindcast," the routes the toys had traveled, producing a map of erratic trajectories that appeared to have been hand-drawn by a cartographer with palsy. Beginning at the scattered coordinates where beachcombers had reported finding toys, the lines wiggled west, converging at the point of origin: 44.7°N, 178.1°E, south of the Aleutians, near the international date line, where farthest west and farthest east meet. The data that Ebbesmeyer's beachcombers had gathered also allowed NOAA's James Ingraham to fine-tune the computer model, adjusting for such coefficients as "windage" (an object with a tall profile will sail before the wind as well as drift on a current).

At first, before they'd sprung leaks and taken on water, the toys rode high, skating across the Gulf of Alaska at an average rate of seven miles per day, almost twice as fast as the currents they were traveling. Among other things, the simulation revealed that in 1992 those currents had shifted to the north as a consequence of El Niño. If the toys had fallen overboard at the exact same spot just two years earlier, according to OSCURS, they would have taken a southerly route instead of a northerly one, ending up in the vicinity of Hawaii. In 1961, they would have drifted along the California coast.

Though with far less certainty, OSCURS could forecast as well as hindcast, and in this respect, Ebbesmeyer and Ingraham were like meteorologists of the waves. Because the weather of the ocean usually changes more slowly than the weather of the skies, they were also like clairvoyants. OSCURS was their crystal ball.

By simulating long-term mean surface geostrophic currents (those surface currents that flow steadily and enduringly, though not immutably, like rivers in the sea) as well as surface-mixed-layer currents that are functions of wind speed and direction (those currents that change almost as quickly as the skies), OSCURS could project the trajectories of the toys well into the future. According to the simulator's predictions, some of the animals that remained afloat would eventually drift south, where they would either collide with the coast of Hawaii in March of 1997 or, more likely, get sucked into the North Pacific Subtropical Gyre.

"*Gyre* is a fancy word for a current in a bowl of soup," Ebbesmeyer likes to say. "You stir your soup, it goes around a few seconds." The thermodynamic circulation of air, which we experience as wind, is like a giant spoon that never stops stirring. To make Ebbesmeyer's analogy more accurate still, you'd have to set that bowl of soup aspin on a lazy Susan, since the earth's rotation exerts a subtle yet profound influence on the movements of both water and air, an influence known to physical oceanographers as the Coriolis force. The Coriolis force explains why currents on the western edges of ocean basins are stronger than those along their eastern edges—why the Gulf Stream, for instance, is so much stronger than the Canary Current that flows south along Africa's Atlantic coast.

Comprising four separate arcs—the easterly North Pacific Drift, the southerly California Current, the westerly North Equatorial Current, and the northerly Kuroshio (the Pacific's equivalent of the Gulf Stream)—the North Pacific Subtropical Gyre revolves between the coasts of North America and Asia, from Washington State to Mexico to Japan and back again.[1] Some of the toys, OSCURS predicted, would eventually escape the gyre's orbit, spin off toward the Indian Ocean, and circumnavigate the globe.

Others would drift into the gyre's becalmed heart where the prevailing atmospheric high has created what Ebbesmeyer christened "the Garbage Patch"—a purgatorial eddy in the waste stream that covers, Ebbesmeyer told me, as much of the earth's surface as Texas. When he is being fastidious, Ebbesmeyer will point out that there are in fact many garbage patches in the world, the one in the North Pacific being simply the largest, so far as we know. For that reason he sometimes refers to it as the Great North Pacific Garbage Patch. Other times, in *Beachcombers' Alert!* and elsewhere, he'll distinguish between an Eastern Garbage Patch lying midway, roughly, between Hawaii and California, and a Western Garbage Patch, lying midway, roughly, between Hawaii and Japan. In fact, both patches are part of what most oceanographers call the North Pacific Subtropical Convergence Zone—a bland term of art made blander still by its initials, STCZ. The scientific community's love for acronyms and abbreviations, rivaled only by that of government bureaucrats, helps explain why Ebbesmeyer has enjoyed much more celebrity in the popular press than he has influence in the scientific community. He possesses a showman's gift for folky coinages, but also, perhaps, a showman's tendency to sensationalize. "It's like Jupiter's red spot," he said. "It's one of the great features of the planet Earth but you can't see it."

He'd never visited the Garbage Patch himself, but he had received eyewitness reports from sailors. "They'd be sailing through there with their motors on—not sailing, motors on," he said. "No wind, glassy calm water, and they start spotting refrigerators and tires, and glass balls as far as you could see."

Anecdotal evidence suggested that similar atmospheric highs had created garbage patches in the five other subtropical gyres churning the world's oceans—including the North Atlantic Subtropical Gyre, which circumscribes the Sargasso Sea, so named because of the free-floating wilderness of sargasso seaweed that the currents have accumulated there. Later, skimming through Jules Verne's *Twenty Thousand Leagues under the Sea*, in a chapter about the Sargasso Sea, I'd come upon a help-

ful explanation for patches of garbage like the one at the heart of the
North Pacific Subtropical Gyre. "The only explanation which can be
given," Captain Nemo says of the seaweed engulfing the *Nautilus*,
"seems to me to result from the experience known to all the world. Place
in a vase some fragments of cork or other floating body, and give to the
water in the vase a circular movement, the scattered fragments will unite
in a group in the centre of the liquid surface, that is to say, in the part
least agitated. In the phenomenon we are considering, the Atlantic is
the vase, the Gulf Stream the circular current, and the Sargasso Sea the
central point at which the floating bodies unite." Nemo's explanation is
mostly accurate, with this one correction: the circular current is the
North Atlantic Subtropical Gyre, of which the Gulf Stream describes
only the north-by-northeasterly arc. *A Sargasso of the Imagination*, I
thought as I listened to Ebbesmeyer describe the Garbage Patch. The
phrase comes from a scene in *The Day of the Locust*, in which Nathanael
West is describing a Hollywood backlot jumbled with miscellaneous
properties and disassembled stage sets.[2]

There is no wilderness of seaweed at the center of the North Pacific
Subtropical Gyre, which circles around the deepest waters on the planet,
which are therefore among the least fertile. It is a kind of marine desert.
If you went fishing in the Great Pacific Garbage Patch, all you'd likely
catch aside from garbage is plankton, a class of creatures that includes
both flora (phytoplankton, tiny floating plants that photosynthesize
sunlight at the water's surface) and fauna (zooplankton, tiny floating animals that live off the tiny floating plants or each other). My *Ocean Almanac* calls phytoplankton "the pasture of the sea" because it is "the first
link in the sea's food chain." Ten thousand pounds of phytoplankton
will ultimately produce a single pound of tuna. The word *plankton* comes
from the Greek *planktos*, meaning "wander" or "drift," because that is
how they get about, going wherever the currents carry them. In going
adrift the castaway toys had become a species of giant ersatz plankton.
So, in a way, had I.

OSCURS's simulations predicted that relatively few of the bathtub
toys would have ended up in the Great Pacific Garbage Patch. The majority would have stayed well to the north, closer to the site of the spill,
caught in the North Pacific Subpolar Gyre, which travels counterclockwise beneath a low-pressure system between the coasts of Alaska and Si-

beria. Smaller and stormier than the North Pacific Subtropical Gyre, the Subpolar Gyre does not collect vast quantities of trash at its center. In counterclockwise gyres of the Northern Hemisphere, the currents don't spiral inward to create convergence zones. They spiral divergently outward, toward shore. Convergence zones tend to collect flotsam. Divergence zones tend to expel it. Floatees trapped in the Subpolar Gyre, Ebbesmeyer's research showed, would have remained in orbit, completing a lap around the Gulf of Alaska and the Bering Sea once every three years, until a winter storm blew them ashore or they strayed through the Aleutians onto one of the northerly currents flowing through the Bering Strait.

There, OSCURS lost them.

Ingraham had not programmed his model to simulate the Arctic. To follow the animals into the ice, Ebbesmeyer had to rely on more primitive oceanographic methods. He went to a toy store and purchased a few dozen brand-new Floatees to use as lab animals in various experiments. Several specimens he subjected to the frigid conditions inside his kitchen freezer in order to find out whether cold would make them crack (it didn't). Others he bludgeoned with a hammer to see what it would take to make them sink (a lot). Even breached and taking on water, they remained semibuoyant.

The toys, Ebbesmeyer concluded, could survive a voyage through the ice. Once beset, they would creep along at a rate of a mile or so per day. How long would it take them to reach the Atlantic? That was hard to say, impossible to say, perhaps. It depends which route they took. Data from other drift experiments, both intentional and disastrous, suggest that flotsam can cross the Arctic in three years, or in six years, or eight, or ten, or more.

Gazing into the indeterminate mists of his climatological crystal ball, Ebbesmeyer nevertheless hazarded an augury, one in which he'd had enough confidence, back in 2003, to put the beachcombers of New England on alert: seven or eight years after the day or night they fell overboard, five or six years after entering the ice pack, some of the toys would escape through Fram Strait and find themselves abob again, this time among icebergs and melting floes. In the North Atlantic some would catch an offshoot of the Gulf Stream and ride it to Northern Europe. Those that strayed west into the Labrador Current would begin the long, two-thousand-mile journey south toward Kennebunkport.

Before flying to Seattle, I'd made a day trip to Maine to visit that beachcombing couple who thought they'd seen a Floatee. Bethe Hagens and her boyfriend Waynn Welton—drawn to each other, perhaps, by the unusual spellings of their first names—had taken me to the southeast end of Gooch's Beach, the scene of their discovery. On that afternoon in 2003, the sun had been shining and the tide had been out. There had been sailboats on the blue water. The day I visited, by contrast, was damp and drizzly. The tide was at full flood. All that remained between the stone seawall and the surf was a narrow runner of sand. A lit Marlboro in one hand and his sandals in the other, rain beading on the lenses of his glasses and in his beard, Waynn Welton strode knee-deep into the sea and sloshed around until, phantasmal in the drizzle, he found what he'd gone wading for. "It was right there!" he hollered, a wave darkening the hem of his shorts. "Right around where that sippy cup is!" I followed his pointed finger. Where at low tide one sunny afternoon a duck—maybe yellow, maybe white—had perched atop the seaweed, a blue sippy cup with a pink lid now floated alone inside a corral of rocks. A wave came in and the cup rolled and bumped around.

I'd assumed that Ebbesmeyer shared my doubts about Hagens and Welton's alleged discovery. The question marks he'd printed on his map suggested as much. Now, however, he told me that there was "no question" in his mind that the couple had indeed seen one of the 7,200 ducks lost at sea. Hagens was a trained anthropologist, after all, with a Ph.D. from the University of Chicago. He drew a courtroom analogy: "It's sort of like, you have an accident at a stoplight. What did you see? Well, there are some good observers and there are some bad observers. Somebody who details what they saw precisely, the jury will listen to that." Besides, Hagens wasn't his only eyewitness. He'd received one other credible report, from "a lady in Scotland" who'd happened on a frog: "Again she didn't pick it up. I said, 'What did it look like?' She said, 'Well, it was kind of buried in the sand.' I sent her a picture, and she said, 'That's it.'"

Hagens had described her duck after hearing Ebbesmeyer describe one on the radio; instead of asking his witness in Scotland to pick her partially buried frog from a lineup of frogs, Ebbesmeyer had sent her a mug shot of the suspect. *Objection!* I thought, but didn't say. Who was I to accuse the learned oceanographer of leading a witness? Of confabulating facts? Of making believe? Besides, like him, like the subscribers to *Beachcomber's Alert!* and to newspapers and magazines the world over, like Eric Carle and his juvenile readers, like Bethe Hagens and Waynn Welton and Big Poppa, I wanted to believe.

Although his library of shoes may suggest otherwise, Ebbesmeyer has not amassed a museum of flotsam in his basement. He collects stories and data, not things. Fat three-ring binders occupy most of the shelf space. They contain "a small portion" of the studies he has conducted over the years. Reading the handwritten labels masking-taped to their spines, I wondered how many of these studies had scientific value and how many were merely the glorified puzzles with which the heavyset Dr. E. had desperately occupied his mind.

I saw binders labeled "Fishing Floats" and "Vikings," "Phytoplankton" and "Drifting Coffins," "Eddies" and "Icebergs." Beside a paperback copy of *The Egyptian Book of the Dead* was an entire binder devoted to Isis and Osiris, the star-crossed Egyptian gods. Ebbesmeyer told me the tragic ending of their tale: "Osiris's brother killed him, put his body in a coffin, put the coffin in the Nile River, and it washed up three hundred miles to the north of Lebanon. His wife, Isis, went to find it, and she did. That's the first documented drift of an object between point A and point B that I know of." It was self-evident to me why oil companies had commissioned Ebbesmeyer to study the eddies swirling unpredictably through the Gulf of Mexico, knocking oil rigs from their moorings, and I knew that peer-reviewed oceanographic journals had published his studies of flotsam. But Isis and Osiris seemed more like armchair archaeology than hard science.

Ebbesmeyer must have sensed my doubts, or else he'd heard them from other skeptics before. In the backyard, seated on the patio, where a string of rubber duckie Christmas lights festooned a grape arbor and wind chimes made mournful noises on the breeze, he waxed ecclesiastical. "There's nothing new around," he said. Take Osiris. Even today, when the Nile floods, flotsam follows that same route. Not even pollution is new. He told me to think of volcanic eruptions, of the tons of pumice and toxic ash an eruption throws into the sea. No, when you studied the history of flotsam long enough you realized that only one thing was fundamentally different about the ocean now, only one thing since the time of the ancient Egyptians had changed. He took a sip of coffee from his mug, which was decorated with a painting of a cat. "See, pumice will absorb water and sink," he said. "But 60 percent of plastic will float, and the 60 percent that does float will never sink because it doesn't absorb water; it fractures into ever smaller pieces. That's the difference. There are things afloat now that will never sink."

Ebbesmeyer went inside and returned a moment later carrying what at first glance appeared to be exotic produce—a new, flatter variety of plantain or summer squash, perhaps. He spread these yellowy lozenges out on the patio table. "Remnants of high-seas drift-net floats," he said. There were four of them, in varying stages of decay. The best-preserved specimens had the hard sheen of polished bone. The worst was pocked and textured like a desiccated sponge that had been attacked with a chisel. Ebbesmeyer picked the latter float up. "This is a pretty cool old one," he said. By "cool," he meant that it told the story of drift-net floats particularly well.

"High-seas drift nets were banned by the United Nations in 1992," his version of this story began. "They were nets with a mesh size of about four inches, but they were, like, fifty miles long. The Japanese would sit there and interweave these for fifty miles. There were something like a thousand drift nets being used every night in the 1980s, and if you do the math they were filtering all the water in the upper fifty feet every year. Well, they were catching all the large animals, and it clearly could not go on."

(I'd heard this part before. As a kid, I'd made my parents buy certified catch-free tuna after reading about how dolphins would get tangled in the nets and drown. The idea of dolphins drowning—dolphins, which spent most of their time swimming around with curvy little smiles on their faces—had made a big impression on me. But then lots of things made big impressions on me as a kid: bloody harp seal pups; the scene in *Close Encounters of the Third Kind* where something you can't see tugs the little boy out the doggy door; a picture in a book of a saber-toothed tiger thrashing in vain as it sank into the La Brea tar; the biography of the psychopathic Reverend Jim Jones that I for possibly worrisome reasons chose to write a book report about when I was twelve, reading therein that in his youth Jones had cut the leg off one chicken and stapled it onto another. My mind was like the moon, cratered with all the big impressions things had made.)

According to Ebbesmeyer, those high-seas drift nets had not gone away, and not only because pirate drift netting still takes place. Before the ban, fishermen had lost about half their nets every year, and because the nets are made of nylon, which can last at sea for as long as half a century, those lost nets were still out there, still fishing. "Ghost nets," they're called.

When he tells stories like these, Ebbesmeyer will sometimes pause dramatically, or whisper dramatically, or punctuate particularly astonishing facts with his eyebrows. He'll say something like, "What happens is, the nets keep catching animals, and then the animals die, and then after a while, the nets get old, and they roll up on a coral reef, and the waves roll it along"—he pauses, leans forward, continues in a stage whisper—"like a big avalanche ball, killing everything in its path." Then his bushy white eyebrows will spring up above his glasses and stay there while he looks at you, wide-eyed with autodumbfoundment.

And killer drift-net balls are genuinely dumbfounding, like something from a B horror movie—so dumbfounding that, smelling a hyperbole, I later checked Ebbesmeyer's facts. A ghost net may not kill everything that crosses its path, but it sure can kill a lot. News reports describe nets dripping with putrefying wildlife. Just three months before I showed up on Ebbesmeyer's doorstep, NOAA scientists scanning the ocean with a digital imaging system from the air had spotted a flock of a hundred or so ghost nets drifting through the North Pacific Garbage Patch. When they returned to fetch them, they found balls of net measuring thirty feet across. "There is a lot more trash out there than I expected," one of the researchers, James Churnside, told the Associated Press. A few years earlier, Coast Guard divers had spent a month picking 25.5 tons of netting and debris—including two four-thousand-pound, fifteen-mile-long high-seas drift nets—out of reefs around Lisianski Island in the Northwestern Hawaiian Archipelago. They estimated that there were six thousand more tons of netting and debris still tangled in the reefs when they left.

In Ebbesmeyer's opinion ghost nets may pose a still greater danger once they disintegrate. While we were conversing on his patio, he handed me the oldest of the drift-net floats. "Hold this a minute," he said. It weighed almost nothing. "Now put it down and look." On the palm of my hand, the float had left a sprinkling of yellow dust, plastic particles as small as pollen grains in which, Ebbesmeyer believed, the destiny of both the Floatees and of the ocean could be read.

Sitting on his patio, I mentioned to Ebbesmeyer my dream of following the trail of the toys from beginning to end.

"It's an expensive thing to do the kind of traveling you want," he said.

I told him I'd travel on a shoestring, roughing it, freeloading, hitch-hiking, crewing on boats, whatever. I was convinced it could be done. Perhaps a shipping line would let me earn my passage to China as a cabin boy. Perhaps a magazine would send me to the Arctic on assignment.

Still skeptical, Ebbesmeyer nevertheless gave me a lead. "Probably you'll want to go with Charlie out on his boat," he said.

Charlie was Charles Moore, captain of a fifty-one-foot catamaran, the oceanographic research vessel *Alguita*. In August of 1997, after com-peting in the Transpac, an annual Los Angeles–to–Hawaii sailboat race, Moore had for no particular reason motored north into the Sub-tropical Convergence Zone, known to sailors as the doldrums. In a 2003 article for *Natural History* magazine, Moore described what he discov-ered during his detour. Approximately eight hundred miles from Cali-fornia, the wind speed fell below ten knots, the water turned glassy calm, and drifts of garbage began to appear. "As I gazed from the deck at the surface of what ought to have been a pristine ocean," Moore wrote, "I was confronted, as far as the eye could see, with the sight of plastic. It seemed unbelievable, but I never found a clear spot. In the week it took to cross the subtropical high, no matter what time of day I looked, plas-tic debris was floating everywhere: bottles, bottle caps, wrappers, fragments."

A year later, Moore, a furniture repairman turned organic farmer turned charter boat captain turned self-trained oceanographer, turned himself into a plastic hunter, sailing back out to the Subtropical Conver-gence Zone, this time equipped with a trawl net and a volunteer crew. They began collecting water samples from the eastern edge of the Sub-tropical Gyre, trawling along a 564-mile loop encompassing exactly one million square miles of ocean. The larger items that Moore and his crew retrieved included polypropylene fishing nets, "a drum of hazardous chemicals," a volleyball "half-covered in barnacles," a cathode-ray television tube, and a gallon bleach bottle "that was so brittle it crumbled in our hands." Most of the debris that Moore found had already disintegrated. Every time he lowered his net he caught in its fine mesh "a rich broth of minute sea creatures mixed with hundreds of colored plastic fragments."

Moore didn't discover this "plastic-plankton soup," as he called it; since before Jules Verne invented Captain Nemo, oceanographers have known that convergence zones collect debris, and since the 1960s they've been worried about the persistence of "pelagic plastic," which they've found in all the oceans of the world, including the Arctic. What Moore did discover were greater quantities of pelagic plastic than anyone suspected were out there. In 2001, he published a paper about his research in a scientific journal called the *Marine Pollution Bulletin*. The undramatic title, "A Comparison of Plastic and Plankton in the North Pacific Central Gyre," belied its dramatic findings. The total dry weight of plastic Moore's samples contained—424 grams—was six times greater than the total dry weight of plankton and half again as much as any similar study had previously found. Moore and his coauthors proposed two hypotheses to explain these results: either the concentrations of plastic in this part of the ocean are aberrantly high, or else "the amount of plastic material in the ocean is increasing over time." Subsequent research has shown that both hypotheses are likely correct: the amount of plastic material in the ocean is increasing, in convergence zones especially.

Out on his front lawn, as I was leaving, I asked the heavyset Dr. E. what he thought of *Ten Little Rubber Ducks*. Despite the ominous future he'd augured in that handful of plastic dust, he thought Carle's cheerful picture book was "delightful," and he hoped that it would "make the ocean fun to kids." He did have one criticism. He couldn't figure out why Carle along with just about everyone else seemed compelled to turn the four Floatees into rubber ducks. Coverage of the story in newspapers and magazines almost always showed a picture of a solitary rubber duck, and usually not even the right kind of duck. What was wrong with the three other animals? "Maybe it's a kind of bigotry," Ebbesmeyer speculated. "Speciesism."

Ebbesmeyer loaned me a set of the toys that had survived his experiments, to be returned when I was done with them. I have been carrying them around with me ever since, and they are at present perched before me on my desk as I write. Monochromatic and polygonal in a Bauhaus sort of way, they bear little resemblance to the rubber ducks in Carle's book or, for that matter, to any other toy animal I've seen. Though blowmolded out of a rigid plastic (low-density polyethylene, I would eventually learn), they look whittled from wax by some tribal artisan.

The frog's four-fingered hands (the left smaller than the right) seem folded in prayer. The limbs of the turtle are triangular stubs, its shell a domed puzzle of hexagons and pentagons. The duck's head, too large for the flat-bottomed puck of a body it sits on, is imperfectly spherical, the flat plane of its beak continuing like a crew-cut mohawk over the top of the skull. Poke an axle through the duck's puffed cheeks and its head would make a good wheel. Wildly out of scale and dyed a lurid, maraschino red, the beaver seems altogether out of place in this menagerie, a mammalian interloper from somebody's acid trip. A seam left by the split mold bisects all four animals asymmetrically, and there's a little anal button of scarred plastic where the blow pin, that steel umbilicus, withdrew.

CUTE NEIGHBORS

"Why do precisely these objects we behold make a world?" Thoreau wonders in *Walden*. "Why has man just these species of animals for his neighbors; as if nothing but a mouse could have filled this crevice?" Since Thoreau's time ecologists have explained why that mouse filled that crevice, and since then Walden Woods has grown far less bewildering. For Thoreau the distinction between the natural world and the man-made one matters less than that between the subjective experience within and the objective world without. For him, both rocks and mice are objects that he perceives as shadows flickering on the walls of his mind. For him, anthropomorphism is inescapable. All animals, he writes, are "beasts of burden, in a sense, made to carry some portion of our thoughts."

The word *synthetic* in its current sense of "chemically unnatural" would not appear in print until 1874, twenty years after the publication of *Walden* and three years after the invention of celluloid, the first industrial synthetic polymer. In its 140-year history, the synthetic world has itself grown into a kind of wilderness. With the exceptions of our fellow human beings and our domestic pets, the objects that make the worlds we behold today are almost entirely man-made.

Consider the following: In nature, there are 142 known species of Anseriformes, the order to which ducks, swans, and geese belong. Of those species only one, the white Pekin duck, a domesticated breed of mallard, produces spotless yellow ducklings. Since the invention of plastic, four known species of Anseriformes have gone extinct; several others survive only in sanctuaries created to save them. Meanwhile, by the estimates of an American sociologist of Chinese descent named Charlotte Lee, who owns the largest duckie collection in the world, the makers of novelties and toys have concocted around ten thousand varieties of rubber duck, nearly all of which are yellow, and most of which are not made in fact from rubber, nor like the Floatees from polyethylene, but from plasticized polyvinyl chloride, a derivative of coal. Why has man just these species of things for his neighbors, a latter-day Thoreau might ask; as if nothing but a yellow duck could perch on the rim of a tub?

Let's draw a bath. Let's set a rubber duck afloat. Look at it wobbling there. What misanthrope, what damp, drizzly November of a sourpuss, upon beholding a rubber duck afloat, does not feel a Crayola ray of sunshine brightening his gloomy heart? Graphically, the rubber duck's closest relative is not a bird or a toy but the yellow happy face of Wal-Mart commercials. A rubber duck is in effect a happy face with a body and lips—which is what the beak of the rubber duck has become: great, lipsticky, bee-stung lips. Both the happy face and the rubber duck reduce facial expressions to a kind of pictogram. They are both emoticons. And they are, of course, the same color—the yellow of an egg yolk or the eye of a daisy, a shade darker than a yellow raincoat, a shade lighter than a taxicab.

Like the eyes of other prey (rabbits, for example, or deer) and unlike the eyes of a happy face, the rubber duck's eyes peer helplessly from the sides of its spherical head. Its movement is also expressive—joyously erratic, like that of a bouncing ball, or a dancing drunk. So long, that is, as it doesn't keel over and float around like a dead fish, as rubber ducks of recent manufacture are prone to do. It's arguable whether such tipsy ducks deserve to be called toys. They have retained the form and lost the function. Their value is wholly symbolic. They are not so much rubber ducks as plastic representations of rubber ducks. They are creatures of the lab, chimeras synthesized from whimsy and desire in the petri dish of commerce.

Apologists for plastics will on occasion blur the semantic lines between the antonyms "synthetic" and "natural." Everything is chemical, they rightly say, even water, even us, and plastic, like every living creature great and small, is carbon-based and therefore "organic." But to my mind the only meaningful difference between the synthetic and the natural is more philosophical than chemical. A loon can symbolize madness or mystery, and a waddling duck can make us laugh. But the duck and the loon exist outside the meanings with which we burden them. A loon is not really mad or, so far as it's concerned, mysterious. A duck is not really a clown; it waddles inelegantly because its body has evolved to dabble and dive and swim. A rubber duck, by comparison, is not burdened with thought. It *is* thought, the immaterial made material, a subjective object, a fantasy in 3-D.

INSIDE PASSAGE

One night, during the twenty-ninth week of her pregnancy, my wife and I attended a practicum in infant CPR. With the other expecting parents, we'd sat around a conference table set with babies—identical, life-size, polyethylene babies, lying there on the Formica like lobsters. The skin of these infantile mannequins was the color of graphite. Even their eyeballs were shiny and gray. Their mouths had been molded agape, so that they seemed to be gasping for air. To dislodge an imaginary choking hazard, you were supposed to lay the baby facedown over your left forearm and strike its back with the heel of your right hand. If you struck too hard, its hollow head would pop from its neck and go skittering across the linoleum. The morning after my visit with Ebbesmeyer, hurtling up the eastern shore of Puget Sound aboard the Amtrak Cascades bound for Bellingham, it occurs to me that "garbage patch" sounds like "cabbage patch," and for a moment I am picturing a thousand silvery, gape-mouthed heads bobbing on the open sea.

The old woman across the aisle, a retired high school chemistry teacher from Montana, tells me that she and her husband are traveling the globe. All they do is travel. She loves every minute of it. They have been to every continent but Antarctica. She teaches me how to say, "I don't have any money" in Norwegian. She tells me about the mural she saw in Belfast depicting a masked man and a Kalashnikov. She tells me about her grandson, who has in fact been to Antarctica. He spent a night dangling from the ice shelf in something like a hammock. *National Geographic* named him one of the top rock climbers in the world, she says. Then he died in an avalanche in Tibet. Left three little boys. She smiles as she says this. In the window behind her, the blue waters of Puget Sound flash through the green blur of trees.

A few seats away, facing me, riding backward, there's a young couple dressed in matching khaki shorts. She is holding an infant. He has a toddler in his lap. A bubble balloons from the toddler's right nostril and pops. She laughs deliriously and slaps the window, leaving snotty little handprints on the glass. "Choo-choo!" she exclaims. "Bye-bye! Woo-woo! I see cows!"

The train groans into a curve. Suddenly there are green and orange and blue containers stacked atop flatbed train cars parked on a neighboring track. The polyglot names of shipping companies speed by: Evergreen, Uniglory, Hanjin, Maersk. Then, at a clearing in the trees, the great brontosaural works of a gantry crane loom up above a Russian freighter loaded with what looks like modular housing. PORT OF SEATTLE, a sign on the crane reads.

We are somewhere east of the Strait of Juan de Fuca—Juan de Fuca, whom I read about in one of the many books I packed into my wheeling suitcase. He was a Greek sailor in the Spanish navy whose real name was Apostolos Valerianos. He claimed to have discovered the entrance to the Northwest Passage at the 48th parallel in 1592. The transit from the Pacific to the Atlantic had taken a mere twenty days, he reported, and the northern lands between these oceans were rich in silver and gold. Despite how familiar this tale must have sounded, for centuries people actually believed him. Although no one knows for certain whether the Greek sailor ever even visited the North Pacific, his de-

scription of the entrance to the passage, then known as the Strait of
Anian, bears a superficial resemblance to the entrance of Puget Sound,
and so the Strait of Juan de Fuca memorializes the pseudonymous per-
petrator of a hoax, and so even our most accurate maps are imaginary.
Looking out at the flashing waters of Puget Sound, I am filled with the
desire to sail out across them, through Juan de Fuca's fanciful strait,
down across the currents of the North Pacific Subtropical Gyre, into its
crowded, lovely heart, the heart of garbage. But I'm short on time. A
ferry ride to Sitka will have to suffice.

At the Bellingham ferry terminal, I find a café table overlooking the
harbor and spend the day reading about the science of hydrography and
the history of the North Pacific. Beside me a bronze seagull the size of
a condor points a wing at the sky, while beneath it a real gull hops
around eyeing my sandwich. Although we are scheduled to embark a
little before dusk, the M/V *Malaspina* is already waiting at its dock.
Viewed from shore, it is a splendid sight, its white decks gleaming, a yellow stripe running the length of its navy-blue hull, its single smokestack
painted in the motif of the Alaskan state flag—gold stars of the big dip-
per against a navy-blue sky. All the motor vessels in the Alaska Marine
Highway system are named for Alaskan glaciers, and the *Malaspina* is
named for the largest, a 1,500-square-mile slow-moving mesa of ice,
which is in turn named for an eighteenth-century Italian navigator,
Alessandro Malaspina, whose search for the Northwest Passage ended
in 1791 at the 60th parallel, in an icy inlet that he christened *bahía del de-
sengaño*, Disappointment Bay.

When I wheel my suitcase down the gangway that evening, the splendor of the M/V *Malaspina* diminishes with every step. The ferry is, I see
upon boarding it, an aging, rust-stained hulk, repainted many times.
Posted in a display case of documents near the cocktail lounge one can read
a disconcerting open letter in which "past and present crew members . . .
bid farewell to this proud ship." Queen of the fleet when it was first
launched in 1962, the *Malaspina*, the letter explains, "will cease scheduled
runs of Alaska's Inside Passage on October 27, 1997." Why the old ferry is
still in service eight years later the documents in the display case do not say.

The solarium on the sundeck, where I will sleep for free in a plastic
chaise longue, is a kind of semitranslucent cave, glazed in panes of
scratched Plexiglas that admit a yellowy view. Electric heaters hang

from the solarium's ceiling, their elements already glowing orange, like those inside of toasters. By the time I drag a plastic chaise to a spot near the back and lock my suitcase to a rail, I am dripping in sweat.

Outside the solarium, in the open air, backpackers are pitching their tents, duct-taping them down so that the wind won't toss them overboard. Soon a rustling nylon village of colorful domes has sprung up. "Tent city," the veteran ferry riders call it. The evening is cool and exhilarating, the sky clear save for a distant, flat-bottomed macaroon of a cloud from which a tendril of vapor rises and coils. The wavelets on Bellingham Bay are intricate as houndstooth, complicated by crossbreezes and by ripples radiating from the hulls of anchored boats.

At last the ferry's diesel engines rumble to life. I am going to sea! Who can resist an embarkation? The thrill of watery beginnings? The dock falls away. On the forested hills of Bellingham, the houses face the harbor. How festive the ferry must look from up there! As the ship turns and slithers toward the horizon, the low sun moves across the windows of the town, igniting them one by one. I stand at the taffrail and think to myself *taffrail*, enjoying the reunion of a thing and its word.

There is something quaintly democratic about the Alaska Marine Highway. Cheap and utilitarian, it exudes a faith in government that, like the *Malaspina* itself, was supposed to have been decommissioned years ago, around the same time that Amtrak went quasi-private and the U.S. Postal Service became a trademarked brand and PBS started licensing Big Bird to toymakers. The ferries have survived, I suspect, because there is no money to be made from what is essentially a maritime municipal bus system connecting the isolated fishing villages and resort towns that dot the islands of the Alaska Panhandle. For most of those towns, the Inside Passage is the only thoroughfare that leads to the outside world, and there are many locals as well as many tourists aboard the *Malaspina* today. The Marine Highway's misleading name suggests what its creators had in mind—a public works project in the spirit of Eisenhower's interstates. And yet travel by ferry no more resembles the solitary confinement of the automobile than these coastal waters resemble a four-lane road.

For one thing, travel by ferry is slow going; the cruising speed of the

Malaspina is sixteen knots, or approximately twenty miles per hour. You can if you are so inclined, as I am, draw pictures of a mountain or an island before it disappears from view, and after several hours, the drone of the engines and the sameness of the scenery induce either boredom or peace of mind, or possibly both. It will take us three days and three nights to travel the 952 nautical miles between Bellingham and Sitka, a distance planes fly in less than two hours.

And then, for another thing, life aboard the public ferry is inescapably communal. It's true that the passengers in the tiny staterooms belowdecks have purchased some privacy, but since staterooms are cheap and the most desirable spaces on the ship are public, there's little sense of economic segregation. We squatters on the sundeck prefer the open air. Everyone eats in the same cafeteria, where the plebeian menu of grayish Salisbury steak and scrambled eggs matches the Cold War–era decor—seats upholstered in vinyl, tables enameled in sparkly Formica. In an America increasingly devoted to the service economy, even the relationship between the paying passengers and the crew feels atypically egalitarian. The crew keeps the *Malaspina* shipshape, but it does not serve. There are no chambermaids or stewards aboard.

If anything divides the ferry's passengers, it's age. Tent city has the feel of a floating youth hostel, or even a floating campground—hence the stenciled sign that prohibits campfires and cookouts (but not, alas, folksy sing-alongs). The retirees tend to congregate on the boat deck in the observation lounge, a sightseeing theater overlooking the bow. Sitting there in the anchored, amply cushioned chairs, it's hard not to feel as though the wraparound windows are movie screens on which footage of the passing scenery plays, though every now and then a passenger outside will walk through the foreground and break the spell. Often as not the passenger in the foreground is a dude in orange-tinted sunglasses and a cowboy hat who seems to be intent on walking a marathon before we make Sitka.

The main source of onboard entertainment is Ed White, a bespectacled "interpreter" employed by the National Forest Service. "Interpreter" is what the Forest Service calls a ranger who is also a tour guide, and I love what the title implies: that a place is like a language. In this case, though, I can't help feeling that something has been lost in translation. Once or twice a day, in the observation lounge, White delivers presentations on topics such as commercial fishing and local wildlife. He informs his audience, for instance, that there are 300,000 hairs on every square inch of a sea otter's pelt. Then he puts his charts and markers away, and the spectators go back to looking at the mountains and the trees, or reading their Carl Hiaasen novels.

For the kiddies, there are daily screenings of family-friendly films on the senescent television in the recliner lounge. My favorite is *Alaska's Coolest Animals*, which features video footage of Alaska's "coolest flyers," "coolest walkers," and "coolest swimmers," accompanied by the voices of children reading from a script. All their lines seem to end in exclamation points. "If a moose doesn't get enough food, it might get starved and covered in snow!" one child says. "Hey, that's a big bear!" says another. Sometimes the narrators use the first person. "I'm sleepy," one says when a bear puts its head on its paws. "I want to go to bed. Bed!"

During the middle of the first night, off the eastern shore of Vancouver Island, the temperature drops, a fog shuts down, and my cell phone loses reception. So much for daily phone calls home. A plastic deck chair, it turns out, makes for a miserable mattress. Cold air seeps between the slats. The government-issue cotton blanket I rented from the ship's purser for a dollar is far too thin. Some of my neighbors in the solarium move inside to sleep like refugees on the carpeted floor of the recliner lounge. I rent a second blanket for the second night, but it hardly makes a difference. The space heaters, too, have little sensible effect. Shivering in a fetal position, I think about that rock climber dangling from the Antarctic ice shelf in a hammock and feel faintly ridiculous. After two nights in the solarium of a cruise ship—a state-operated, poor man's cruise ship, but a cruise ship nonetheless—I have already had my fill of adventuring.

The Alaskan stretch of the Inside Passage snakes through the Alexander Archipelago, a chain of one thousand or so thickly forested islands, some small as tablecloths, some large as Hawaii. These are, in fact, the tops of underwater mountains, part of the same snowcapped range visible on the mainland to the east. Most rise steeply from the water and soar to cloudy heights. Before going there I expected Southeast Alaska to feel like a giant outdoor theme park—Frontierland—and the shopping districts of the resort towns where the gargantuan cruise ships dock confirm my worst expectations. Cruise ship companies now own many of the businesses in those districts and may soon be able to "imagineer" (as the folks at Disney call it) every aspect of your vacation experience. But the backwaters of the Inside Passage, too narrow and shallow for the superliners to enter, are something else. They contain lost worlds.

In the narrowest of the narrows, it feels as though we are motoring down an inland river—some Amazon of the north—rather than along the ocean's edge. Although this is the Pacific, the water doesn't look, smell, or sound like the sea. Neither waves nor flotsam gets past the outer islands to the placid interior. In the summer, the rain and the streams of glacial melt together make the channels so brackish they seem fresh enough to drink, and in places the minerals those streams carry turn the channels a strangely luminous shade of jade. The forested banks sometimes loom so close you could play Frisbee with a person standing on shore. Hours go by when we see no other ships, or any sign of civilization besides the buoys that mark the way among the shoals.

Early in the morning, fog rises here and there from the forests of hemlock, cedar, and spruce. It is as if certain stands are burning, except that the fog moves much more slowly than smoke. In some places it forms tall, ghostly figures, and in others, it spreads out horizontally like wings. On the far side of one mountain, a dense white column billows forth like a slow-motion geyser that levels off into an airborne river that flows into a sea of clouds. I've begun to notice currents everywhere, a universe of eddies and gyres. Phytoplankton ride the same ocean currents that carried the Floatees to Sitka. Zooplankton follow the phytoplankton. Fish follow the zooplankton. Sea lions, whales, and people follow the fish. When, at the end of their upriver journey, salmon spawn and die en masse, their carcasses—distributed by bears, eagles, and other scavengers—fertilize the forests that make the fog, which falls as rain, which changes the ocean's salinity. All deep water travels along what oceanographers, when speaking to laymen, call the "conveyor belt," which begins in the North Atlantic, where surface currents, warmed by the tropical sun, made salty by evaporation, carried north by the Gulf Stream and the North Atlantic Drift, upon arriving in these cold latitudes, chill enough to sink below the comparatively fresh water spilling from the Arctic.

There, off both the east and west coasts of Greenland, the sunken Atlantic water creeps slowly south, through the abyss, beneath the Gulf Stream, over the equator, into the Antarctic Circumpolar current, which carries it to the South Pacific, where it begins creeping north. After a thousand years—a millennium!—some of that deep water conveyed by the conveyor ends here, in the North Pacific, where the ancient, life-giving element wells up, carrying nutrients with it, nutrients that fertilize the Alaskan fisheries. Much of what oceanographers know about where deep water goes they learned studying radioactive isotopes released into the sea as fallout from nuclear tests. I'm becoming a devout driftologist. The only essential difference between rock, water, air, life, galaxies, economies, civilizations, plastics—I decide, standing on the *Malaspina*'s deck, totally sober, watching the fog make pretty shapes above the trees—is the rate of flow.

We round a bend, and an inlet comes into view. Protruding from the forested banks beside a waterfall is an abandoned sawmill. The windows are shattered and one corner of the building has collapsed. Any second the whole structure might tumble into the sea. The trees on these islands are part of the seventeen-million-acre Tongass National Forest. Fifty years ago, the timber industry was booming here, but in the last twenty years nearly all the sawmills and pulp mills have shut down.

There are ghost towns and ruins all over the islands of the Inside Passage, vestiges of its long history of extractive industries gone bust. In the 1800s the Russian fur trade made Sitka, then the capital of Russian Alaska, the largest city on the entire west coast until San Francisco eclipsed it in 1849—the Paris of the Pacific, some hyperbolist dubbed it. Then sea otters and fur seals grew unprofitably scarce. In the 1870s, after the Russians had sold off their exhausted North American hunting grounds, the world acquired a taste for canned Alaskan salmon, which in the age of refrigeration it has largely lost, though there are a few canneries left.

Today the only thriving industry here besides fishing is tourism. In towns that the cruise ship lines have not yet tarted up, you can sense what the local economy would be like if the cruise ships left. By the docks at the gold rush town of Wrangell, for instance, one wall of a wooden fishing shack has been shingled with old drift-net floats, and three school-age girls sit at card tables in a parking lot selling garnets chiseled from nearby Garnet Ledge. They appear to be competitors, not friends, each sitting stoically behind her outspread wares, the prices handwritten on tags of masking tape. They go to sleep, perhaps, reading Milton Friedman's *Free to Choose* under their covers with flashlights. I buy a five-dollar rock from a girl named Tiffany, who has punctuated her xeroxed sales brochure with smiley faces and illustrated it with a hand-drawn geological diagram of strata, beginning at ground level (there's a little house) and descending sixty feet to a lopsided circle labeled "Earths core." Dots at ten feet represent garnets. Tiffany has also drawn a maze, at the entrance to which stand two stick figures with lanterns on their heads, rays of light emanating like tentacles. "Help the Miners find Their way to Garnet Ledge," her instructions read.

Fifteen minutes from the ferry dock are Wrangell's most famous treasures, a collection of petroglyphs depicting birds, fish, and sea mammals carved in the same geometric style as the Tlingit and Haida totem poles, thickets of which can be found all along this coast. The petroglyphs are only visible at low tides and in certain casts of light. I want badly to go see them, but the *Malaspina* is making a brief stop. Late passengers will be left behind.

Everywhere they look, archaeologists find them—buffalo sprayed with pigments onto the walls of caves, killer whales cut from cedar or stone, horses molded from gutta-percha or plaited out of straw. Our primal fear of predators and our hunger for prey cannot alone account for this menagerie. Three thousand years ago in Persia, someone carved a porcupine out of limestone and attached it to a little chassis on wheels. Four thousand years ago in Egypt, someone sculpted a mouse and glazed it blue. Why blue? Who ever heard of a blue mouse? Is this the forebear of the red beaver and the yellow duck?

In fact, many of the figurines that look to us like toys turn out to have been totemic gods or demigods, used in religious ceremonies or funerary rites. To make the archaeological record all the blurrier, some totems in some cultures were given to children as playthings once the festivities had ended. One thing in the archaeological record is clear:

animals held an exalted position in the lives of both children and adults. Even after the missionaries came and cleansed them from the temples, the animistic gods survived, adapting to the altered cultural landscape. In Europe of the Middle Ages, the most popular book after the Bible was the bestiary, a kind of illustrated field guide to the medieval imagination, wherein the animals of fable and myth were reborn as vehicles of Christian allegory. From the bestiary came the idea that after three days a pelican could resurrect a dead hatchling with her blood, and from the bestiary we learned that only a virgin girl can tame a unicorn. Even Aesop, that pagan, remained a favorite with old and young alike well into the seventeenth century.

Gradually, as allegory gave way to zoology, we decided that animals—as the children in the old Trix commercial inform the envious, sugar-addicted rabbit—were for kids. "Children in the industrialised world are surrounded by animal imagery," John Berger notes in his essay "Why Look at Animals?" Despite the antiquity of zoomorphic toys and the "apparently spontaneous interest that children have in animals," it was not until the nineteenth century that "reproductions of animals became a regular part of the decor of middle class childhoods—and then, in [the twentieth] century, with the advent of vast display and selling systems like Disney's—of all childhoods." Berger traces this phenomenon to the marginalization of animals, which the age of industrialism incarcerated as living spectacles at the public zoo, treated as raw material to be exploited, processed as commodities on factory farms, or domesticated as family pets. Meanwhile, "animals of the mind"—which since the dawn of human consciousness had been central to our cosmologies—were sent without supper to the nursery. Animals both living and imaginary no longer seemed like mysterious gods. They seemed like toys.

Go bird-watching in the preindustrial libraries of literature and myth and you will find few ducks, which is puzzling, considering how popular with the authors of children's books ducks have since become. Search, for instance, the fields and forests of Aesop, whose talking beasts are the ancestors of both Chanticleer the Rooster and Walter the Farting Dog, and you will meet ten cocks, a cote of doves, several partridges, a caged songbird, ten crows, two ravens (one portentous, the other self-loathing), a dozen or so eagles, four jackdaws (one of whom wishes he were an eagle), many kites, flocks of cranes, two storks, three hawks, a cote of pigeons, three hens, a sparrow with a bad case of schadenfreude, four swallows, many peacocks, a jay who wishes he were a peacock, many swans, two nightingales, two larks, an owl, a gluttonous seagull, a thrush ensnared in birdlime, and nary a single duck.

Aesop's fables exhibit considerable ornithological knowledge, but their primary aim is to transmute animal behavior into human meaning—to burden them, as Thoreau would say, with some portion of our thought. The closest thing to a duck in Aesop's fables is the famous goose, the one who lays the golden egg and then succumbs to the carving knife. In a Kashmiri version of the same tale, Aesop's barnyard-variety waterfowl becomes the Lucky Bird Huma, a visitor from the magical avian kingdom of Koh-i-Qaf. A Buddhist version of the tale replaces the egg-laying goose with one of the only mythical ducks I have found, a mallard plumed in gold, which turns out to be a reincarnation of the Bodhisattva. (The birds of myth, as Leda learned, are often divinity in disguise.)

In all three versions of the fable, the human beneficiaries sacrifice their magically profitable waterfowl on the altar of their greed. The farmer kills the goose, cuts it open, and finds no eggs. Dreaming of rupees, a Kashmiri woodcutter accidentally asphyxiates the Lucky Bird Huma while carrying him to market in a sack. A family of Brahmin women decide to pluck out all of the Bodhisattva's golden feathers at once; they turn into the worthless feathers of a crane. Unlike the others, the Buddhist version tells the fable from the bird's point of view, and for that reason it is peculiarly affecting. Both Aesop's fable and the Kashmiri one show us the folly of human desire, and it is satisfying, reading them, to watch our wicked, bumbling protagonists endure dramatically ironic reversals of fortune. The Buddhist fable shows us the folly of human desire, but it also makes us experience that folly's cost, the debt of suffering our appetites can incur. The tone of the final sentences is more sorrowful than ironic. Trying to escape, the once golden mallard stretches his plucked wings but, featherless, finds he cannot fly. His captors throw him into a barrel. With time, his feathers grow back, but they are plain white ones now. He flies home, never to return.

Not all the passengers aboard the *Malaspina* are transported, or even entertained, by the Alaskan scenery. There is, for instance, a teenager vacationing with his parents—vacationing, I'm inclined to guess, against his wishes. He is almost always alone, wandering or sitting around, lost in adolescent thought. He wears the same outfit every day, jeans and a black T-shirt on which appears the cryptic, presumably ironical phrase IRON CHEFS ARISE.

When the rain lets up, which it rarely does, he adjourns to the outer deck to practice martial arts. On the last night of my passage to Sitka, I watch him one deck below striking poses of graceful ferocity in the shadowy deck light. He kicks his leg out high and holds it there an impressively long time. He stiff-arms an imaginary foe. The clouds have blotted out the stars and moon. No lamps burn onshore. Beyond the *Malaspina*'s rails, the only light is the shine the ferry casts on the black water—that and the green and red twinkling of the buoys.

More than the sublime scenery we've passed through, more than the charismatic megafauna we've seen both on-screen in the recliner lounge and live, this is the scene I will remember best from my ferry ride, I feel certain—this karate kid with his black ponytail and his ironical shirt, out there shadow dancing in the deck light as we thread our way brightly and noisily among green and red beacons, past quiet islands we sense but cannot see.

BEACHCOMBING THE PACIFIC

On the morning I disembark, Tyler and Dean Orbison are just returning from a two-week, three-hundred-mile beachcombing expedition to Lituya Bay and back. They go on such expeditions every summer, traveling farther and farther afield every year, poking around in bunkers abandoned at the end of World War II, walking beaches where the only footprints in the sand are animal tracks. They have a cabin cruiser big enough to sleep in and an aluminum skiff for going ashore. From the cruiser, they look for V-shaped coastlines that funnel the tides, and they look for "jackstraw"—driftwood logs jumbled like a pile of pick-up sticks—and, most important of all, like prospectors panning in the tailings, they look for "good color," their term for plastic debris visible from afar. Where there's some color, there's sure to be more. Their style of beachcombing is by necessity a tag-team affair. One person has to stay in the skiff to keep it from foundering on the rocks while the other person wades in and combs. They take turns. Dean prefers to hunt high up, in the purple fireweed, where storms will throw objects out of the reach of tides. Tyler, Dean's son, is "a digger." Like a human metal detector, he's learned to divine the location of buried objects by reading the terrain.

This year for the first time Tyler and Dean started combing in seaside caves where tangled driftwood will form a kind of flotsam trap. It's dark in the caves. You have to beachcomb with a flashlight. It's also cold, but the labor of log-lifting keeps you warm. The effort's worth it. Every cave the Orbisons search contains a farrago of wrack—a Dawn dish detergent bottle, glass fishing floats, Floatees. Half a water pistol turned up in one cave, the other half in another. By far the most common objects the Orbisons find are polyethylene water bottles. They have begun keeping the screw tops, cataloging the varieties. On this last trip they identified seventy-five brands, many of them foreign in origin. Up in Lituya Bay they saw a live black wolf and the bones of a whale, and they picked wild strawberries, and when their cooler ran out of ice they floated alongside a glacier and broke off a chunk.

Now, at the end of my first day ashore, they've fetched me from my hotel. "Growing up here, I mean, there's nothing," Tyler tells me from the backseat of his father's truck while we're waiting for his parents to emerge from Sitka's only supermarket. "I mean we don't even have a mall. So I took to the outdoors pretty hard."

It is clear that Tyler has never given much thought to the marginalization of animals. You wouldn't either if you'd grown up in Southeast Alaska, where bears make off with household pets, and ravens alighting on transformers cause power outages, and bald eagles sometimes come crashing through dining room windows. If anything, it's the people who occupy the margins here. Just look at a map: Sitka perches on the coastal brink of Baranof Island, wedged between mountainous wilderness to the east and watery wilderness to the west. Sitkans share their island with an estimated 1,200 grizzly bears—more than are found in all the lower forty-eight states combined. In May and June, eagles and ravens—the supreme deities in the pantheon of the native Tlingit—wheel overhead. In July and August, the creeks grow dark with spawning sockeye and chum. In November, the whales and the whale watchers arrive. People like me may feel sorry for the 1.2 million sea otters that the Russian American Company parted from their pelts in the early 1800s, but since the Endangered Species Act protected them in the early seventies, otters have repopulated Sitka Sound with such procreative gusto that local fishermen now regard them as pests—crop-thieving, net-wrecking vermin of the sea.

Tyler's arms and face are partly tanned, partly sunburned, and like the karate kid of the M/V *Malaspina*, he is wearing jeans and a black T-shirt, only his T-shirt is not in the least cryptic or ironic. Decorated with bicycles and boats, it commemorates Sitka's annual triathlon. Twenty-three years old, he is currently earning his teaching credentials at the University of Alaska in Fairbanks, a city of fifty thousand. More than five times the size of Sitka, Fairbanks is too urban and too populous for Tyler's taste. "I don't do cities very well," he says. As soon as classes let out, he hurries home. This summer, he's been working part-time on commercial fishing boats and part-time teaching a summer school class on mammals to middle schoolers. With what time remains, he goes hunting and beachcombing with his father, whom he calls by his first name.

Dean used to be an engineer at the pulp mill. After it shut down, he started working for the power company, running the hydroelectric plants that electrify all Sitka. He's semiretired now, which means he has plenty of time to play outside with his son. They are, by all appearances, best friends. They are also two of Curtis Ebbesmeyer's most devoted disciples.

When Tyler's parents finish their grocery shopping, we drive off to have a look at the salvaged flotsam piled on their front porch. Tyler rummages around, pulling out item after item and keeping up a running curatorial commentary. There are plastic buoys; a tightly sealed tin can of air; something that looks like the lid of a blender; a plastic housing with Russian characters on it. About one object on their porch Tyler and his father disagree. Tyler thinks it's a boat muffler. Dean thinks it's the cover off an underwater cable. "We're not sure what that is," Tyler concludes, "but it's pretty skookum." The haul also includes a message in a plastic bottle.

Even the most skeptical of travelers, upon discovering a message in a bottle, must experience a frisson of wonder. Fishing the scrolled parchment out, you can't help but hope that the words scrawled across it will spell something disastrous or mysterious, like, "Of my country and of my family I have little to say. Ill usage and length of years have driven me from the one, and estranged me from the other" (the opening sentences of Poe's "Ms. Found in a Bottle"[3]).

These days messages in bottles seem to be sent mainly by schoolchildren or by drunken tourists, which would explain their characteristic style—more like seaborne graffiti than poetry. Among drunken tourists, the shopworn SOS-from-a-deserted-island-or-sinking-ship conceit is a perennial favorite. Schoolchildren tend to be more matter-of-fact. In 2003 an octogenarian beachcomber in Australia found a bottle. "My name is Harmony," the message inside read. "It's my birthday. I am nine years old today. If anyone finds this, please ring me." From a bottle they dug up on a New Zealand beach in 2002, two boys withdrew the following memorial, dated December 2, 1912: "This note is to commemorate the enjoyable experience by two Nelson College boys, and also in memorance of our notable land-mark." When his ship, the USS *Beatty*, was torpedoed off Gibraltar in 1943, all a sailor—who actually was gazing into the mortal abyss—could think to say was, "Our ship is hit and sinking. Maybe this message will reach the U.S. someday." In a sense, mortality is the theme of all messages in bottles, which are addressed to distant times as well as distant shores. They are little time capsules, escape pods, sea beans of memory, loosed on the waters of oblivion.

Like most beachcombers of the Pacific Rim, the Orbisons started out collecting Japanese fishing floats, the glass balls that you sometimes see hanging in nets from the ceilings of seafood restaurants, or decorating the window displays of maritime boutiques. The popularity of glass floats owes partly to their delicate, soap-bubble beauty, partly to the Kuroshio Current that sweeps them across the Pacific and bowls them up the beaches of the American West Coast, and partly to Amos L. Wood, an aeronautics engineer and beachcombing enthusiast whose books *Beachcombing for Japanese Floats* and *Beachcombing the Pacific* have become to beachcombers what Audubon guides are to bird-watchers.

A century and a half ago, beachcombers tended to be transcendental weirdos like Ellery Channing and Henry Thoreau. Back then, much of New England's shoreline was as wild as Alaska's is today and more treacherous to passing ships. Just before Thoreau arrived at Provincetown in 1849, a ship carrying Irish immigrants sank off Cohasset. The bodies of the drowned lay strewn along the beach, torn asunder by the surf and fish. "The Gulf Stream may return some to their native shores," Thoreau later wrote, "or drop them in some out of the way cave of Ocean, where time and the elements will write new riddles with their bones." Even where no shipwrecks had occurred, a Cape Cod beach in 1849 was "a wild rank place" littered "with crabs, horse-shoes and razor clams, and whatever the sea casts up—a vast *morgue*, where famished dogs may range in packs, and crows come daily to glean the pittance which the tide leaves them."

Still a recent coinage, the word *beachcomber* in 1849 meant approximately what we mean by "beach bum": it evoked a character like the narrator of Melville's *Omoo*, a transient ne'er-do-well who had fled from civilization hoping to sample tropical women and tropical fruits and loaf around beneath the blowsy palms. "Idle, drunken, vagabond," one Australian author wrote in 1845, "he wanders about without any fixed object, cannot get employed by a whaler or anyone else, as it is out of his power to do a day's work; and he is universally known as 'the beach-comber.'" The local Cape Codders whom Thoreau met on his seaside rambles usually took him for a traveling salesman. What other explanation could there be for a vagabond with a walking stick and a knapsack full of books?

By the 1980s, when Amos Wood published his how-to manuals, American beaches had become seaside playgrounds frequented not by dogs or crows but by the sun-worshipping masses. As for our vagrant beachcomber, he had become, in Wood's definition, "any person who derives pleasure, recreation, or livelihood by searching ocean, lake, and river shores for useful or artful objects." Inclusive as that definition sounds, Wood's books are intended not for "the casual visitor to the beach" strolling barefoot along the shore plucking up pretty pebbles and seashells as keepsakes, but for "the serious beachcomber," a mercenary, methodical prospector of the sands. The casual visitor to the beach "flits about" without any "specific purpose," Wood writes, whereas a "serious searcher plans his hike, selects the tide and wind conditions that are favorable, prepares for an extended trip, and has a particular objective in mind."

Thoreau's rambling style of beachcombing—*extravagant sauntering*

he would call it—appeals to me far more than Wood's forensic treasure hunting does. If I tried to follow Wood's advice, I wouldn't last a
weekend before retiring my metal detector to that cabinet of fleeting en-
thusiasms which also contains various musical instruments, a teach-
yourself-Russian CD-ROM, and a guide to bicycle repair.

The Orbisons have read Wood's books and have followed some of his
advice, though much of what he says about beachcombing in California
and Washington does not pertain to the shores of Alaska, where glass
fishing floats tend to shatter on the rocks. Although they probably know
more about beachcombing in Alaska than anyone, I doubt that either
Tyler or Dean would consider himself a serious searcher. They are outdoorsmen. Their beachcombing grew out of their hunting. They hunt
animals and in the intervals between hunting seasons they hunt shipwrecked junk, and the way they talk about spotting a plastic duck and
killing a bear makes it seem like there's no great difference between the
two. "When I was younger, Dean would find a glass ball before me,"
Tyler says, "and I would get so mad. And when you're mad you walk
right over them. To me it's a lot like hunting: You'll do a lot better if
you're chilling and hanging out."

In the Orbisons' living room are stacks of pelts, a glass cabinet containing, as of that summer, 120 Floatees (the largest collection Curtis
Ebbesmeyer knows of), another containing Japanese fishing floats, and
another of skeletons and skulls. I ask Tyler to identify the skulls on one
shelf. "Bear, bear, gorilla, bear," he says, his finger bouncing from skull
to skull. "The gorilla we bought, but the bears, we shot those. That's an
elk. Birds, all sorts of birds down there. We do a lot of trapping too, so
we have mink, marten, rabbit, otter, fox."[4]

In the summer of 1993, when they began discovering bath toys, the
Orbisons gave up collecting for driftology. Tyler was just twelve at the
time, but he was the one to find their first toy, one of the beavers, and he
remembers the moment vividly. "We were on Kruzof Island, looking for
glass balls," he says. "We didn't really know what else to look for. It was
beautiful weather; the reason we went to Kruzof is because it's really
hard to get ashore, and that's where we go when the weather's good. We
were up beyond the high tide line. It was in the drift a ways. It had been
there for a little bit. And I thought, This is cool. It was bleached out, ex-

actly like the beavers we find now. I would say it had been there since the winter storms."

They assumed that their beaver was a solitary castaway, but when they arrived back in town, talk of the mysterious invasion was in the air. Dean and Tyler went looking for more plastic animals, and found them. They started keeping meticulous records, treating the Floatees as data, which they eventually reported to Ebbesmeyer. A year or two later, the oceanographer began publishing *Beachcombers' Alert!* and the Orbisons were among his first subscribers. They own every issue. "Curt tells us what to look for, and we go out there and find it," Dean explains. This year, at Ebbesmeyer's bidding, they searched for and found a computer monitor, Japanese surveying stakes, hockey gloves, "antisandals" (a sheet of rubber from which flip-flop–shaped blanks have been stamped), part of a naval sonobuoy, and six new Floatees, including a turtle they chiseled from the ice. After cataloging the junk on their porch and showing it to Ebbesmeyer, they take most of it to the dump.

QUACK, QUACK

We make plans to go beachcombing together the next day so that Dean and Tyler can show me how it's done, and I return to my hotel day dreaming of an extravagant saunter along some wild shore, but that night, I come down with a nasty fever that I must have caught aboard the *Malaspina*. I spend the weekend alone, shivering and sweating in the smallest, cheapest hotel room in Sitka. On the walls are framed paintings of snowy New England farms. The window offers a view of a neighboring rooftop where two beer cans lie crumpled in a puddle and, beyond, appearing and disappearing from within a veil of fog, of the green slopes of Mount Verstovia. It rains. I sleep. Hours pass. I hear voices from the bar below. The window curtains brighten and dim. My wife calls the hotel—my cell phone is still out of range—to tell me what she learned at this week's obstetric exam. Her cervix is "partially effaced," and the baby's head is low. He or she is ready to be born. Does this mean I should fly home? There's no telling, no sure sign; the due date is as unpredictable as ever, and I have yet to set foot on an Alaskan beach, yet to go hunting for flotsam on the sand. I am of two minds, and of two hearts, and feverish—it feels—with ambivalence.

When my fever breaks for good, I don my yellow slicker and my quick-dry nylon pants ("Adventure" pants, the label said) and flee into the long Alaskan dusk. After two days indoors, the rainy air feels blessedly cool and smells both of the ocean and of the green mountains soaring up from behind the onion minaret and steeple of St. Michael's, Sitka's historic Russian Orthodox cathedral. Made of wood, painted gray and white, St. Michael's is pretty but looks to me more like an ornate barn than a cathedral. Nonetheless, a cathedral it is. I slurp down some udon at a sushi place and saunter extravagantly around town.

Although the bars are open, and loud, the streets of Sitka are strangely quiet, and I realize why: almost no one's driving. Tourists like me are mostly on foot and few locals appear to be out and about. In the darkened windows of the gift shops mannequins wear G-strings upholstered in fur from which plastic eyeballs gaze. BALEEN FOR SALE, reads a hand-lettered sign in the window of a Native arts cooperative. Fronds of the hairy stuff are on display, as are instructions telling you what to do "if you come across a dead stranded marine mammal." Down at Crescent Harbor, gray, molten slivers of Sitka Sound shine like tines between the dark spars of outrigging and masts.

To the northwest, neighboring Kruzof Island shelters the harbor from winter storms and heavy surf, but to the southeast the sound opens wide onto the Pacific, which explains why this spot has become a mecca for beachcombers. The only major town in Southeast Alaska situated on the windward side of the Alexander Archipelago, Sitka lies in the path of transoceanic waves, waves that may have originated in the coastal waters of Russia or Japan.[5] Peering from a dock, I search the water for my elusive quarry. No luck.

The residents of Sitka all seem to know the story of the bath toys, which they invariably refer to as "the duckies" or "the ducks." The one person I meet who has not heard of the ducks is a young commercial fisherman named Fred who came to Sitka a year or so ago. I run into him one rainy night at the Pioneer Bar.

The "P bar," as locals call it, caters to commercial and recreational fishermen, who keep odd hours. It therefore opens early and closes late. The window blinds almost never come up. Rows of black-and-white photographs, mostly of fishing vessels, hang on the walls above the bottles. I find Fred curled around the bar like a C-clamp, leaning on his elbows, the sleeves of his sweatshirt pushed up, dangling a beer from his fingers and nursing at the foam. Emboldened by whiskey, I introduce myself. I tell him my story, the one about the rubber ducks, and he tells me his. He used to be a corporate copywriter for GMAC in Dearborn, Michigan, and made a decent but anesthetizing living. On top of his job, he had to take care of his brother who has "a psychological problem" and is "basically a dependent." Then, a year ago, Fred cut loose. He gave his brother the keys to his house and fled to Sitka to seek his fortune on the seas. Sketching little diagrams on napkins, he teaches me about the different kinds of commercial fishing—purse seining, gill netting, longlining.

He buys one round, I buy another. The mood turns confessional. I feel guilty about my pregnant wife in New York. Fred feels guilty about his poor brother in Michigan. There's also the girlfriend he broke up with when he left. So far things haven't worked out as planned, he confides. He wants a spot on a longline trawler because those guys earn six figures. It's hell while you're out there on the Bering Sea, but you can cash in after a year. Problem is, a spot on a longline trawler is hard to get, so maybe he'll take up day trading, or maybe he'll write a novel, he has a great idea for a novel, or maybe he'll go back to Michigan and settle down. He thinks he'd be a great father, if only he could earn enough money to take care of both his family and his brother. Suddenly, he's on the verge of tears, and I feel obliged to offer comfort or advice.

What do I, a quixotic duckie hunter, possibly have to say to this wayward prospector fallen on hard luck? I tell him that it sounds like he has a lot of choices, that he'll probably be happier once he decides, at which suggestion Fred grabs my arm and leans close, his moist eyes glittery with freaky excitement. "That's it, you're so right," he says. "I've got to decide." He lurches from his stool. "I want you to meet someone." He leads me down the bar to an old woman in a black leather motorcycle jacket. She is smoking a long cigarette and staining the filter with lipstick. How and to what degree Fred knows this woman is not clear. Neither is her age. Neither is my mind. I am, I realize, totally shitfaced.

Fred wants me to tell this woman about the voyage of the ducks, but she already knows the story.

"Who the hell gives a rat's ass about that?" she asks, rotating on her stool to appraise me. Then, without warning, she reaches around my shoulders, snaps the yellow hood of my slicker up over my head, and barks, "You *look* like a rubber duck." She and several other patrons of the P bar are laughing at my expense. Their faces blur around me, like the faces of carousel horses. But the old woman in the motorcycle jacket is not satisfied just yet. She wants to hear me quack.

An attentive calm falls over the patrons of the P bar. Everyone—the bartender, the fishermen in their quilted vests—is listening, watching, waiting. Or so it seems to my foggy mind.

"Quack," I say.

"Again," the old woman says, with sadistic glee.

"Quack," I say. "Quack, quack."

BAHÍA DEL DESENGAÑO

Standing at the helm of the *Morning Mist*—a white twin-engine troller with outrigging as tall as flagpoles and orange floats the size of beach balls dangling like ornaments from its rails—is Larry Calvin, a spry, thin, white-haired fisherman in suspenders. On Calvin's black ball cap a fish leaps above the words ABSOLUTE FRESH. Back in the *Morning Mist's* cabin are a dozen-odd academics. Out the windows of the *Morning Mist* are mountains and water and, so far as I know, though I can't see them, whales and yellow ducks.

A self-employed left-wing entrepreneur who subsidizes his fishing with profits earned in the building supply business, Calvin embodies an old brand of American individualism that seems to flourish in the strange demographic conundrum that is maritime Alaska, a place both rural and coastal, both red and blue, Western and Tlingit, industrial and aquacultural and wild. In a magazine rack in the ship's cabin I spot the latest issue of the *Nation* and a copy of David C. Korten's antiglobalization jeremiad *When Corporations Rule the World*. The bridge of Calvin's ship looks like a cross between a cockpit and an office. Above the big silver steering wheel and the ball-tipped throttle levers there flickers an array of computer monitors connected to the antennae bristling from the roof. Beyond the monitors, out the windshield, the bright water stretches west to the horizon. With our course for Kruzof Island set, Calvin doffs his cap, drops it on the dashboard, steps halfway out the door, and stands there, left hand on the throttle, face in the sun, white hair blowing in the wind.

The academics aboard the *Morning Mist* include oceanographers, archaeologists, anthropologists, linguists, historians—all of whom have come to Sitka for the annual Paths Across the Pacific Conference, a symposium that coincides with the Beachcombers' Fair. Curtis Ebbesmeyer is here wearing his sea-bean necklace and a baseball cap decorated with stickers from Seattle coffeehouses. So is Dean Orbison, dressed in his customary plaid woodman's shirt and a pair of Sitka sneakers—the local name for knee-high rubber boots. The organizing theory of the conference is this: people started crossing the Pacific Ocean by boat tens of thousands of years ago, far earlier than was previously thought. Some Asian immigrants sailed to America by mistake when blown astray. Some came on purpose, paddling along the coast of the Beringian ice shelf, traveling a little farther east with every generation. Little is known about these ancient migrations, and on the way to Kruzof Island, an oceanographer named Thomas Royer tells me one reason why. Sea levels have risen so much since the last ice age that the earliest settlements in Alaska are now one hundred meters underwater.

Another passenger, an archaeologist, interrupts to contest Royer's figures. Sedimentation adds about a centimeter a year to the ocean floor, this archaeologist says, which means you'd have to dig deeper than one hundred meters. You'd have to dig about four hundred meters to find any artifacts.

Most of the chitchat aboard the *Morning Mist* is similarly interdisciplinary and esoteric. When did people first begin using boats—forty thousand years ago? Fifty thousand years ago? What does it take to start a migration—a critical mass, or a few individuals with the itch to explore? Why did people migrate in the first place—the profit motive? Hunger? Exile? Chance? Computer models like OSCURS may help these marine archaeologists reconstruct the routes of those transoceanic migrations, the history of which is inseparable from the history of global climate change. In other words, it is humanity's own past and future that the oceanographers are scrying in the tangled drift routes of the toys.

Looking out from the cabin of the *Morning Mist*, Royer teaches me how to read the surface of the sea. "You see that smooth area there?" he asks. "Either temperature or the salinity will change the surface tension of the water, so the same wind will ruffle the water in one spot, but not in another." All variations in the surface are the effects of hidden causes. What to me looks like an homogeneous expanse is in fact a kind of shifting, aquatic topography.

In the half hour or so it takes us to get to Kruzof, centuries seem to recede. Sitka disappears into a blur of blue horizon. The world out here would not have looked much different a millennium ago, I think. It resembles the opening verses of Genesis. There is only the land, the water, the trees. And then, in my peripheral vision, an orange figure swoops and dives. It's a kite. A kite shaped like a bird, and there on the beach below it are three figures, a father and his children, dressed in colorful swimsuits. They have rented one of the two rustic cabins that the National Forest Service maintains on Kruzof Island. The cabin itself becomes visible now, tucked back into the trees. Here, in the forest primeval at the foot of a dormant volcano, is a scene from the Jersey Shore. Larry Calvin anchors the *Morning Mist* well away from the rocky shallows and ferries us in several at a time aboard his aluminum skiff. The father flying the kite hollers hello, the children eye us warily. Ebbesmeyer hands out white plastic garbage bags in which to collect our riddles and discoveries. Dean Orbison will lead one beachcombing party to the south. I join Ebbesmeyer's party, headed north.

I try to remember what the Orbisons and Amos Wood have taught me. Up ahead, where the beach curves and tapers into a sickle, there's lots of jackstraw and even a little color—a fleck of blue, a daub of red. To get there we have to cross Fred's Creek, which spills down through the trees over terraces of rocks before carving a delta of rivulets and bluffs through the sand. The delta is perhaps a dozen yards wide, and those of us without Sitka sneakers have trouble getting across. I manage to leap from rock to rock. Ebbesmeyer, who ambles effortfully along, is in no state to go rock jumping. He hikes up into the trees and crosses where the creek narrows. Reuniting on the far shore, we make our way down the beach spread out in a line, scanning the sand. Ebbesmeyer launches into one of his litanies of facts. Bowling balls float, he informs us, or rather the nine-, ten-, and eleven-pounders do. Heavier ones sink. And did you know that the valves of clams are not symmetrical? A colleague of his once surveyed the clamshells along a mile of beach. "At one end of the beach, it's mostly rights, and the other end it's mostly lefts." The currents can tell the difference.

For the first time since I entered Alaska aboard the *Malaspina* a week ago, the rain clouds have cleared. A strong breeze is blowing inland across the sparkling waves. To our right, there is the sound of the surf, to our left the soughing of the trees. Peering into them I see only shadowy depths of green. The beach here is more gravelly than sandy. It's like walking over peppercorns. Our boot soles crunch, and I fall into a kind of trance.

No matter how crappy the pittance the tide leaves, no matter how ominous the riddles in the sand, beachcombing has its delights. There is pleasure in setting your senses loose. At the sight of something half-buried, the eye startles and the imagination leaps. At the edge of the waves flickers a silver flame. A hundred yards off, from beneath a pile of driftwood, glows a small, fallen sun. Then, at the moment of recognition, there is a kind of satisfying latch. The silver flame? An empty bag of Doritos, torn open. The small sun? A red, dog-chewed Frisbee. The strange becomes suddenly familiar once again, though never quite so familiar as before. The beach can strip familiar objects clean of their usual associations, a wild beach especially. And occasionally the strange remains strange. Occasionally the object you've inspected is unrecognizable or exotic or mysteriously incongruous. Occasionally that surf-tossed bottle turns out not to have been left by a camper but jettisoned from a Malaysian shrimp boat crossing the Andaman Sea.

Today, there are no messages in bottles, no computer monitors or Nikes, no toys. According to Ebbesmeyer, the beachcombing this year has not been good. It all depends on the winter storms. Then, too, there hadn't been a good container spill that he knew of since a batch of hockey gloves went overboard in 2003. I clamber over the jackstraw, finding there a predictable assortment of water bottles and spent shotgun shells, but also a polystyrene ice cream tub, a plastic length of hose, nylon nets, huge cakes of Styrofoam, all of which I dump into Ebbesmeyer's bag.

"Aw, man," Ebbesmeyer says of the Styrofoam. "That'll break up into a billion pieces. Aw, man. That's the worst stuff. In Seattle you can't recycle Styrofoam. Pisses me off. So what do you do with it? See all those little cells? The irony is, it's made of polystyrene, which sinks, and they foam it to make it into something that floats. That's what I think of when I see that stuff, all the windrows of Styrofoam, coffee cups with barnacles growing on them. You say you'd love to get it off the beach, but there's no way." He tells me about a container that spilled a shipment of filtered cigarettes. "There are about ten thousand polymer fibers per butt—that's, what? Ten to the order . . . about ten billion fibers for just one container." His eyebrows spring up above his glasses.

Luckily for them, none of Eric Carle's ten rubber ducks runs into a Laysan albatross. The encounter would not be pretty. The Laysan albatross is probably the most voracious plastivore on the planet. Three to four million cigarette lighters have been collected from seabird rookeries on Midway, and naturalists recently recovered 252 plastic items from the carcass of a single Laysan albatross chick. During a lecture he gave yesterday at Sitka's Rotary Club, Ebbesmeyer showed a slide of them, those 252 items. At first the photograph calls to mind stained glass. Then as you look closer, you start spotting familiar objects strewn amid the shrapnel. Two cigarette lighters and a dozen-odd bottle caps appear to be good as new. Somewhere among those 252 items may be the remains of a Floatee.

At the edge of the ebbing surf, we come upon the fresh footprints of a bear. The beach ends, the shoreline giving way to a labyrinth of wave-washed boulders into which the footprints continue. "Stonehenge for bears," says Michael Wilson, a Canadian geoarchaeologist who later this week will deliver a lecture titled "Natural Disasters and Prehistoric Human Dispersal: The Rising Wave of Inquiry."

Wilson, in Sitka sneakers and jeans, follows the footprints into the boulders, talking loudly. The wind is behind us, and we assume that the bear will keep its distance, but you can tell that Ebbesmeyer's feeling nervous. I am too. We both start glancing into the trees. Wilson's spotted something, something big and blue, and runs ahead to see what it is, the red windbreaker tied around his waist trailing like a cape. Now we're following two pairs of footprints, Wilson's and the bear's. His discovery turns out to be an empty plastic barrel with the word "toxic"

printed on the lid—an empty drum of boat fuel, most likely. It appears
to be watertight. Wilson thumps it like a bongo then hoists it up above
his head and roars like one of the apes in *2001: A Space Odyssey*.

We'd like to take the barrel back with us, but, deciding that the
damn thing is just too big, end up leaving it where we found it, among
the rocks. As we turn to retrace our steps I think of Wallace Stevens's
anecdotal jar: *The wilderness rose up to it, / And sprawled around, no longer
wild*.

Some of the archaeologists in our beachcombing expedition have
studied the midden heaps of shells that prehistoric seafarers left around
the Pacific Rim. Garbage often outlasts monuments, and if a millennium or three from now, archaeologists come looking for us, they may
well find a trail of plastic clues.

I have yet to reach the end of my own trail of clues. The toys first
made landfall on Kruzof Island in the autumn of 1992. So far only a
thousand or so of them have been recovered by beachcombers. The rest
are still out there, circling the Gulf of Alaska, or riding an ice floe
through the Arctic, or lying under wrack and sand in some out-of-the-
way cave of the ocean. And at least one or two could be stranded in the
Great Pacific Garbage Patch, circling around among ghost nets and
fishing floats and refrigerator doors. Ebbesmeyer believes that patch be-
tokens nothing less than the "end of the ocean." If he's right, then the
yellow duck I've been chasing is not only an icon of childhood but a genuine bird of omen, and the legend of the rubber ducks lost at sea is not
merely a delightful fable suitable for children but a cautionary environmental tale. I don't yet know whether to believe Ebbesmeyer's auguries.
There are more riddles to solve. For now, though, I've followed the
trail—and pushed my luck—far enough.

In 1827, returning from another failed attempt to find the Northwest
Passage, Lieutenant Parry, upon learning that he was to become a father, sent a letter home to his pregnant wife: "Success in my enterprize,"
he wrote, "is by no means essential to our joy, tho' it might have added
something to it; but we cannot, ought not to have *everything* we wish."

On Kruzof Island, for the first time since Bellingham, my cell phone
is picking up a signal. Certain that I'll be able to catch up with the rest
of our search party, or at least with the slow-moving Dr. E., I fall back
and call Beth. Just to be safe I've decided to fly home from Alaska three

days sooner than planned, I tell her, gazing out at lapping waves. A week after my return, following a difficult, 30-hour labor, she'll give birth to a son, the sight and touch of whom will dispel my usual, self-involved preoccupations and induce a goofy, mystical, sleep-deprived euphoria. Holding our pruny, splotchy, meconium-besmirched, cone-headed son, she'll cry, and when she does, so will I. These will be tears of joy, of course, but also of exhaustion and awe and, truth be told, of sadness. Holding my son for the first time, I will feel diminished by the mystery of his birth and by the terrible burden of love, a burden that, requiring hopefulness, will feel too great to carry, but which I will take up nonetheless.

In the meantime, back on Kruzof Island, there is Fred's Creek to cross.

Loaded down with our plastic bag of scavengings, Ebbesmeyer is standing at the creek's edge, contemplating the water and the rocks, looking for the way. It's shallow, Fred's Creek, but he'd rather not get his feet wet. Nor would I. To his right, the creek comes spilling out of the forest, descending terraces of rocks, flowing under mossy, fallen trees. To his left it widens into a miniature delta, carving lilliputian canyons in the sand, flowing around rocky islands, before emptying into the Pacific. He places his foot on a partially submerged stone, which shifts beneath his weight. Reconsidering, he heads once again to the sylvan narrows. I spring from rock to rock, enjoying the acrobatic challenge, making it across the creek's mouth in a few hops, boot soles wet but the cuffs of my quick-dry Adventure pants dry. By now, the rest of our party is out of both earshot and sight. Up in the trees, balancing, a scavenged walking stick in one hand, white bag of plastic trash in the other, blue ball cap visible through the branches, Ebbesmeyer is still midstream, having a hard go of it. "Throw me the bag," I call to him, and he does. It lands with a splash at the creek's edge. I met the oceanographer in person only a week ago, but for the moment I feel oddly protective of him, oddly filial. I watch the trees for bears. Finally he's made it to the other side. Together we walk to the landing on the beach and wait for Larry Calvin to come for us.

THE SECOND CHASE

The transition is a keen one, I assure you, from schoolmaster to sailor, and requires a strong decoction of Seneca and the Stoics to enable you to grin and bear it.

—Herman Melville, *Moby-Dick*

A KEEN TRANSITION

All I knew about Chris Pallister when I decided to go to sea with him was that he ran some sort of nonprofit called GoAK (pronounced *GO-ay-kay*), an acronym for Gulf of Alaska Keeper, and that this group was doing some sort of beach cleanup at some place called Gore Point, and that during the course of this cleanup a number of Floatees had been found—the first new finds I'd heard about since returning from Kruzof Island two years before. Plastic frogs had been found, Pallister told me when I called him from Manhattan, plastic turtles had been found, beavers, and, yes, ducks too. His "big boat" was already at Gore Point, he explained, serving as a sort of floating bunkhouse for GoAK's crew, but that weekend he would be heading there in his "little boat." If I could get my ass to Anchorage by Friday, I was welcome to tag along. He could use a deckhand, he said, and it would be nice to have the company.

I looked for Gore Point in my *Atlas of the World* and failed to find it. A little research revealed why. Gore Point (population o) is one of the wildest places left on the American coastline, and one of the last places on the planet you'd expect to have a garbage problem. Unegkurmiut Indians used to spend their winters there, and supposedly you could still find signs of them (a scar on a tree trunk where some hungry soul had scraped the bark away to get at the nourishing cambium beneath; the faint concavity a house pit had left in the mossy earth). But the last of the Unegkurmiut—their numbers decimated by smallpox, the survivors lured off by ill-paid work in the canneries of the Alaska Commercial Company—vacated the premises more than a century ago. There had been no permanent residents since. Gore Point was now part of Alaska's Kachemak Bay State Wilderness Park. Despite the pretty scenery and the charismatic megafauna to be seen there, few nature lovers bother to visit the park's outer coast, which can be reached only by helicopter, floatplane, or boat, and then only when the weather permits, which it often does not.

In *Beachcombing the Pacific*, Amos L. Wood divides North America's Pacific coastline into twelve regions, providing "a detailed analysis" of each. As beachcombing territory, the region encompassing Gore Point "ranks high." It ranks high because the Gulf of Alaska's divergent currents unload lots of flotsam onto those wild shores, but also because few casual beachcombers are dogged enough, or reckless enough, to go there. The region is, Woods writes, "about the most inaccessible of any around the North Pacific Rim, and thus is the least traveled." Gore Point also happens to lie six hundred miles downcurrent from Sitka— six hundred miles farther, in other words, along the trail I was following.

At the time Pallister extended his invitation, I was in no condition to go beachcombing on any Pacific beach, accessible or otherwise. I was, in fact, recovering from back surgery—minor, routine, noninvasive back surgery, but back surgery nonetheless; a microdiscectomy to be exact, necessitated by an acutely herniated disc that had impinged agonizingly on my sciatic nerve, and even more agonizingly on my not-so-well-laid plans. Happenstance, it turns out, is a terrible travel agent. The previous spring, after many months of preparation, I'd quit my teaching job in order to continue the journey I'd begun. I'd asked the Orbisons to take me with them on their annual beachcombing trip, and they'd agreed. But on the eve of their departure, I found myself in a Manhattan hospital, back-wrecked on a gurney, listening to an anesthesiologist explain to my wife that he had no idea whether or not his specialized services were covered by my insurance plan, then adding, under his breath, while searching for a vein, that he could really use a drink. The surgery had nevertheless gone well. For the previous ten days, the pain shooting from my spine to the tips of my toes had been so great, I'd been unable to walk or stand or sit. When the anesthesia wore off, aside from the aching suture, my pain was gone. I could walk. I could stand. I could beachcomb.

After a week of postoperative convalescence, while the Orbisons were scavenging in sea caves, I'd begun making calls in hopes of finding some other Alaskan willing to give me a free ride. I wasn't inclined to be picky. My frantic queries eventually led me to Marilyn Sigman, director of the Center for Alaskan Coastal Studies in Homer, a fishing village on Cook Inlet. Sigman led me to GoAK and Chris Pallister. It is no doubt ill-advised to volunteer as a deckhand while recovering from back surgery, even routine, noninvasive back surgery. But I was desperate. Pallister was embarking for Gore Point that weekend with or without me. I'd already gone adrift. There was no turning back. Like it or not, I was a professional duckie hunter now. I wasn't about to spend the summer convalescing in Manhattan, not when there were Floatees to be found.

At my postoperative examination, I told my surgeon that I had some "urgent business" in Anchorage, leaving out the part about a voyage to inaccessible coasts. Inspecting his handiwork, which was healing nicely, my surgeon said he guessed it would be okay for me to travel, so long as I didn't lift anything heavier than ten pounds, doctor's orders. No bending from the waist. No sitting for prolonged periods of time. For the next six weeks, there was an elevated risk of reherniation, he said.

"Promise me you'll ask for help with your luggage," he said. I promised. "Promise me you'll get up and walk around during the flight to Anchorage. Book an aisle seat." I promised. He wrote me a new scrip for Vicodin and another for anti-inflammatory steroids, just in case. The following morning I stuffed a pair of brand-new Sitka sneakers into a brand-new ergonomic backpack with wheels, asked the cabdriver if he'd mind loading said backpack into the trunk, kissed Beth and our almost-two-year-old son Bruno good-bye, and flew, hell-bent, to Anchorage.

RESURRECTION BAY

South of Gore Point, where tide rips collide, the rolling swells rear up and sharpen into whitecaps. A moment ago, Chris Pallister was rhapsodizing about the miraculous anatomy of dolphins, which he'd read about in *Discover.* "The metabolic calculus wasn't enough to account for their speed in the water!" he hollered. "They've got all kinds of physical traits and adaptations for diving at depths! They've got a cortex that's kind of like a sponge!" Then we slammed into a steep ten-foot wave that sent Pallister's toilet kit flying from the dash, a tube of toothpaste and a disposable razor scuttling across the cockpit floor. I scurried to retrieve them. Now the second ten-foot wave hits. I stumble back to my station in the copilot's seat, port side, behind the busted windshield wiper, and ransack my shoulder bag for yet another one of the ginger candies I've brought along as an over-the-counter remedy for seasickness. Quiet with concentration, Pallister decelerates from fifteen miles per hour to eight, strains to peer through a windshield blurry with spray, tightens his grip on the wheel, and like a skier negotiating moguls coaxes his little home-built boat, the *Opus*, through the chaos of waves. Our progress becomes a series of concussions punctuated by troughs of anxious calm. In this, I have begun to gather, it resembles the rest of Pallister's life.

Speaking with him on the phone from Manhattan, I'd pictured Pallister as the sort of all-American nature guy you see in advertisements for energy bars or camping equipment, clad in Gore-Tex or a flannel shirt, portaging a mountain bike or kayak through rugged terrain, or mesmerizing with his border collie beside a campfire while a gazillion digitally enhanced stars twinkle overhead. Instead, eleven hours after kissing my wife and kid good-bye, I was met in baggage claim at Anchorage International by a small fifty-five-year-old self-employed insurance attorney in khaki Adventure pants and a grass-green windbreaker emblazoned, where the breast pocket would go, with the yellow logo of the University of Alaska–Anchorage. His dark brown hair was cut, monkishly, in a straight line across his forehead as if with the aid of a carpenter's level. He had a crooked smile, the left corner of his mouth sagging so that he seemed to be scowling even when he wasn't. The chrome frames of his glasses looked difficult to break. Through them he squinted up at me and, in reference to the self-portrait I'd e-mailed him, said, "Yeah, I guess that says *FS*."

To make myself easy to identify, I'd worn a red ball cap bearing the initials of the high school where I had until recently taught, initials that were artfully but somewhat illegibly entangled, like those on a Yankees cap. "Yeah, I guess you're pretty tall," Pallister added, still squinting, his right eye squinting more than his left so that he reminded me of Popeye. "And yeah, you do look kind of young for your age."

This is an understatement. In my twenties, I was mistaken for a teenager; now, in my midthirties, for a college undergraduate. In coat and tie

I bring to mind a parochial schoolboy. My condition—*chronic juvenilia*, it should be called—owes mainly to a congenital shortage of facial hair and a congenital excess of cheek. I have the face of a schoolboy and the back of a middle-aged man.

I'd forewarned Pallister about my recent surgery, and my doctor's orders. He'd promised to do all the heavy lifting. He was well-acquainted with back trouble, he told me, loading my ergonomic backpack into the backseat of his estranged wife's Toyota. He'd had spinal surgery seven times—for sciatica, stenosis, and injuries he'd suffered in a traffic accident. The surgery had left a white scar that wormed out of his shirt collar up the back of his neck and disappeared, at the base of his skull, into his hair. "I'm five-foot-eight but I used to be five-foot-nine," he'd later tell me. "All that surgery chopped me down an inch."

It was almost 1 A.M. in New York. Here at the northerly latitudes of Anchorage it felt like late afternoon. Peering up through the windshield as we drove south along a six-lane boulevard, past strip malls and office parks, I watched a single floatplane fly north toward the mountainous horizon, its pontoons printing an equal sign on the bright, papery, overcast sky.

Sciatica and stenosis weren't Pallister's only troubles, I soon learned. Several years prior, while recovering from that traffic accident, he'd had to give up his law office and start working from home. Then, a year before I met him, he'd lost the lease on the house in which he and his wife, Jane, had raised three sons. The following winter, six months before I met him, after three decades of marriage, Jane left him for reasons he didn't understand. Dispossessed, deserted, he was living like a castaway in a duplex condominium. He'd offered to put me up there so that we could get an early start fitting out.[6]

Aside from the kitchen, the only room in Pallister's condo that seemed recently inhabited was the home office, which served as the headquarters of both GoAK and his struggling law practice. Here, the bookcases sagged with law volumes and three-ring binders. "Navigability," the spine of one of the binders read. More law volumes were stacked in precarious towers on the floor, and drifts of papers covered almost every flat surface. Post-it notes like yellow petals circled a computer monitor on which, in screen-saver mode, a slideshow of nature photography played. Pallister had taken the photographs himself on his many hunts and hikes: mountains, tundra, sunsets, streams, more mountains, more tundra, more streams. The wall above the desk was papered in maps and charts of Alaska's outer coast.

"Okay, so we're here," Pallister said, pointing out Anchorage on one of the charts. "Gore Point's way out here." He moved his finger to the southern tip of the Kenai Peninsula, a wing of land the size of Connecticut that extends 160 miles from the south-central coast of the Alaskan mainland into the Gulf. On the northwestern, leeward side of the Kenai Peninsula is Cook Inlet, a long shallow tidal channel. On the eastern side of the peninsula is Resurrection Bay. Between the inlet and the bay are several hundred thousand acres of wilderness, some federally protected, some protected by the State of Alaska, and on the outer coast of the state-protected wilderness, at the southernmost tip of the peninsula, is Gore Point.

"There's this big, high peak," Pallister said. All that connects the peak to the mainland is a narrow crescent-shaped isthmus—"a witch's finger of land," he called it—that gets in the way of the prevailing winds and currents. The Alaska Coastal Current, I'd learned, flows north along the Alaska Panhandle, past Kruzof Island, past the sea caves the Orbisons like to go foraging in, past Juneau. Colliding with barrier islands, the current makes a sharp left, following and hugging the coastline west until it reaches the Kenai Peninsula, where, ricocheting off Gore Point, it bears south by southwest and continues on toward the Aleutians, becoming the Alaskan Stream. "The isthmus is barely above sea level," Pallister said. "On the west side of it the forest is pristine, but on the east side all the lower branches are stripped off. You can tell that hellacious winter storms have pounded the crap out of it."

The windward shore of that isthmus is what's known to beachcombers and oceanographers as "a collector beach." According to the *Anchorage Daily News*, of the 10.8 million gallons of oil that spilled from the *Exxon Valdez* in 1989, more ended up on the windward shore of Gore Point than on any other beach in Alaska. In a single month workers there had filled six thousand plastic bags with toxic goulash—"oily sand and gravel, patties of emulsified crude, tar coated flotsam and jetsam, and the oil coated carcasses of birds and sea otters," the *Daily News* reported at the time. These bags the workers loaded onto a landing craft, which carted them off to the nearest landfill, eighty nautical miles away, in Homer. The same currents and winds that brought the oil bring flotsam, both man-made and natural. Wave action and strong flood tides have built up a berm of pebbles and driftwood ten feet high. Those hellacious storms throw flotsam up over the berm and into the forest beyond, where it remains. Unlike the oil spill, the incoming flotsam never ends. Every tide brings more. Over the course of the last several decades, ever since the dawn of the plastics era, a kind of postmodern midden heap has accumulated there. When Pallister first set foot on Gore Point, the floor of that forest was "covered in every conceivable type of plastic," he said. There was colorful debris a hundred yards back into the trees.

The single bag of trash Ebbesmeyer and I had collected during our day trip to Kruzof Island was nothing, Pallister assured me. All along Alaska's convoluted outer coast were shores littered with debris. Most of that debris was plastic, and much of it—the Asian, Cyrillic, and Scandinavian characters printed on bottles and fishing floats suggested—was crossing the Gulf of Alaska to get there. "Go out to Kodiak Island, or Kayak Island, or Montague, those first barrier islands," Pallister said. "They have an unbelievable amount of plastic trash." Thomas Royer, the oceanographer I'd met aboard the *Morning Mist,* had done research on the windward side of those barrier islands, and he later confirmed for me Pallister's description.

Before founding GoAK, Pallister and a charter boat operator named Ted Raynor, now GoAK's field manager, helped organize an annual volunteer beach cleanup in Prince William Sound. Over the course of four summers, working their way east from Whittier, the volunteers managed to scour approximately seventy miles of rugged shoreline. At that rate, Pallister and Raynor calculated, it would take two hundred years to clean Prince William Sound just once. The annual cleanup began to seem like a symbolic gesture at best, at worst a Sisyphean exercise in futility. It would take far more than three days a year of volunteerism to clean up Prince William Sound. It would take months. It would take "professional remediation contractors," as well as volunteers. It would take logistics of an almost military complexity. And it would take money. In 2005, Pallister, Raynor, and a NOAA oceanographer named John Whitney chartered GoAK as a 501(c)(3) nonprofit and started soliciting donations.

According to its grandiose mission statement, GoAK's purpose is to "protect, preserve, enhance, and restore the ecological integrity, wilderness quality, and productivity of Prince William Sound and the North Gulf Coast of Alaska." In practice, the group has done little else besides clean up trash from beaches. In the lower forty-eight, beach cleanups like those organized by the Ocean Conservancy tend to involve schoolchildren scouring the shore for candy wrappers and cigarette butts left by recreational beachgoers. GoAK's cleanups, by contrast, are costly expeditions into the wild. The group's volunteers must be eighteen or older, and everyone, myself included, must sign a frightening waiver in which they agree not to hold the organization liable for such perils as "dangerous storms; hypothermia; sun or heat exposure; drowning; vehicle transportation and transfer; rocky, slippery, and dangerous shorelines; tool and trash related injuries; bears; and"—in case that list left anything out—"other unforeseen events." Pallister, damaged as he was, seemed almost astrologically condemned to endure such events.

In the summer of 2006, the group's first summer in action, GoAK and a hundred or so volunteers—some traveling by kayak, some by bush plane, some by fishing boat—bivouacked along the Knight Island archipelago at the entrance to Prince William Sound. When the volunteers went home, Raynor, Pallister, and a team of several remediation contractors—friends and family of GoAK's founders—kept going. In all, they managed to clean some 350 miles of shoreline, picking up enough trash to fill forty-six Dumpsters, an accomplishment that earned GoAK the 2006 Outstanding Litter Prevention Award from Alaskans for Litter Prevention and Recycling (ALPAR)—which is, it should be said, not an environmental group, as its name misleadingly suggests, but a charitable organization whose board, in Pallister's own words, represents "the who's who of big business in Alaska."

Neither, I was surprised to learn, does Pallister consider GoAK to be an "environmental group." To me, he confided that he is "a greenie through and through," but publicly he calls GoAK "a conservation group." Why? Because "in Alaska conservation plays better." What's the difference? In Alaska people tend to think of environmentalists as tree-hugging, anti-hunting "animal welfare types," he explained, whereas conservationists are avid outdoorsmen who love nature but don't make trouble.

I'd assumed that the Gulf of Alaska Keeper was part of the Water-keeper Alliance, the network of environmental watchdogs that Robert F. Kennedy Jr. helped found. It isn't, and I'd later learn that Water-keeper officials had objected to GoAK's use of their brand. Although he still hoped to apply for membership in the alliance, Pallister refused to change GoAK's name. Waterkeeper's objections are without legal merit, he insists. He knows. He's checked. "They've trademarked 'riverkeeeper,' 'soundkeeper,' 'baykeeper,'" he'd tell me, "but not 'Alaska keeper.'"

An enthusiastic hunter, Pallister has little patience for animal-welfare types, but he idolizes Robert Kennedy Jr. as well as John Muir, to whom he believes he is distantly related. He is, in effect, a closet en-vironmentalist. In his thirties, after a decade or so working construction to support his family, Pallister went back to school, eventually earning a JD from Lewis and Clark College, in Portland, Oregon. Later, in hopes of becoming, like Kennedy, an environmental lawyer, he acquired a cer-tificate in enviromental and natural science law. Later still, on a NOAA Sea Grant, he'd gone to Washington as a staffer for Alaska's Republican senator Frank Murkowski. The first day he reported to duty, Pallister told me, Murkowski assigned him the task of rifling through the En-dangered Species Act for loopholes. Though disillusioning, his year in D.C. had been for him an education in the expediencies that politics in Alaska, perhaps more than elsewhere, requires.

He would make more trouble if he had the deep legal pockets to do so, he says. But in Alaska, the loser in a lawsuit pays all the legal fees, and when you take on an oil company or a multinational mining con-glomerate, the legal fees can be ruinously steep. In Alaska, only the best-endowed environmental groups dare to litigate. Which is why, even though he's read Edward Abbey's *The Monkey Wrench Gang* more than once, Pallister insists in public that GoAK's work is "not political." He couldn't afford to be political, he said. He depended on the generosity of corporate benefactors, as well as on the largesse of pro-development politicians. "We don't care if donating burnishes your image," he told me. "At least you're doing something for the environment."

The Gore Point cleanup project was far more ambitious—and far costlier—than any GoAK had so far undertaken. It was also their first mission paid for in part with federal funding, funding made available by the 2006 Marine Debris Research, Prevention, and Reduction Act, one of the few pieces of environmental legislation that President George W. Bush ever signed and the latest in a long line of federal actions that have, over the past quarter century, failed to turn back the rising tide of marine debris.

The act authorized $5 million of the annual federal budget to help the Coast Guard better enforce anti-littering laws and another $10 million for a new Marine Debris Program to be administered by NOAA. In addition to conducting its own research, prevention, and reduction efforts, the Marine Debris Program is charged with disbursing matching grants to "any institution of higher education, nonprofit organization, or commercial organization with expertise in a field related to marine debris." The Alaska Republican senator Ted Stevens, a coauthor of the act and then chairman of the powerful Senate Appropriations Committee, had made sure that a disproportionately large sum would be directed to his home state.

In the winter of 2007, Pallister applied for one of the grants. By then GoAK certainly had acquired the requisite expertise. Despite all that the group had accomplished in its first summer, Pallister was unsatisfied. It wasn't enough to clean beaches near coastal communities. Alaska has 33,000 miles of coastline, most of which is wild and remote. "GoAK's goal," his successful grant application explained, "is to remove the plastic MD"—the plastic marine debris—"from as much of the Prince William Sound and Gulf of Alaska shoreline as possible."

The shoreline Pallister thought about most, the one that had become for him a kind of Everest, was the outer coast of Montague Island, a hundred-mile-long femur-shaped bar of mountainous land that stretches across the entrance to Prince William Sound. "There are forty miles of beaches that are covered in plastic" but no place to safely anchor a boat, he told me in his home office. "We'll have to put people on that shoreline by airplane, and land them on the beach, and then support them by airplane."

For now the costs and logistics of such a "massive undertaking" were more than GoAK could handle. Before attacking Montague, Pallister had determined, he and his crew needed to practice on a remote, heavily fouled beach where airplanes weren't necessary; where it would be safe for even a small supply vessel like the *Opus* to anchor. On the east side of Gore Point was a tranquil lagoon and a sheltered pebble beach, the perfect spot for a base camp from which it would be a short walk across the isthmus to the debris-strewn windward shore. The Gore Point cleanup would be a sort of pilot project, an experiment in logistics that, if successful, Pallister hoped to repeat on a larger scale. What would it take, he wanted to know, to clean up one wild beach?

I had my own questions that I hoped to answer at Gore Point. To beachcombers, that midden heap of flotsam had made Gore Point a happy hunting ground, one of the best places in Alaska to find exotic oddities. To Pallister it had turned a wilderness park into an accidental dump. To me it sounded like a kind of wonder, akin to the Mammoth Caves or Stonehenge or the Mesa Verde cliff dwellings, except that the Gore Point midden heap was the collaborative work of both nature and man, an unforeseen marvel that the ocean had wrought with the raw material we'd provided it. It also sounded like an unsolved environmental mystery—unsolved and possibly unsolvable. Who, if anyone, could be held accountable for all that plastic trash? What did it really forebode—for us, for the sea?

Another trough of anxious calm. Another concussive wave. Yet to fly from the dash are Pallister's wristwatch, a ziplock bag of venison jerky he made himself from a blacktail deer he shot himself, a bag of trail mix purchased yesterday morning in the Anchorage branch of Costco, and a spiral notebook serving as captain's log. As if playing the nautical equivalent of Whac-a-Mole, I manage to keep these items in place. In the spiral notebook are the entries Pallister made the last time he attempted to take the *Opus* out. The notebook is open to a page inscribed with the following cautionary reminder: "BEAR COVE / Keep close to large rock on port side. / Do not go between shoaling rocks and port side." Accompanying this message is an alarmingly cartoonish hand-drawn map of Bear Cove and its large rock. That last voyage ended badly but not, though it might have, disastrously: Way out on Prince William Sound, the ancient, rebuilt, 120-horsepower four-stroke Volvo inboard/outboard overheated and stalled. A fishing boat came to Pallister's rescue, towing the *Opus* back to safe harbor.

Ready to come to our rescue today is the *Patriot*, a charter fishing boat two or three times the size of the *Opus*, captained by Cliff Chambers, who may well weigh twice as much as Chris Pallister. A sweet, mustachioed bon vivant, when we met up at Seward Harbor this morning, Chambers had encased his prodigious gut in a T-shirt conveying the message that at Hog Heaven could be enjoyed three things: BIKES, BABES, AND BIG FISH. He enjoys all three, but about the babes he appears to have discriminating tastes. As an in-kind donation to GoAK, Chambers agreed to ferry volunteers to and from Gore Point, free of charge, so long as Pallister buys the fuel; the volunteers he's ferrying today—a mother and grown daughter from Alaska's North Slope who answered Pallister's televised call for volunteers less out of do-gooderism than out of a desire for a cheap vacation in the famously beautiful Kenai Fjords— are in Chambers's estimation the inferior variety of babe.

"Have you met these girls?" he asked Pallister as we were preparing to embark. "They're kind of a step down from Isabelle. They're nice"— "nice" here implying all that they were not.

Isabelle, I learned, was a vacationing French babe who'd recently spent a couple of weeks as the *Patriot*'s solitary deckhand, and although Chambers's transactions with her were strictly professional, those two weeks seem to have been among the best in his career as a charter-boat captain. What left the biggest impression wasn't the beauty of Isabelle, though that left an impression, nor her exotic Frenchiness. What left the biggest impression was the cooking, about which Chambers also exhibits discriminating tastes. Isabelle cooked him omelettes and crepes he remembers still. His current deckhand, a blond ponytailed dude of so few words one wonders whether he might be mute, is also a step down from Isabelle, Chambers discreetly informed us, when said deckhand was doing something out of earshot involving ropes.

Not so discreetly, upon first catching sight of the *Opus* tied up at the docks of Seward Harbor, Chambers informed us that he esteemed it to be of the inferior variety of boat. "Holy shit!" he exclaimed, addressing me, within earshot of Pallister. "You're going to ride on that? Did you go to church last Sunday?"

Unfortunately, although its discriminating captain and its nice inferior babes and its mute ponytailed deckhand are willing to come to our rescue, the *Patriot*'s traveling speed is five miles per hour slower than the fifteen sustained by the *Opus*, and Cliff Chambers is several knots more prudent than Chris Pallister. For hours, we've been on our own.

Back in the 1980s, you may recall, the specter of fouled beaches was one of America's most frequently recurring collective nightmares. The Jersey Shore was awash in IV bags and used syringes. New York's garbage barge haunted the world, trailing windblown flotsam in its wake. On the approach to Kennedy Airport, the protagonist of *Paradise*, a late Donald Barthelme novel, looked out his airplane window and saw "a hundred miles of garbage in the water, from the air white floating scruff."

Then, like the ozone hole, all that floating scruff seemed to go away. Perhaps we tire of new variations on the apocalypse the same way we tire of celebrities and pop songs. Perhaps all those syringes and six-pack rings, no longer delivering a jolt of guilt or dread, were carried off by an ebb tide of forgetfulness. By the end of the 1980s, we had newer, scarier nightmares to fear. Who could be bothered about seabirds garroted by six-pack rings when Alaska's shores were awash in Exxon's crude? Who could be bothered about turtles tangled in derelict fishing nets when the ice caps were melting and the terrorists were coming?

Or perhaps—even a well-informed landlubber might be forgiven for thinking—this particular ecological nightmare, like the nightmarish ozone hole, really had been laid to rest. In the mid-1980s New York's sanitation department began deploying vessels called TrashCats to hoover up scruff from the waterways around the Fresh Kills landfill. Elsewhere mechanical beach sweepers did the same for the sand. In 1987, the federal government ratified Marpol Annex V, an international treaty that made it illegal in the territorial waters of the signatory countries to throw nonbiodegradable trash—that is, plastic—overboard from ships. Later that same year (the same year, it so happens, that members of the United Nations signed the Montreal Protocol, the treaty that banned ozone-destroying chemicals outright), the garbage barge returned to New York after four months at sea and had its cargo incinerated. The good news for the ocean kept coming: In 1988, Congress passed the Ocean Dumping Reform Act, which forbade cities to decant their un-treated sewage into coastal waters. In 1989, the Ocean Conservancy held its first annual International Coastal Cleanup, which has since grown into the largest such event in the world.

But beautification, it seems, can be deceiving. Although many American beaches—especially those that generate tourism revenues—are in fact a whole lot cleaner these days than they used to be, the oceans are another matter. In *Garbage Land*, her book-length exploration of the American waste stream, the journalist Elizabeth Royte reports that in the mid-nineteenth century, officials in New York cleaned dead horses from the streets by tipping the carcasses into the river along with household refuse. How the times have changed, you might think. At least today there aren't equine carcasses putrefying beneath the Brooklyn Bridge, or feral pigs scavenging slop from the gutters of lower Broadway, gutters that during those unplumbed, horse-drawn times flowed with shit, equine and porcine and human.

Yet since the mid-1800s, not only has the volume of New York's household refuse grown, to four million tons a year; its chemistry has changed. Depending on where they sample, oceanographers have found that between 60 and 95 percent of today's marine debris—the preferred bureaucratic term for flotsam and jetsam—is made of plastic. Despite the Ocean Dumping Reform Act, according to a 2004 EPA report, the United States still releases more than 850 billion gallons of untreated sewage and storm runoff every year, and in that sewage are what the Environmental Protection Agency charmingly calls "floatables"— buoyant, synthetic things: Q-tips, condoms, dental floss, tampon applicators.

The tide of plastic isn't rising only on Alaska's uninhabited shores. In 2004, oceanographers from the British Antarctic Survey completed a study of plastic dispersal in the Atlantic Ocean, north and south. "Remote oceanic islands," their survey showed, "may have similar levels of debris to those adjacent to heavily industrialized coasts." Even on Spitsbergen Island, in the Arctic, the survey found on average one plastic item every five meters. Farther south, in the mid-Atlantic and the Caribbean, at the edge of the Sargasso Sea, they found five times as much— one plastic object every meter. And on "the best-studied and more remote Atlantic shores" the historical data "suggest unabated increase."

Not even oceanographers can tell us exactly how much floating scruff is out there; oceanographic research is simply too expensive and the ocean too vast. A 1997 paper, citing studies conducted in the 1970s, estimated that globally 6.4 million tons of garbage were set adrift from ships every year; later estimates are higher still. In 2002, *Nature* magazine reported that debris in the waters near Britain doubled between 1994 and 1998; in the Southern Ocean encircling Antarctica the increase was a hundredfold.

The *Encyclopedia of Coastal Science*, about as somniferously clinical a source on the subject as one can find, predicts that plastic pollution will grow worse in the twenty-first century because "the problems created are chronic and potentially global, rather than acute and local or regional as many would contemplate." The problems are chronic because, unlike the marine debris of centuries past, commercial plastics persist, accumulating over time, much as certain emissions accumulate in the atmosphere. The problems are global because the sources of plastic pollution are far-flung, but also because, like emissions riding the winds, pollutants at sea can travel.

Though no oceanographer, Pallister has a theory about where all the flotsam and jetsam on Alaska's outer coast is coming from. "There's a weather phenomenon we have here called the Pineapple Express," he told me in his Anchorage condo. "That's a winter low that sets this prevailing wind pattern from the Hawaiian Islands right toward the Gulf of Alaska, and it will just funnel this way for days on end if not weeks on end. That wind is blowing right across that bunch of plastic out there." He was talking, of course, about the Great Pacific Garbage Patch.

Like Cliff Chambers, when I first laid eyes on the *Opus*, what I saw bobbing there among the steel outriggers and fiberglass motorboats gave me pause. The twenty-four-foot-long, eight-foot-wide wood-and-epoxy hull was painted dark red. The plywood cabin, painted gray, resembled the turret of an antique tank. The see-through, snap-on plastic curtains enclosing the tiny deck made the boat look like a floating gazebo—fit for a picnic on a lily pond, perhaps, but not for a nine-hour resupply mission around the Kenai Peninsula's outer coast. Pallister conceded that the *Opus* might be "the ugliest boat in the harbor," but he'd take his homebuilt cabin cruiser over "a cheap, mass-produced" fiberglass Bayliner any day.

"*Opus* as in *magnum opus*?" I asked, noticing the name stenciled in gray block letters over the transom.

"*Opus* as in the penguin," said Pallister. "You know, *Bloom County*? Berkeley Breathed? Man, that guy was great!"

Despite appearances, the *Opus* is perfectly seaworthy, Pallister assured me. Back in the eighties, he built it with the help of a naval-architect friend according to a "Coast Guard–approved" design. Furthermore, Pallister himself is in the Coast Guard Auxiliary. Members of the Coast Guard Auxiliary, he explained, are boat-owning civilians trained to assist in search-and-rescue operations.

All this sounded pretty reassuring. Then I learned that there are no life jackets aboard the *Opus*. Then that there is one survival suit on board but only one, and it is in Pallister's size, not mine, and Pallister is a little man, wiry and short, so that with his pants tucked into his Sitka sneakers he brings to mind a jockey. At least there is a lifeboat aboard, one of those inflatable skiffs that I now know are generically referred to by the brand name Zodiac. Then again our Zodiac isn't outfitted with a motor, only a pair of plastic oars, and its very presence atop the roof of the *Opus* brings to my schoolteacher's mind Stephen Crane's "The Open Boat," a masterful short story based on Crane's harrowing, true-life survival, in 1896, of a shipwreck off the coast of Cuba. "Many a man ought to have a bath-tub larger than the boat which here rode upon the sea," Crane wrote. The same could be said of our Zodiac and, almost, of the *Opus*.

Then I learned that not only are there no life vests aboard, nor a survival suit in my size, nor a motor for the Zodiac; there is no water maker with which to make potable water, just a big plastic five-gallon bladder that Pallister filled up in Seward Harbor with a hose. Nor is there what we seafarers call a "head," only a Porta Potty stowed in a little hatch beneath the planking of the deck, planking currently obstructed, like most of the deck, by boxes and coolers of provisions purchased yesterday morning at the Anchorage branch of Costco. In these boxes and coolers are, for instance, pillows of spinach, vats of mayonnaise, packages of hot dogs the size of vibraphones. If the outboard should stall, and if we should run out of fresh water, at least we have plenty of food and a few cases of Pepsi and Dr Pepper to drink. Pallister is opposed to bottled water, for reasons both thrifty and ecological, as well as to coffee, for reasons neurochemical and moral; he considers coffee an addictive drug and therefore a substance abused by weak people like me. But for inexplicable reasons all varieties of Pepsi and Dr Pepper—bottled, canned, diet, sugary, and, yes, caffeinated—are in Pallister's weird book A-OK.

Even if I could easily access it, the use of the Porta-Potty would require a new level of intimacy between me and my commanding officer, a level of intimacy that, despite the alpine altitudes of intimacy to which we've already climbed in the past forty-eight hours, I don't particularly wish to explore, suffering as I intermittently and unpredictably do from that variety of sphincteral stage fright that men are prone to experience—a surprisingly large number of men, I suspect, considering the demure metallic and to my mind excellent barriers that one increasingly finds protruding between the urinals of public restrooms.

When he needs to relieve himself on a long voyage aboard the *Opus*, Pallister will simply pull into the sheltered lee of some island or peninsula, put the outboard in neutral, clamber nimbly onto the swim deck, and offer up a steamy contribution to the hypothermically cold Alaskan sea. Affixed to the stern, the swim deck of the *Opus* is about the size of a park bench. When last Pallister stopped to tinkle, a couple of hours ago, back in the shallows of Wildcat Pass, I gave the swim deck one look and concluded I could wait.

"Suit yourself," Pallister said with the characteristic note of head-shaking disapproval with which he greets the mostly idiotic behavior of his fellow hominids.

Thus, on top of gingery queasiness and an aching postoperative lumbar scar, I am now, as we go bucking and crashing through the tide rips off Gore Point, also experiencing serious urological distress, not to mention serious misgivings about having impulsively accepted a free ride to a wild isthmus I'd never heard of, from a heartsick conservationist I'd never met.

Armed with a brand-new GPS gizmo, familiar with the route, confident of his seamanship, Pallister has neglected to bring along a copy of *United States Coast Pilot* volume 9, *Pacific and Arctic Coasts Alaska: Cape Spencer to Beaufort Sea*. The *Coast Pilot* is the navigator's most trusted guide to America's territorial waters, which radiate from the land like a blue nimbus two hundred miles thick. Anyone worried that the Image has vanquished the Word can seek solace in its pages. All nine volumes are marvelous documents, each paragraph distilling centuries of first-hand observations made by both sailors and scientists. They are like literary atlases, those nine volumes, literary Google Earths, translating the great big mysterious world into detailed descriptive prose.

In addition to harbors and landings and facilities and interesting geographic features, the *Coast Pilot* alerts mariners to assorted perils of the sea, and if Chris Pallister or I had brought along volume 9, and if we'd turned to chapter 4, page 197, we would have come upon a note warning us about the perilous "tide rips with steep, short choppy seas . . . 3 to 5 miles S of Gore Point."

Other navigational descriptions of Gore Point are similarly cautionary. According to *Exploring Alaska's Kenai Fjords: A Marine Guide to the Kenai Peninsula's Outer Coast*, a copy of which Cliff Chambers keeps aboard the *Patriot*, "Mariners unfamiliar with this unforgiving area tend to pass through without stopping to sightsee." One such mariner, Charles Clerke, captain of the *Discovery*, companion ship to Captain Cook's *Resolution*, put it more blasphemously and memorably. Coasting in the vicinity of Gore Point in 1778 on a search for the Northwest Passage, the *Resolution* encountered thick fog and gale-force winds. "This seems upon the whole," Clerke wrote in his journal, "a damn'd unhappy part of the World."

Poor Charles Clerke. He was more damn'd than he knew. From Alaska, the *Resolution* and *Discovery* would retreat to the warm, tropical shores of Hawaii. On a previous stop, they'd found the island to be a garden of earthly delights inhabited by gift-giving, scantily clad, erotically generous, if somewhat kleptomaniacal natives. The idyll they dreamed of while weathering Alaskan storms would prove, like most idylls, a fleeting illusion. For historically obscure reasons—possibly because the Hawaiians mistook Captain Cook for a fertility god named Lono, possibly because the English overstayed their welcome, draining the resources of their hosts, but mostly because Cook attempted to punish Hawaiian thieves—the natives turned hostile. While attempting to take their king hostage, Cook would meet his death at the hands of an angry crowd. Charles Clerke would replace him as the expedition's commanding officer, but not for long. Off the damn'd, unhappy coast of Siberia, he too would succumb, not to natives but to tuberculosis, leaving in command one John Gore, the namesake of Gore Point.

At least I can't complain about the weather. All day, the sea conditions have been perfect—lots of sunshine, a breeze out of the northeast that is refreshingly cool but not cold. When we embarked from Seward this morning, there were lots of other boats out on Resurrection Bay—tour boats taking tourists for a peep at Bear Glacier, charter boats full of fishermen hoping to land a halibut, little speedboats full of day trippers celebrating the beauty of the Kenai Fjords with coolers of beer.

Over the shining, turquoise water we went, past Bear Glacier, past the tour boats full of tourists admiring Bear Glacier, past Fox Island, where the artist Rockwell Kent spent the winter of 1918 roughing it with his nine-year-old son, Rockwell III, an experience Kent later wrote about in a book, titled *Wilderness*, that is, mostly on account of Rockwell III, quite possibly my very favorite of the many, many books—transcendental, survivalist, sociological, satirical—that outsiders have written about their Alaskan adventures.[7]

Personal circumstances have no doubt biased me in favor of Kent's book. I've entirely failed to reconcile fatherhood with adventuring. When I decided to quit my job and resume the chase, Beth and I recognized that my intermittent but prolonged absences would be hard on me, harder on her, and especially hard on our son, Bruno. But we'd told ourselves that he would profit vicariously from my travels, my discoveries. I would come home with souvenirs and pictures and stories that would fill his little mind with wonderment. I would bequeath to him a patrimony of curiosity—about Alaska, the ocean, the mysterious land on the other side of the planet whence most of his possessions came—and instill in him a sense of responsibility for the natural world. And perhaps that is how it will turn out in the end. For now, though, all I seem to have bequeathed to him is timidity.

Bruno, about to turn two, is this summer afraid of fire engine sirens; of the skeletons at the natural history museum; of Spider-Man, whose face to his eyes resembles that of a skeleton at the natural history museum; of SpongeBob SquarePants, whom he calls "that funny snail"; of the Wild Things in *Where the Wild Things Are*. Even *Sesame Street* characters scare him (he doesn't like their "googly-eyes"). He is also, Beth told me when I called her one last time this morning from Seward just before Pallister and I shoved off, afraid of the sea. Hearing this, I thought with affectionate distress, *He is his father's son.*

Although in adulthood I've learned to bluff, I am an exceptionally fearful person, ill-suited to the role of journalist, or adventurer, or errant duckie hunter. The greatest among my many fears—greater than a fear of heights that makes me faint-headed if I stand too close to a plate-glass window in a skyscraper; greater, even, than my fear of mysterious contagions, or of terrorists, or of bankruptcy, or of disgrace—is my fear of sharks and therefore of the sea. For that fear I hold neither myself nor my parents responsible.

Those whom I hold responsible are the feckless counselors of the summer camp to which my parents sent me at the impressionable age of seven or eight. Those counselors, coaches and college kids mostly, would subject us campers to a morning routine of desultory sport involving pink rubber kickballs. In the afternoon, they would gather us around a television on a wheeling cart and plunk into the boxy maw of a VHS player a film.

One afternoon the film of choice was Steven Spielberg's *Jaws*. There I sat, cross-legged on the linoleum, trying as best I could to be a well-behaved child, a *good* child, and there on the screen coeds lounged around a campfire on a beach, smoking something. And there on the screen one of the female coeds ran naked into the moonlit waves. And on the screen, as she swam, there came a sudden jerk, and a look on her face of awful yet somehow comical surprise. And there on the screen the naked coed disappeared beneath the waves. Across the water spread a slick of blood. And on the beach, the following morning, there appeared a dismembered forearm. And although I once dreamed of becoming a marine biologist, I have been afraid of the sea ever since—so afraid that even now on trips to the seashore I'll wade in waist deep but no farther.

I mentioned none of this to Beth when I spoke to her this morning. Friends of ours had invited Beth and Bruno to spend the weekend at a beach house in Delaware. The house was beautiful, she told me, and Bruno was having a grand time playing with the daughter of their hosts. All the same Beth was beginning to wish they'd stayed home, on account of Bruno's fears. Yesterday, when they went to the beach, he'd refused to go near the waves. Today, he was refusing to go to the beach at all. She'd tried coaxing him, goading him, coercing him, but the very mention of the beach sent him into screaming, sobbing paroxysms of dismay.

I told her what parents always say when trying to reassure each other or themselves—"It's just a phase." I told her this, but in truth I wondered if my absence might be partly to blame, my absence and my genes. Perhaps, I suggested, Bruno would like to hear his father's voice? Beth summoned him to the phone, and at the mention of my name, he came running, running so excitedly he tripped. I heard it happen. The crash, the wail, Beth rushing to comfort him. Standing there on the Alaskan coast, gazing out past the marina at Resurrection Bay, listening to my son wailing inconsolably somewhere on the coast of Delaware, in a house I couldn't imagine, I felt like a truant, a deadbeat dad. Like a serial deserter who'd dressed up his restlessness in the trappings of a quest.

Still sniffling, Bruno finally took the phone, or more likely his mother held it to his ear. I could almost see them, him slumped in her cross-legged lap. "I slipped," he said in his tiny toddler's voice, which on the phone seemed tinier still. "I fell on the floor."

"I'm in Alaska," I said, hoping he'd think of the picture book I'd given him, called *Alaska*, in which appear illustrated mountains, illustrated eagles, illustrated sled dogs, illustrated bears. If he was thinking of such marvels, he gave no sign, so I tried again: "Last night, I slept on a boat."

"I slipped," Bruno repeated.

"Are you going to be okay?"

No answer. Probably he was nodding, yes, forgetting that I couldn't see him nod. He had only just begun to learn the magic of telecommunication, the trick of speaking to a disembodied voice. Into the silence between us came the cosmic hiss of satellites, the static of the spheres. If he were nine years old instead of two, perhaps I could have brought him with me, and perhaps it would have done him some good. In *Wilderness*, as the Rockwell Kents, younger and elder, are crossing Resurrection Bay, braving a storm in their open boat, Rockwell III, showing "a little panic," says to his father, "I want to be a sailor so I'll learn not to be afraid."

At his first glimpse of his hermitage on Fox Island, Kent the elder had said to himself, "It isn't possible, it isn't real!" Which just about sums up how I felt motoring across Resurrection Bay in the copilot seat of the

Opus. So many books have been written about Alaska, and so many cruise ship commercials shot there, that the place, even when you're in the midst of it, can seem a symbolical mirage.

In his late twenties, in pursuit of that mirage, Pallister persuaded his high school sweetheart, at the time pregnant with their first child, to move with him to Anchorage because of a map—a map of Prince William Sound that he'd happened upon in the back of *Field & Stream*. Growing up in Montana, he'd fallen in love with the mountains of the American West. He and his brothers hiked in them, camped in them, hunted deer and mountain goats in them. Then, as a teenager, Pallister read Joshua Slocum's *Sailing Alone Around the World* and dreamed of going to sea. What he glimpsed in *Field & Stream* seemed to him, as Fox Island had seemed to Kent, a kind of paradise, an American wilderness that was maritime and mountainous both. Thirty-five years later, the spell had yet to break, though the place of Pallister's dreams has receded into the blue, nostalgic distances of remembrance. To his great dismay, there are plenty of other people in the world who want in on his paradise.

Although Pallister scorns organized religion, considering it the enemy of reason, there is something puritanical about his brand of conservationism, which is in large part a crusade against idiotic hominids. Like many conservationists in Alaska, he dates the beginning of his activism to March 24, 1989, the day the *Exxon Valdez* ran aground on Bligh Reef. What troubled Pallister the most wasn't the spilled oil, however; it was the crowds—the volunteers, the cameramen, the news anchors, the oilmen, the politicians. "All of a sudden there were literally thousands of people in places where I'd never seen people before," he told me. "I thought to myself, 'Holy Christ! This is on the national news. Everybody's going to see how beautiful this place is. It's going to be a tragedy for Prince William Sound.'"

What he feared had come to pass. On the eve of our embarkation, when we were towing the *Opus* over the Kenai Mountains, the traffic on the Seward Highway had been thick as on the Long Island Expressway at rush hour. So thick it felt as though we were participating in some sort of exodus or pilgrimage. It was the weekend, a weekend in the middle of July, at the height of Alaska's tourist season. The salmon were running and the creeks were crowded with fishermen scooping up silvers with dip nets, so many fishermen and so many silvers (the water in places was dark with them), you might have thought it was actual silver the fishermen were prospecting for.

All these travelers in all these cars and trucks and recreational vehicles were joining the rush, heading out with their boat trailers and tackle and kayaks and mountain bikes to worship recreationally on whatever vestiges remain of the so-called Last Frontier. To drink beer beside a campfire. To make an offering of wild salmon to some wild, American god. Pallister and I were heretics in their midst. Or maybe *heretics* isn't the right word; maybe I was an apostate in their midst, and Pallister a fanatic. In any case, we both had nonrecreational reasons for making the pilgrimage to Resurrection Bay, and Pallister's reasons were different from mine. I was on a possibly quixotic quest to get my hands on a hollow plastic duck. Pallister was on a possibly quixotic quest to regain a paradise lost.

This morning, after a night on the *Opus*, the first of several to come, after gassing up, it was thrilling to be out on the water, far more thrilling than my slow, soggy, fever-inducing ferry ride to Sitka two years ago, for much the same reason that riding a motorcycle or horse is more thrilling than riding a Greyhound bus. The *Opus* is so small, even the short, six-foot seas of Resurrection Bay made the boat indulge in nearly every one of the six degrees of freedom. The copilot's seat is so close to the water, I could put my hand out the Plexiglas port window and feel the spray. When a pair of Dall's porpoises started racing alongside us, cavorting, braiding their wakes around ours, they seemed near enough almost to touch. Beyond Fox Island we emerged from Resurrection Bay and into the Gulf of Alaska. Before us the North Pacific stretched all the way to Sitka, all the way to Hawaii, all the way to the Great Pacific Garbage Patch, all the way to Hong Kong and Guangzhou.

On we coasted, on a southwesterly bearing, among the forested tops of partially submerged mountains, past the point where the tour boats ventured, on and on, until there were not even any fishing boats to be seen, including the *Patriot*, only wildlife—orcas, Dall's porpoises cavorting at metabolically improbable speeds through submarine canyons, puffins dragging their football bodies over the waves as they flap-flap-

flapped their little wings. White mountain goats speckled the sorrel-green lower slopes of snowcapped granite peaks. A pair of black dorsal fins described momentary arcs then vanished, their owners, a humpback mother and calf, having slipped below the surface without breaching or turning fluke, much to my disappointment.

It was hard to carry on a conversation over the rebuilt outboard's roar. That Pallister is slightly deaf—or as he puts it, "deaf as a stump"—made it harder still. Mostly he shouted and I listened. He shouted about the metric system, how crazy it was we hadn't adopted it. He shouted about overpopulation, how it was the root of all environmental evils. He shouted about exercise, to which he objected almost as strenuously as he did to overpopulation. "People go walking out in the road with weights in their hands!" he exclaimed in disbelief. "Why don't you build a wall or dig a hole in the ground?" That's one reason he likes cleaning up beaches: "It makes me feel like I'm doing something that's real important! It gives me a reason for being alive!"

Between these bursts of opinion, we stared in silence at the horizon, Pallister gnawing on jerky, I popping ginger candies, the outboard sucking down boat diesel—$200 worth. Pallister would aim the *Opus* at some speck of land in the distance, and for what seemed like hours we'd watch the speck grow into the forested top of a partially submerged mountain, until finally we were alongside it, and then, suddenly, beyond it, where yet another expanse of ocean would open up before us, and far ahead there would be a new speck of land to aim for.

Sometimes Pallister steered us into narrow passes where the shallow currents ran riotous as river rapids over hidden rocks that sent our electronic depth finder into a beeping tizzy of alarm. "That's one thing about this boat!" he shouted as we were transiting one narrow pass. "I've been driving it around for so damn long that I can take it places that I wouldn't take a big boat! I can go in damn shallow water! Cliff doesn't want to go through these smaller passes!"

Overhead on the cabin roof, ten white tubes—scrolled up NOAA charts with place names scrawled on them in marker—kept working their way out of their wooden rack. Every so often, Pallister would shoot a glance in their direction, curse under his breath, and, with a flick of his left hand, swat them back into place, where they would immediately begin working loose again. And every so often, the half of the windshield with the broken wiper would get so scummy with salt spray that I, the solitary deckhand, would fill an empty Gatorade bottle from our five-gallon bladder of fresh water, stick my hand out the port window, and douse the windshield clean.

At last, the speck of land on the horizon was Gore Point, which, even more than Kruzof Island, resembled the edge of some new world. While Pallister was rhapsodizing about the radial musculature of dolphins, I peered through the starboard Plexiglas at Gore Point's windward shore, searching the jackstraw for the abundance of color Pallister's descriptions had led me to expect. All I could make out at that impressionistic distance were the white hem of the breaking surf, a gray stripe of rocky beach, a bone-white stripe of driftwood, a green band of trees. Then the tide rips reared up and we slammed into the first of many steep, choppy, ten-foot waves. Then the second. Then the third.

"There are some hellacious forces in nature, aren't there?" Pallister shouted.

FOREBODING

In the lee of Gore Point, the ten-foot whitecaps subside as abruptly as they arose. Pallister pilots the *Opus* into the tranquil lagoon he's told me about, our wake unzipping widening V's behind us. The landscape that now reveals itself: a gray crescent beach that stretches between the mainland and Gore Peak, the tip of which, according to the *United States Coast Pilot*, "can easily be mistaken for an island" when viewed from sea.[8] At its northern end, the gray crescent of pebbles terminates at the foot of a great cliff five hundred feet tall, an insurmountable granite palisade that has at this late hour cast a third of the lagoon into bottle-green shadow. Behind the pebbles stretches the forested isthmus. The vertiginous geography of the Kenai Fjords—cliffs plunging into underwater canyons—is the work of two geological forces: tectonics and glaciation. Sixty-five million years ago, colliding plates crumpled the land into valleys and peaks. Two million years ago, in the Pleistocene, advancing glaciers buried the Kenai Peninsula under an ice sheet that extended all the way to the continental shelf, gouging valleys in the land above water, trenches on the seafloor below, all the while grinding mountains down into boulders, boulders into rocks, rocks into pebbles and gravel, which—around twelve thousand years ago, as the Ice Age ended and the ice sheets (by then four thousand feet thick) went into retreat—sedimented into moraines, providing the tides and currents with the raw material for a pebble beach. Geologically as well as historically, this is indeed a comparatively new world, a work still very much in progress—or at least very much in flux.

A few dozen yards from shore, in the shadowy shallows of the lagoon, floats Pallister's big boat, the *Johnita II*, a fifty-foot Murray Chris-Craft twin-engine yacht. It is in this gray-green world an incongruously harlequin sight, at once shabby and luxurious, white fiberglass radiant above the dark water, colorful laundry fluttering from chrome rails. Atop the uppermost deck rises a structure resembling a treehouse, an unpainted wooden scaffolding enclosed with plastic tarps. Pallister pulls the *Opus* alongside the yacht, cuts the engine, balances himself on the port gunwale, springs at exactly the right moment onto the *Johnita II's* swim deck, and lashes the two boats together. When he's finished, the *Opus* looks like the yacht's sidecar. I crawl from little boat to big boat on hands and knees, voice recorder and digital camera hanging by lanyards from my neck.

The *Johnita II* was clearly intended to be a pleasure craft, not a floating bunkhouse. It has three decks, four cabins, two lounges furnished with armchairs and sofas, a mess the size of a breakfast nook, a well-appointed galley well-stocked with Pepsi, a neglected wet bar, a washer, a dryer, and, praise be, two heads, one of which I retreat to posthaste.

Several owners ago, the yacht allegedly belonged to Ed McMahon of *The Tonight Show*—an allegation that I have not been able to confirm, though the early-eighties decor (burnt-orange curtains, faux-wood paneling) lends some credence to the claim. "He and Johnny Carson had parties on this boat," Doug Leiser, GoAK's crew manager, tells me when I emerge from the head. For the past two weeks, Leiser has been cohabiting the *Johnita II* with his and Pallister's sons, employed for the better part of the summer as GoAK's professional remediation contractors, earning $200 a day apiece—a not inconsiderable sum that Pallister, a bit defensively, insists is fair. "These are all rugged guys," he told me, rugged and hardworking and far more efficient than volunteers; volunteers are good for "community outreach"—nonprofit-speak for "public relations"—but are unpredictable, not to mention expensive to insure. Furthermore, his paid crew would be out here in the wilderness for a month or more, longer than most volunteers can stand, effectively on the job 24/7, with no shore leave for R & R, at least not shore leave of the sort they'd like.

"The boys," as everyone at Gore Point calls GoAK's five remediation contractors, are all approximately college age, give or take a year, so it's no surprise that after two weeks their close quarters have come to resemble a frat house. There are dirty dishes in the galley, *Playboys* on a couch, fragrant sports sandals piled among rubber boots beneath a coatrack heavy with wet-weather gear. There is even, we will later discover, a secret stash of beer that Pallister's sons have hidden from their teetotaling father.

I notice a tattered copy of *Tortilla Flats* beside a rumpled sleeping bag. It belongs to Keiler Pallister, the eldest and tallest of the boys. Of Steinbeck's novel Keiler will later remark, "They sure drink a lot of wine in that book!" From Steinbeck he'll move on to Hemingway's *The Old Man and the Sea*, of which Erik Pallister, the next eldest, will remark, "That the one about the marlin? The whole book's about that?"

Pallister's youngest son, nineteen-year-old Ryan, is the one I find most sympathetic, in part because he reminds me of my more likable students (he wants to know everything I have to tell him about the chemistry and history of plastic), and in part because Pallister is far harder on Ryan than he is on Keiler and Erik. When I mention my son Bruno's fear of sea bathing, Doug Leiser says, "You just got to make him do it."

"And if he doesn't want to," Ryan says sardonically, "pressure him."

These days the *Johnita II* belongs mainly to John Cowdery, a Republican state senator in the Alaskan legislature. Cowdery, per Pallister, isn't much of a "boat guy." During the winter of 2001, while the *Johnita II* was tied up at the docks in Whittier, so much snowmelt gushed from a leaky drainage hose into the hull that the yacht sank to its upper decks. "I guess that pickled it pretty good," says Leiser.

Once a salvage team managed to float it to the surface, Cowdery had the yacht towed around the Kenai Peninsula to Anchorage. Pallister bought a 49 percent share in the wreck for a song and then spent the next five years making it shipshape. He put in new water tanks, rewired the electrical system, replaced the rotten carpeting, built that weird wooden treehouse thing above the exposed upper deck, and cleaned oily bilgewater from the furniture, gray streaks of which remain. He finished his restoration, most of it, last fall. This is the yacht's first summer in action.

Along a shelf under a window the boys have arrayed an exhibit of castaway curiosities scavenged from Gore Point's windward shore: a thimble-size likeness of R2-D2, a hollow plastic witch riding a hollow plastic broom, a weird turtle with a hole in it like a doughnut, a bottle labeled HANDS CLEARNSER [sic], and—lo and behold!—Floatees: three frogs, two beavers, one turtle, one duck. Although I've already seen the Orbisons' impressive collection, seeing these, out here in the wild, at the scene of their recent discovery, feels different, as though I'm getting closer and closer to one of the X's marked on my map. As expected, the frogs are still green, the one turtle still blue, whereas the formerly maraschino beaver has faded to a milky beige, and the duck—the duck has turned the yellowy white of banana flesh.

Although Leiser and Pallister are old friends, it's clear that in the field Pallister is boss. Like a general visiting his troops, he stomps about inspecting the yacht, assessing damages, exclaiming "Holy Christ!" in response to the mess the boys have made, barking commands ("Someone had better get up here and pin down these clothes before they blow overboard!"), all the while demanding intelligence from Leiser.

GoAK's crew has worked fast, faster than expected, Leiser tells us. We are almost too late. The Gore Point midden heap is almost gone. In the past two weeks, ten workers—Leiser, the boys, field manager Ted Raynor, and three volunteers from Homer—have already filled around 1,200 garbage bags weighing, on average, fifty pounds each. In my notebook, I do the math. That's sixty thousand pounds of trash collected along a single half-mile beach. By comparison, consider this: volunteers participating in the Ocean Conservancy's 2006 International Coastal Cleanup picked up on average eight thousand pounds of trash per mile from the coast of New York State. "There's probably a day's work left out here, one good swath," Leiser says.

Impressive as Leiser's numbers are, the success of GoAK's rescue

mission remains in doubt. Pallister still doesn't know how in the hell
he's going to get all that trash off that windward shore once his crew fin-
ishes bagging it. The original plan was to load the bags onto six-wheel
all-terrain vehicles, drive them across the isthmus to the protected lagoon, and transfer the bags onto an amphibious barge. As during the
cleanup of the *Valdez* oil spill, the barge would ferry the bags eighty
nautical miles to the landfill in Homer. But archaeologists with the
State Parks Department, worried that the ATVs could damage the Unegkurmiut house pits, recently told Pallister, no six-wheelers. So now
how is he supposed to get those bags across the isthmus? Sweat equity?
Zip lines? Helicopters?

Nor does Pallister know how he is going to cover his multiplying
costs even if he works out the logistics. He still hasn't raised all the cash
and in-kind donations that his $115,000 matching grant from NOAA
requires, and he won't see a penny of federal money until he does. Last
spring, Pallister campaigned hard to convince Alaska's state legislature
to chip in. He sent letters accompanied by photographs of debris. This is
state land, he reasoned, so it only makes sense that the state should help
clean it up. When his appeals seemed to fall on deaf ears, he did what
any citizen well educated in the ways of American politics would do: he
hired a pair of lobbyists. The lobbyists began persuading pro-development
representatives in Juneau that GoAK isn't "just another environmental
group"—that, in other words, the group poses no threat to anyone's
profitable interests.

The lobbyists delivered, winning a $150,000 allocation for GoAK by a
narrow margin. But at the last minute, Alaska's new Republican
governor—the once obscure, now infamous Sarah Palin—vetoed the allocation. When he first put out his call for volunteers on the local evening
news, Pallister described the "super camp" that GoAK intended to build
out here. There'd be an electrified bear fence, he said. There'd be tents.
There'd be food. More than a hundred Alaskans answered his call. But
without that state allocation, he had to slash his budget. The super camp
was downsized to an unfortified, bring-your-own-food-and-tent, dump-
your-own-shit volunteer ghetto. Instead of a hundred volunteers, he
ended up with five, each of whom had enlisted for a single ten-day tour.
Scrambling to come up with the matching funds his NOAA grant required, Pallister turned to corporate donors, notorious polluters among

them: Princess Cruises, the Alyeska Pipeline, and British Petroleum, whose sunflower logo decorates most of GoAK's garbage bags.

Still falling more than $60,000 short of his estimated costs, with the end of the cleanup drawing near, he began entertaining the unthinkable: hitting up Exxon, an act regarded in Alaska's environmental community as tantamount to signing a contract—in North Slope crude and otter blood—with Satan. What choice did he have? A chartered helicopter would run him approximately $2,000 an hour, the amphibious barge $4,000 a day, and already he was bouncing checks.

After we've unloaded the *Opus*, Ted Raynor and the boys pull on their knee-high rubber boots and ferry us ashore, Raynor's brindled pit bull Bryn perched like a figurehead at his rubber skiff's prow. Pallister's sons are everything I preconceived their father to be. They seem to have stepped from the pages of an outdoor adventure magazine. They carry around carabiners, which they refer to as "beaners." They wear striped knit caps and lots of polar fleece but no life jackets. During their time at Gore Point, they've grown beards. They enjoy punching each other in the arm. Earlier today, to access a beach just north of here, they free-climbed a fifteen-foot cliff. And, I now discover, they can launch and land a Zodiac with acrobatic grace. As we approach the breaking surf, Erik, in a pair of mirrored sunglasses, cuts the outboard. As if on cue, Keiler and Ryan leap out, splashing up onto the pebble beach, hauling the Zodiac with me in it behind them.

In the long midsummer twilight, glacial pebbles clattering beneath their boots, their shadows tall as spruce trees, the boys toss around the new Nerf football that I and Pallister have brought for them, doing their best to keep it from Bryn, who popped their last football with a single chomp. Eyes on the ball, Raynor's pit bull runs barking from boy to boy to boy. "That damn dog just bit my ass!" one of them shouts.

Meanwhile, two volunteers, Bree Murphy, a researcher from the Center for Alaskan Coastal Studies, and Michael Armstrong, a reporter for *Homer News*, are playing bocce ball with plastic fishing floats on a sandy course they've groomed clean of pebbles. Looming over the scene is a shipwreck. Rusted, sundered, half-sunken into the pebbles, it is in an advanced state of decay, but its name—*Ranger*—

can still be read, written in white block letters on the remnants of its black bow.

Pallister and I are anxious to have a look at the cleanup site before dinner. Ted Raynor leads the way, Bryn racing on ahead, sniffing the ground for marmots and bears. Raynor is a forty-something bachelor, with close-cropped hair turning from orange to gray, but like me he still has boyish cheeks, cheeks so ruddy with sunburn and burst capillaries, he seems to be blushing even when he isn't. As we walk, he rants. He rants against "the idiot" who wrecked the *Ranger* and didn't bother to clean it up. He rants about the archaeologists and their precious "culturally modified trees"—the ones modified by Unegkurmiut. "If I take my chain saw and cut down a tree tomorrow, four thousand years from now they're going to be worshipping me because I culturally modified a tree," he rants. "I, Ted Raynor, modified it."

In response to this remark I think and feel complicated thoughts and feelings. When you embed as a journalist, no matter which subcultural group you embed in, it's hard to resist the tug of sympathy, which is perhaps why organizations from the U.S. armed forces to the Gulf of Alaska Keeper invite strangers like me into their midst. Strong as gravity, the tug of sympathy can pull you out of your detached point of view into the foggy atmosphere of relativism. It's kind of like the Stockholm syndrome, or what happens in *Heart of Darkness* to Mr. Kurtz. Before you know it, you're rooting for your sources even if, under other circumstances, under the influence of different sources, you might root against them.

I want GoAK to succeed. I also sympathize with the archaeologists. The Unegkurmiut didn't merely alter trees when they were hungry. These spruce forests were for them what whale oil was in Melville's time and what fossil fuels are in ours—the resource on which the entire economy, the entire material culture, depended. Spruce bark served as siding for Unegkurmiut houses and steam baths. With the pitch they patched the skin hulls of their kayaks and bidarkas, waterproofed their wooden bowls, started their fires. They chewed spruce pitch chewing gum, and applied spruce pitch bandages to their wounds. A culturally modified tree is to an Alaskan archaeologist what an oil field or strip mine or, for that matter, a collector beach would be to Raynor's hypothetical archaeologist of the future.

Most vehemently, Ted Raynor rants against Sarah Palin and all the blind indifference that she in his mind has come to represent: "You'd think that since this is a state wilderness marine park, the only one we have, the state would take some responsibility, but we got our funding cut by the state, by our governor." A few hundred yards south of the bocce court, just past a big color wheel that one of the volunteers has assembled out of fishing floats on the gravelly sand, we turn single file through the surfgrass onto a narrow trail. "If state land is not a state responsibility, then is it the Indonesian government who's responsible?" Raynor asks. He is clearly fond of this line about the Indonesian government. Over the next few days, I will hear him deliver it multiple times. I don't yet have an answer to Raynor's question, but Sarah Palin did, an answer both simple and simplistic: since much of the debris comes from elsewhere, it is neither Alaska's fault, nor Alaska's responsibility.

By the end of that summer, I will get to know Raynor pretty well. In addition to delivering screeds, he likes to imagine that his actions are divinely ordained, but his preferred deities, unlike Palin's, are Mother Nature and his dead pit bull Codi—Saint Codi, Raynor calls him now. For good luck, he keeps a gold-plated locket of Saint Codi's ashes on the dashboard of his eighteen-foot tin fishing boat, the *Cape Chacon*, anchored across the lagoon from the *Johnita II*. "Mother Nature likes us," I will hear him say once. "We call what we do 'wiping Mother Nature's ass.'"

Raynor also likes to put on authoritative airs, as if he were in charge of an actual military operation. Instead of answering yes-or-no questions "yes" or "no," for instance, he'll say, "affirmative" or "negative," as when, after a volunteer mentions that Japanese fishermen no longer use glass balls as floats, Raynor replies, correctly: "Negative. They still use them." Formerly a charter boat captain like Cliff Chambers, a couple of years ago he got sick of chartering. Now being field manager for GoAK's summer beach cleanups is his only job. Last year he worked six weeks. This year he is going to work ten. "Hey, I'm improving," he says. The rest of the year he trains for marathons by running up mountains. According to Pallister, Raynor has run eight hundred miles in the past nine months.

A favorite topic of conversation is which actresses he would and wouldn't "do." He finds it amusing to pretend that the actress Jessica

Alba is his girlfriend. Sarcastically, he's given to using what he thinks is fashionable slang—"I'm down with that," or "You the man." He's also given to rhapsodizing, with heartfelt sincerity, about the beauty of the Alaskan wilderness in general and of Gore Point in particular. It is, he feels, one of the most beautiful places on earth. The Alaskan wilderness and his pit bull—the living one—are the two things he cares most about in this world. He tried settling down once, with a kayaking guide. Then one morning she woke up and said, "I thought you might be the one. But you're not."

The trail dips and meanders along the boundary between a spruce forest and a meadow where wildflowers are in bloom. Raynor identifies them for me by name: purple fireweed, lupine, wild geranium, chocolate lily. "Know why they call it a chocolate lily? 'Cause it smells like chocolate. Take a sniff." I bend the somewhat purple, vaguely chocolate-colored flower to my nostrils. It doesn't smell like chocolate. It smells like a flower. "Gotcha," Raynor says, with a big self-satisfied grin that makes it easy to imagine what he was like as a teenager.

The trail curves into the forest. We pass an Unegkurmiut house pit, and a culturally modified tree. I begin to notice the branches Pallister had mentioned, the ones sheared off by the winter storms. They're furry with moss. On the forest floor beneath them grows a dense understory of salmonberry and devil's club, the latter of which Pallister warns me not to touch. Pinching the edge of one of the big flat leaves, he gingerly lifts it up. The underside bristles with thorns that are, he says, "damn hard to get out."

Ravens caw in the treetops, and Bryn, AWOL in the understory, barks in reply. Here and there shafts of sunlight penetrate the canopy, gilding the trunks of some of the trees but not others, dropping pretty pools of shine that make the devil's club fairly glow. As clouds pass overhead, the pools dim and brighten, brighten and dim. In the distance, trash bags, some yellow, others white, flash between the trees. "We used to be walking by garbage already," Raynor says. We are at present exactly four hundred feet from the ocean. Raynor knows. Michael Armstrong, the reporter for the *Homer News*, "GPS'd it." If you listen closely, you can barely hear the faint huzzah of the surf crashing on the windward shore. The huzzah grows steadily louder as we walk.

"Holy crap," Pallister says. Our file comes to a halt. Snaking across the forest floor is what appears to be a black plastic anaconda but what is in fact some sort of PVC duct, at least a dozen feet long, pumped full of Styrofoam. "Haven't seen anything like that before," says Pallister. "Foam must be to keep it from crushing at depth probably. Christ almighty." A little farther he spots a hollow stalk of bamboo and shouts, "Hey! Bamboo!" Bamboo, of course, is not native to Alaska. The stalk of it lying there in the devil's club almost certainly traveled here from Hawaii or Asia. My hopes rise. This second beachcombing expedition promises to be far more fruitful than my day trip to Kruzof Island. I am nearing, I feel certain, the terminus of one thread in my tangled trail.

I hoped that in following my trail I'd attain some variety of enlightenment, but here, on this wild isthmus, approaching this trail's end, I am if anything increasingly confused, increasingly, well, bewildered. Maybe Alaskan travelers like John Muir and Rockwell Kent have spoiled me to Gore Point's wild beauty, leading me to expect too much.[9] The wild beauty—and "purity" and "sublimity"—of such Alaskan places have been remarked on so often that Gore Point's wild, pure, sublime beauty seems to me unremarkable. The usual words and thoughts and metaphors, the comparisons between a forest and a cathedral, or between the wilderness and the Louvre, have been worn out from overuse, the Alaskan scenery wrung dry of its power to astonish or exhilarate or enrapture. I can't help wondering whether we need to reimagine the meanings of *wilderness* yet again, or perhaps abandon the word altogether. Perhaps its contradictory associations have worn the word out. Perhaps it signifies so much, it signifies nothing. Perhaps the more we worship it, preserve it, memorialize it, manage it, the more humanized, the more man-made, the more iconic, the more synthetic the wilderness becomes. Gore Point isn't simply an isthmus in the wilderness, after all; it's an isthmus in a state-protected wilderness park. Can a park be wild?[10]

The trail bends 90 degrees as it nears the shore, and then we are there, before a great cairn of trash bags, pilgrims before a shrine. Among the bags are spherical fishing floats strung like beads the size of melons onto loops of rope. Pallister is dumbstruck, rapt with admiration. "I'm amazed you guys got this cleaned up this fast," he finally says. "I'm completely amazed." Surveying the scene, we can see other heaps of bags, distributed every few dozen yards along the length of the beach near the forest's edge. There are more bags piled about back in the forest. Here and there, clustered in the grass, are loose objects too big or heavy for bags—the wheel of a car, a microwave oven, a television screen that, shorn of its cabinet, looks naked somehow, like a brain without a skull. Crusty sucker rings left by gooseneck barnacles dot its glossy screen. The barnacles themselves are long gone, scraped away in the surf, perhaps, or plucked off by scavenging birds.

A hundred yards farther we leave the trail and wade through the devil's club, thorns catching at our pants, to the last acre yet to be cleaned up. Behind the moldering trunk of a fallen spruce, a great drift of flotsam has collected, like water behind a dam. As we approach, the mossy earth begins to crackle and crunch underfoot. How strange! What is that sound? And then I recognize it, a sound one does not expect to hear in an old-growth forest: that of crumpling plastic bottles, a whole stratum of them, buried under humus and moss. This is what the entire shore was like two weeks before, Raynor says, and given the bagged evidence, I believe him.

I climb down into the remaining drift and rummage around. At the surface, drift-net floats of the sort Ebbesmeyer showed me on his patio are the most abundant item; polyethylene water bottles, the second-most, many of them embossed with Asian characters.

"An amazing amount of crap, isn't it?" Pallister says.

It is. Lost in a crap-inspired reverie, the three of us marvel for a moment in silence. Off in the understory Bryn is still barking at the ravens. Somewhere nearby a creek is flowing over rocks. The sunlit leaves of the devil's club brighten and dim. Along with the wind in the branches, you can hear the unseen waves. And I'm sitting atop plastic. I unearth a flip-flop, and then, a few moments later, an empty container of Downy, the fabric softener. Surely, somewhere in this great drift there is yet to be found a Floatee.

Pallister had answers to some of my doubtful questions about the value of a pristine wilderness. He thought the Gore Point midden heap was unsightly, but he objected to it on ecological as well as aesthetic grounds. This waste, he believed, was genuinely hazardous. "They're pulling plastic out of that gyre and analyzing what's on the surface of it," he told me back in Anchorage, "and they're finding concentrations of persistent or-

ganic pollutants a million times higher than in the surrounding seawater." By "they" Pallister mainly meant the legendary Charlie Moore and
his crew of sailors and researchers.

The Great Pacific Garbage Patch is not merely a cosmetic problem,
Moore like Pallister contends, nor is it, he believes, merely a symbolic
one. No one knows exactly how many marine mammals and sea turtles
and seabirds die when they entangle themselves in debris or ingest it.
One widely cited if dubiously round estimate puts the toll of casualties
at 100,000 animals a year. "Entanglement and ingestion, however, are
not the worst problems caused by the ubiquitous plastic pollution,"
Moore wrote in his *Natural History* article. Plastic polymers, as any toxicologist can tell you, have the peculiar propensity to adsorb chemicals
colloquially known as POPs—those "persistent organic pollutants" Pallister was talking about.

Such substances are "hydrophobic," meaning that water molecules
repel them. In the ocean, they collect at the surface, much as olive oil
collects atop vinegar. POPs are also lipophilic, which means that oily
substances attract them—or "adsorb" them, to use the term of art. In
chemistry, when one substance is *ab*sorbed by another it permeates it.
An *ad*sorbed substance merely coats the surface, which is what POPs do
to oily solids such as petrochemical plastics. The propensity of plastics to
adsorb such substances is so well-known, in fact, that in the lab, toxicol-
ogists will sometimes dip a wand of sterilized plastic into a water sample
to see what it collects. POPs also happen to include many of the villain-
ous poisons made famous by Rachel Carson's *Silent Spring*—PCBs, dioxin, DDT. Now controlled in most developed nations but less so
elswhere, such toxins are surprisingly abundant at the ocean's surface.
Even in the U.S., where they've been banned, PCBs continue to leach
from dumps—on coastlines, on abandoned military bases in the
Aleutians—into the watershed. By concentrating these free-floating
contaminants, Moore worries, even microscopic particles of plastic
could become "poison pills."

Such fears are not new. In 1972, long before Moore went trawling in
the Subtropical Convergence Zone, Edward J. Carpenter of the Woods
Hole Oceanographic Institution found between one and twenty plastic
particles in every cubic yard of Long Island Sound. A few years later, he
found elevated levels of plastic and tar at the heart of the Sargasso Sea.

Other oceanographers followed Carpenter's lead. If you're awake enough, and have a long enough attention span, you can read dozens of scientific papers published in the 1970s on, for instance, plastic containers discarded on the beaches of Kent, on the "evidence from seabirds of plastic particle pollution off Central California," on concentrations of DDT and PCBs in the western Baltic. The list goes on and on. Over the course of the 1980s, the bibliography of scientific studies on the impact of plastic pollution grew to encyclopedic lengths.

Carpenter, like many of the scientists who followed him, also worried about plastic's environmental impact. Particles could encourage bacterial growth, he speculated, or block the intestines of fish, and like Moore he worried about toxicity. It wasn't the contaminants the particles had adsorbed that troubled Carpenter most. It was the toxins in the plastics themselves, a concern Moore, like many of the scientists who came before him, shares. Once fish or plankton ingest those poison pills, Moore speculates, the poisons both in and on the plastic may enter the food web. And since lipophilic toxins concentrate, or "bioaccumulate," in fatty tissues as they move up the chain of predation—so that the "contaminant burden" of a swordfish is greater than a mackerel's and a mackerel's greater than a shrimp's—these poison pills could be poisoning people too.

Far-fetched as the theory sounds, the bioaccumulations of mercury in seafood provide a well-established precedent for the plastic-poisoning hypothesis. According to the "National Report on Human Exposure to Environmental Chemicals" put out in 2007 by the Centers for Disease Control (CDC), all sorts of synthetic substances with scary, polysyllabic names are getting into the average American's bloodstream, including both POPs and ingredients in plastic—ingredients like the endocrine-disrupting, gender-bending phthalates that resin manufacturers add to PVC to make it more pliable. As early as 1978, a paper published in the prestigious journal *Science* identified "phthalate ester plasticizers" as "a new class of marine pollutant."

The sources of these contaminants are usually impossible to determine, which is why the plastic poisoning hypothesis is so difficult to prove, or redress. When it comes to plastic, the "single biggest 'smoking gun,'" in Moore's opinion, is bisphenol A. First concocted in the laboratory as a synthetic form of estrogen, bisphenol A was later polymerized and is now found in everything from baby bottles to compact discs to the linings in tin cans—linings invented to protect us from botulism. Bisphenol A turned up in 93 percent of the urine samples tested by the CDC.

The mere presence of a toxin does not make it a health risk, of course, as the precedent of mercury illustrates once again. Various factors— from dose size to developmental stage (fetuses and small children are most vulnerable) to genetic predispositions—can negate or amplify the danger, and the effects of endocrine disruptors like bisphenol A, which can play out in subtle ways over the course of a lifetime, are especially difficult to determine. High doses of bisphenol A, for instance, have been found to cause weight loss in rats. Counterintuitively, however, small doses administered in utero led to abnormal weight gain later in life. Some scientists now suspect that fetal exposure to substances like bisphenol A may be partly to blame for soaring obesity rates and declining sperm counts in people, and to rising rates of hermaphroditism in populations of frogs and fish that inhabit polluted waterways.

The trade group that represents the chemical companies and resin plants and extruders that together constitute the plastics industry was until recently known as the American Plastics Council. It is now known as the American Chemistry Council; *plastics*, apparently, is a dirtier word than *chemistry*. Plus, with its new name, the trade group is easy to confuse with the American Chemical Society, an independent scientific guild that, among other things, credentials university chemistry departments. The ACC has called the studies on plastic poisoning "fascinating" but inconclusive, and even worried scientists agree: there are no conclusive findings yet. No conclusive findings, just some disturbing ones. Although a PVC duck in the bathtub may well be harmless to your child, no one really knows how the postconsumer plastics that escape the landfill are altering the chemistry of the environment. The accidental experiment, which began a century or so ago, is ongoing.

"Holy Christ, Ted!" Pallister exclaims while I'm rummaging around among the drift-net floats. "We're going to have to find some clearings for the helicopter!"

Last week, Pallister contacted a helicopter pilot in Homer, who'd assured him that timber companies regularly airlift logs out of forests as dense as this. So long as GoAK loaded the debris into Super Sacks, and so long as the weather wasn't too foul, the forest would pose no problem. Super Sacks are giant, white, rip-proof plastic bags, the size and shape of a balloonist's gondola, that the shipping and construction industries use to sling cargo—up to four thousand pounds of it—through the air. Peering into the canopy of branches, imagining the airlift, Pallister has his doubts. My gaze follows his. As the spruce trees rise, they seem to draw into a twiggy funnel at the distant end of which is a gray asterisk of sky.

EUREKA

That night eating halibut tacos around the campfire I learned that the seven Floatees on display aboard the *Johnita II* weren't the only ones GoAK had collected in the previous two weeks; others had been secreted away in backpacks and duffel bags, destined in most cases for the eBay auction block, where they could fetch, rumor had it, anywhere from a few dollars to a few hundred. Nor were toys the only items mentioned in *Beachcombers' Alert!* that the currents had delivered to Gore Point. Hockey gloves had been found. So had a computer monitor. So had numerous toddler-size Nike sports sandals, a shipment of which had fallen overboard in 2003. Gore Point was a driftological mother lode, and yet—aside from Michael Armstrong, a loyal follower of Curtis Ebbesmeyer who the next morning would be returning to Homer with a duffel of derelict treasures—no one besides me seemed particularly interested in the riddles the tides had written with flotsam and jetsam on Gore Point's windward shore.

No room for us aboard the *Johnita II*, Pallister and I retired that night to the *Opus*—he sleeping astern, atop the engine hatch, mendicant-style, without a blanket or pillow; me in the V-berth, which was a kind of crawl space, small as a pup tent, inside the bow. The V-berth was where Pallister's boys would sleep on hunting trips when they were young, all three of them, snuggled in there like cubs in a den. It was padded with a thin foam mattress, and furnished with flannel blankets, and cozy as could be. There was a little reading light, and fishing rods velcroed to the bulwarks. Going to sleep, I felt the way I used to feel camping in the backyard as a kid.

Late in the night, a storm blew up. A big swell came rolling in from the Pacific. The waves sloshed and slapped against the plywood hull all around me, and the little boat banged and thumped and knocked against the yacht. Waking in the brief predawn dark, I felt more like Jonah tempest-tossed on the Mediterranean than a kid in a tent. I assumed that this was what it was usually like, sleeping in a V-berth in the maritime wilderness. The next morning the skies were mostly clear, the waves mostly calm. That, Pallister told me when I boarded the *Johnita II*, was one of the roughest nights at sea he could remember.

After breakfast, Cliff Chambers, having delivered his shipment of inferior babes, prepared to ferry the previous rotation of volunteers back to Homer. As the *Patriot* was about to raise anchor, Chambers's silent, ponytailed deckhand stripped down to a pair of black bikini briefs, climbed onto the starboard rail and did something astonishing.

With Pallister, Leiser, Raynor, the boys, and me looking on from the *Johnita II*, the deckhand balanced on the *Patriot*'s starboard rail as if at the tip of a diving board, arched up onto the balls of his feet, raised his arms into a wedge, and plunged fingers first—a pale, headlong flash—into the frigid, bottle-green lagoon. Pallister, Leiser, and Raynor shook their heads and jeered. "He seemed like a nice guy," Pallister said. "But I don't think he knew very much about boats the way he was yanking on that Evinrude." The boys and I clapped and whooped. The deckhand surfaced, dog-paddled around to the ladder mounted on the *Patriot*'s stern, pulled himself aboard, and, without comment or a smile or a bow or any other acknowledgment of the derision and admiration with which his astonishing dive had been received, shook his long hair, flinging droplets around in the sun. Moments later, the *Patriot* departed, Pallister disappeared belowdecks with a toolbox, and the rest of us headed off to Gore Point's windward shore.

"It was a few feet down, buried under all sorts of stuff," Raynor said, beaming with pride, when I came running. He was sitting on a log way back in the forest, his T-shirt tucked into his jeans, his jeans tucked into his boots. There was something prim about him. For an outdoorsman, he was surprisingly kempt. After two weeks in the wilderness, his hair was miraculously short-cropped. Perched beside him on the log was his unearthed prize, a hollow plastic beaver. "Must have been there for years," he said. How many years—fifteen, ten, six—was impossible to say.

Although I, like Pallister, am no oceanographer, it occurred to me that here was a flaw in Curtis Ebbesmeyer's scientific method. In the data that the Orbisons and other beachcombers collected, Ebbesmeyer scried oceanic patterns. The Orbisons found more toys in some years—1992, 1994, 1999, 2002, and 2004—than in others. Those peak years, Ebbesmeyer reasoned, must correspond with the laps the toys were taking around the North Pacific Subpolar Gyre, and their laps must correspond, he postulated further, with the gyre's orbital period. Therefore, the North Pacific Subpolar Gyre must complete a revolution "about every three years."

Never mind that the intervals in the Orbisons' records were erratic. Some peak years had occurred two years apart, some five years apart. Averaging the intervals to "about three years" didn't exactly seem scientifically precise. Now, at Gore Point, I'd discovered something else fuzzy about Ebbesmeyer's logic: the date of Ted Raynor's latest discovery—July 16, 2007—indicated nothing about when this particular beaver had washed up. All it indicated was that on this date, today, a foul-mouthed, pit-bull-owning, surprisingly well-groomed nature lover had come along and liberated that beaver from the wrack. The dates of the Orbisons' discoveries were similarly ambiguous. Tyler Orbison, a digger, had chiseled toys from ice, excavated them from beneath pebbles and jackstraw and sand. Ebbesmeyer's data, in other words, had been corrupted by happenstance.

By the time Raynor started hollering about his discovery, his voice echoing through the woods, in that strange way that voices echo through woods, glancing off tree trunks, perhaps, muted by all the rustling forest noises, I'd made a discovery of my own. GoAK's labors, I'd discovered, were even more Herculean than I'd imagined. Cleaning up debris turns out to be slow, mind-numbing, back-straining work, especially if, but not only if, you happen to be recovering from a microdiscectomy. Trying my best to obey my doctor's orders, I'd sit down atop a pile of debris and shovel everything in reach into a garbage bag emblazoned with the sunflower logo of BP. When nothing remained in reach, I'd move over a little and start shoveling again. The other members of GoAK's crew worked all around me, several yards apart, some sitting, some stooping and rising, gleaners harvesting surreal produce—plastic gourds, fungi of foam. Even more than gleaners, what they resembled were rag-and-bone pickers, as nineteenth-century trash scavengers were known. All over the developing world, rag-and-bone pickers are still at work, some of them in colossal landfills from which they somehow scrape a living, and in which they seek out not rags and bones but plastic and aluminum recyclables.

Not all Asian flotsam and jetsam washes out from landfills, however. Now illegal in most of the developed world, the dumping of trash at sea is still widespread in the developing world—Africa, Latin America, Asia. Go to the collector beaches of Europe and you'll find great payloads of flotsam and jetsam originating from the coasts of the Caribbean. Here then is another way in which plastic pollution resembles airborne pollutants: the developed world burned fossil fuels indiscriminately while their economies were maturing and only belatedly decided to mitigate the environmental impacts. So go the coal plants, so goes pelagic plastic. You might then be inclined to condemn those developing countries. But for which markets do Chinese coal plants burn? To whom do we send our virtuous recyclables? We Americans consume and throw away far more per capita than does the developing world. In the early years of the new millennium, shipping containers that delivered Asian goods to American shores often returned carrying postconsumer waste—recycled paper, plastics, metals ferrous and nonferrous. For a while there, postconsumer waste was California's leading export. The Chinese market for raw material had by 2005 driven the value of polyethylene water bottles up to around twenty-five cents a pound—piddling compared with what a castaway toy could supposedly fetch on eBay, but not chump change when you're living, as the Filipino rag-and-bone pickers of Payatas do, on less than two dollars a day.

Out here on Gore Point, what to GoAK's gleaners looked like an unsightly blight would have looked to a rag-and-bone picker, as to a driftologist, like the mother lode. Then, too, on Gore Point, the gleaning, though back straining, was far more pleasant than it would have been in a Filipino landfill. Here, the weather was lovely and the cool forest smelled mostly of sea breezes and spruce needles and damp earth and tree falls undergoing their slow-motion combustion, returning to dust. Then too, the paid rag-and-bone pickers here were earning a hundred times what the average Filipino earns.

Every now and then someone would find something remarkable—a bottle with Arabic writing on it, a toddler-size sports sandal, a Russian vacuum tube—and hold it up for the rest of us to see, before pocketing it or, more often, dropping it into a bag with the other trash. When you stepped back to examine your progress, the difference would hardly be noticeable. But the hours and bags added up, and by late in the day, it was clear that the last of the Gore Point midden heap would be bagged by nightfall.

I was beginning to despair of finding a Floatee myself, and the more I dug and shoveled and gleaned, the more intense my determination grew. I kept seeing them: *There! Under that drift-net float, something green! A frog?* I'd inevitably unearth a disappointment: an empty bottle of dish detergent, say, or yet another float, embossed, often as not, with Asian characters. To make my defeat taste all the more bitter, when I recounted the legend of the bath toys lost at sea to the North Slope volunteers, one of them breezily announced that she'd found a plastic duck a few hours back and dropped it thoughtlessly into her bag. I considered trying to locate and tear open the bag in question. By then we'd piled up dozens of them. It was hopeless. That duck was lost. Who knew what other archaeological treasures, and what other driftological data, and what other stories those bags contained? It was a shame really that, if GoAK succeeded in its rescue mission, all these artifacts would end up along with banana peels and coffee grounds in the Homer landfill before an archaeologist of the ordinary had a chance to examine them illimitably long.

Late in the afternoon, my luck turned. Squatting amid the prickly leaves and stalks of devil's club, plucking kernels of Styrofoam out of the humus, I spotted something. Was it? The familiar tail? Teardrop shape, waffled in cross-hatched lines? I bent back the devil's club with the toe of my rubber boot. Sure enough, there, bleached pale, perched atop dead leaves in the shadow of a spherical fishing float, was a plastic beaver. Not a duck, but good enough.

By dinnertime all that remained on that forest floor were a few sprinklings of Styrofoam—"bear messes," Ted Raynor called them. "You can see the huge rake marks they make with their claws," he told me. "They find something with any sort of smell at all, they just rip into it, they just"—he pretended to be a bear—"*raaar!* The Styrofoam messes we find it's always bears. *Raaar!*"

Our day's hard labor done, we climbed, some of us less nimbly than others, over the slippery, shifting bone-white logs of the driftwood berm, built a fire on the windward beach, and gathered around it to celebrate. A case of Milwaukee's Best and another of Molson emerged from the secret stash. Out on the water the salmon were jumping. Pallister's three sons fetched fishing rods and stood at the edge of the surf, casting and reeling in, until one of them snared a dolly, not as tasty as a silver, but it, too, was good enough.

They gutted it, wrapped it in foil, and baked it among the cinders "barbarian style," and when it was done, they passed the foil package around, and we plucked out steaming morsels of the pink flesh with our fingers, and when there was none left we cooked hot dogs on sharpened alder branches, and Ted Raynor gave a speech about making Mother Nature happy. We'd restored this awesome place to the way Mother Nature had intended it, he said. Although I found his pantheism sentimental and suspect, the mood of beery triumph was infectious. Far, far away, in the lower forty-eight, green consumers were shopping for carbon offsets and energy-efficient lightbulbs. Meanwhile, this merry band of rugged eco-mercenaries or conservationists or remediation contractors or whatever the hell they were had ridden out into the wilderness to do battle against pollution, and the impact of their actions could be reckoned in tons.

We stayed late, tossing the empties into the fire, where the labels sizzled away and the aluminum whitened. Night didn't fall, but the windward shore faced east, and as the sun sank below the mountains behind us, the ocean took on the dull gray sheen of a pencil rubbing, and a weird sort of shadowy darkness gathered in the cloudy sky, a darkness like that of night scenes in old movies. As it gathered, the mood began to turn.

The boys were passing around slices of watermelon. "None for me," Raynor said. "It would make me feel . . . "—he leaned in a little, smirking in anticipation of the funny crack he was about to make, his face flushing an even brighter red from beer and firelight, and lowered his voice conspiratorially. "It would make me feel . . . darker. You know what I mean?"

"No, I don't," Erik Pallister said, but he did know. We all did. The silence grew long and awkward.

"You don't?" Raynor asked with what seemed genuine if drunken bewilderment. *Doesn't everyone know,* he seemed to be thinking, *who likes to eat watermelon?*

"No," Erik persisted, righteously. "Missed that one."

Raynor retreated a little from our sociable circle, toward the edge of the firelight, the edge of the darkness, into the safety of his solitude, and gazed out across the leaden Pacific. Beside him on the pebbles, nose resting on her paws, Bryn had curled up on somebody's sweatshirt, possibly mine. "Sometimes I think I'd like to head out to that horizon until I ran out of fuel," Raynor said now to no one in particular. "And then just pull the plug and have a life-changing experience, if you know what I mean." Once again, we knew what he meant. "A real life-changing experience, pull the plug. There'd be something beautiful about that, you know? To die in isolation."

I said something then that, after I'd said it, sounded more cruel than I'd intended. "That's kind of why Ishmael in *Moby-Dick* went to sea," I told Raynor. "So he wouldn't shoot himself in the head."

"I understand that," Raynor said. He polished off his fourth or maybe it was his fifth beer. Then he tossed the crumpled can into the embers. We'd overstayed the campfire's welcome. Boozy triumph had soured into boozy estrangement. We were no longer merry comrades-in-arms, but a band of shitfaced *isolatoes.*

Preoccupied by the airlift, and by his wife's desertion, in no mood to celebrate, especially when alcohol was involved, Chris Pallister had retired early, gloomily, to the *Johnita II.* The broken water maker wasn't the only technical difficulty afflicting the yacht. There was also, among other things, a broken showerhead: when you turned it on, the showerhead and hose would pop from the plastic stall and fizz around like a demented snake.

Pallister was anxious to finish all the repairs and return to Anchorage. He had people to call, donations to raise, an airlift to arrange. Out here in the wilderness, we were off the grid, incommunicado. The yacht's sat phone was too expensive to use except in emergencies, and then, too, for the past few days, it had been on the fritz. Every extra day his crew spent out here cost Pallister more money. He was also racing against the seasons. Fall comes early to the Kenai Peninsula's outer coast. By mid-August, the purple fireweed would finish blooming, and on the upper slopes of the Kenai Mountains the tundra would be tingeing red. The weather could turn for good. (The first time Thomas Royer, the oceanographer I met aboard the *Morning Mist*, attempted to visit the Kenai Peninsula's outer coast in winter, his research vessel encountered fifty-foot seas.) By mid-August the southeasters could start howling in off the Pacific, buffeting Gore Point's windward shore, making waves surge up into driftwood, stripping branches, hurtling plastic beavers a hundred yards back into those trees. If that happened, you could forget about your airlift. If that happened, Ted Raynor and the boys would have to lash down the heaped bags with cargo nets and pray they survived the winter.

THE CRYING INDIAN

Three days later, the morning after Pallister and I depart, an unfamiliar fishing boat appears at the mouth of the leeward lagoon. Aboard are a man and a boy, father and son. The son is twelve years old. For the last three weeks they've been boating and beachcombing along the outer coast. On their return to Homer, while waiting for a favorable tide, they decided to make one last stop at the best beachcombing spot they knew of: Gore Point. After breakfasting at anchor, they motor ashore, cross the isthmus on the familiar trail, and discover that their happy hunting ground has been replaced by a desecration—garbage bags, great cairns of them piled twenty feet high, twenty yards apart. On the perpetrators of this driftological crime the father unleashes an earful of expletives.

"An idiot named Brad Faulkner came ashore and gave us crap about cleaning beaches," Raynor would write that night in his logbook when he retired with Bryn to the *Cape Chacon*. "I insulted him as often as possible. Then he left."

Three days later, at the Driftwood Inn, in one of the cheapest hotel rooms in Homer, I receive a call from Chris Pallister. "Apparently some jerk stopped by Gore Point," he says. "Brad Faulkner. Some sort of fisherman, I take it. Apparently he tore into the guys for the cleanup, for a lot of various reasons. I'm not sure what kind of bee he has in his bonnet." Then, to his credit, Pallister adds this: "I thought you might want to talk to him."

I'd driven to Homer in an old GMC Jimmy, owned by Doug Leiser, that Pallister had commandeered on my behalf. It hadn't been my plan to drive an old GMC Jimmy to Homer. It had been my plan to continue voyaging on my own, following the currents west, perhaps all the way to the Aleutians, returning to New York by the end of the month, in time for Bruno's second birthday. But I was improvising, surrendering to happenstance, riding the drift, and with every passing day the drift was leading me into wilder waters.

Pallister had convinced me to stick around for the Gore Point airlift. It hadn't taken much convincing. I didn't want merely to learn whether or not GoAK succeeded or failed in its rescue mission. I wanted to witness the denouement for myself. The day after we returned to Anchorage, Pallister managed to land a $50,000 grant from an outfit in Juneau called the Marine Conservation Alliance Foundation. To his great relief, panhandling from Exxon wouldn't be necessary. Now all he had to do was line up the helicopter and the amphibious barge. The airlift could take place as early as next week, he estimated, and if I stuck around, I could be there to witness it.

In the meantime, he proposed a reconnaissance mission to the outer coast of Montague Island. He'd been meaning to conduct a survey of that coast anyway, he said, and was certain the amount of "plastic crap" out there would make an even bigger impression on me than what I'd seen at Gore Point. I'd never ridden in a bush plane, the mention of which summoned from the depths of my mind the theme song of *Raiders of the Lost Ark*. We'd fly out in the morning, fly back in the afternoon. That was the plan. Then the skies darkened, and the rain began to fall on Anchorage, pouring from the eaves of Pallister's condo in a curtain that played a dispiriting drumroll on the little wooden balcony.

It was stormy out at Gore Point, too. "Boy how the weather can change," Raynor noted in his logbook on July 21, the day after Brad Faulkner paid his unwelcome visit. On July 22, Raynor wrote, "Got rocked good last night. The current kept us broadside to the waves all night. Got up at 7:45 A.M. just as the first raindrop hit." The first of many raindrops, it turns out.

In Anchorage, every bush pilot Pallister contacted gave him the same bad news: the weather conditions prohibited a flight to Montague's windward coast. Tired of biding time in Pallister's lonely condo, I accepted the keys to Leiser's SUV as if they were keys to some Alcatraz or Elba of the mind. Nominally, I was going to Homer as a favor to GoAK, delivering four empty fuel drums to Cliff Chambers so that Chambers could, on his next run to Gore Point in the *Patriot*, refuel the Zodiacs and the *Johnita II*. But I had my own, ulterior motives for going.

From Michael Armstrong I'd heard that Homer was a strange place: part fishing village, part tourist trap, part hippie town, part Alaskan satellite of Seattle. According to my *Lonely Planet* guide, Homer was "a mythic realm, like a northern Shangri-La, which bestows itself upon the faithful only after a long, difficult pilgrimage to get there." Homer was also "the arts capital of Southcentral Alaska" and home to a colony of Russian Old Believers, the Puritans of the Orthodox Church, and it boasted more "wonderful eateries than most places 10 times its size," a dubious travel book claim. After three days roughing it at Gore Point, and another two subsisting on Subway sandwiches and Pepsi in Pallister's condo, such enticements were to my ears irresistible. I also wanted to speak to an environmental lawyer named Bob Shavelson.

Director of the Cook Inletkeeper, which unlike GoAK is indeed part of the Waterkeeper Alliance, Shavelson is Chris Pallister's great and true nemesis. Alaska is a big place geographically, but a small place socially and politically. The two men had known each other, and disliked each other, a long time. Towing the *Opus* back to Anchorage, Pallister had told me a story, his narration of which sounded suspiciously one-sided. Before founding GoAK, he'd spent five years trying to start a new Waterkeeper chapter—the Prince William Soundkeeper, it was to be called, and Pallister had expected to serve as the group's first executive director. But something went wrong. An unforeseen event occurred. Another director was chosen in his place, a woman who, Pallister alleged, had compromising ties to commercial fishing. He blamed this coup on Bob Shavelson.

The Cook Inletkeeper's headquarters are in Homer, at the end of a winding unpaved road, in a two-story building that the group shares with the local chapter of the World Wildlife Fund. The place on the sunny morning I visited seemed brand-new, stylishly decorated in an eco-contempo sort of way: dark green carpeting, natural wood trim. A big Calderesque mobile made of driftwood and sea glass spun slowly around in the lobby, above shelves of Cook Inletkeeper T-shirts made in America from organically grown cotton. I noticed on one wall, among save-the-beluga-whale posters, a sign informing me that I had entered a COKE AND PEPSI FREE ZONE.

Accompanied by his border collie, Shavelson—stocky, forty-something, dressed in a button-down, pen tucked into his breast pocket, his graying hair receding into a widow's peak—gave me a tour of the premises, which included a well-equipped laboratory where that morning a summer intern with a newly minted bachelor's degree from Evergreen College was titrating water samples to be tested for pollutants. Like Pallister, Shavelson was concerned about plastic pollution, but he considered it just one among many man-made environmental threats to Cook Inlet—ominous, certainly; grave, possibly; the gravest? Far from. The dizzying list of contaminants that could be found in this watershed belied the pristine illusions peddled in tourist brochures. There were the pharmaceuticals people were pissing into the waste stream, the depleted uranium leaching from ordnance fired during military exercises, the heavy metals from mining tailings, the pesticides carried here by breezes and currents, the oily water dumped back into Cook Inlet by a Chevron production facility at Trading Bay, the "gray water" discharged from superliners, not to mention the invasive species that stowed away on oceangoing ships.

"I've got very strong differences of opinion with Chris Pallister," Shavelson told me after we'd adjourned to his office, where a bumper sticker on a file cabinet commanded me to BOYCOTT CONOCO & CHEVRON. Shavelson objected to GoAK because marine debris was their "sole focus," and because people had confused the Gulf of Alaska Keeper with the Cook Inletkeeper, and because Pallister so indiscriminately accepts and promiscuously advertises donations from known polluters like Princess Cruises and BP, and because one of GoAK's directors, John Whitney, was an administrator with the federal agency, NOAA, charged with disbursing marine debris grants in a supposedly impartial way (later that summer, because of such complaints, Whitney would resign from GoAK), and because Pallister was paying some of that NOAA grant money to his own three sons. This "appearance of a conflict of interest" was "the kind of thing that can hurt organizations like mine that don't operate that way," Shavelson said.

I mentioned to him what Pallister had told me in his defense: how he'd like to expand GoAK's focus but didn't have the deep legal pockets to do so; how he planned to apply for Waterkeeper certification; how volunteers were expensive to insure and less efficient than his three sons. Shavelson conceded that it was hard for environmental groups to litigate against polluters in pro-development Alaska. But he felt that GoAK was only making the problem worse, draining public resources that might be put to better use and in the meantime giving polluters the opportunity to remediate their polluted reputations. In his opinion, the Gore Point cleanup was essentially a boondoggle verging on eco-graft. Beach cleanups could teach the general public to be "good stewards" of the environment, so long as you worked with local communities and enlisted lots of volunteers, which is something that the Cook Inletkeeper tried to do, but cleanups alone did little to solve the problem once and for all. To do that, you had to stop it at its source.

The evidence on this point seems to be unambiguously on Shavelson's side. Year after year, equipped with garbage bags and good intentions, hundreds of thousands of volunteers participate in the Ocean Conservancy's International Coastal Cleanup (ICC), and year after year, the tonnage of debris is greater than before. Seba Sheavly, a marine biologist who ran the ICC until 2005, will happily admit that the Ocean Conservancy's cleanup "has never been about curing the problem of marine debris." It has always been, she told me when I called her at her offices in Virginia, "a public awareness campaign." Sheavly is now a private consultant who lends her expertise on marine debris to such estimable clients as the U.S. Environmental Protection Agency and the UN Environment Programme. She's also worked for the American Chemistry Council, whose public relations department eagerly gave me her phone number. Sheavly considers the 2006 Marine Debris Act "the best chance we've had in years to make real progress." Other environmentalists I spoke to regard the act as merely the latest in a long line of toothless legislative actions that have failed. "If you look at how much plastic is out there," says Shavelson's boss, Steve Fleischli, president of the Waterkeeper Alliance, current federal policy seems, "well, rather comical." In the opinion of both Sheavly and the ACC, marine debris is mainly a local littering problem, and the primary value of coastal cleanups lies in the lesson they teach volunteers—"that what they're picking up comes from them."

As I'd learned from Pallister, and seen for myself at Gore Point, only a fraction of the debris washing onto Alaska's outer coast comes from local litterbugs. On much of Alaska's 33,000-mile-long shoreline, in fact, there are no local litterbugs. On most Alaskan shores, as on those of the Northwestern Hawaiian Archipelago, or Spitsbergen Island in the Arctic, there are no people at all. But there is plastic. I for one wasn't sure what edifying lessons to draw from all the flotsam and jetsam I'd helped bag. I hadn't thrown those water bottles into the ocean, or lost those derelict fishing nets, or sent containers of toys and shoes tumbling overboard.

"I've never known a six-pack ring to take itself off of a stack of cans, go out the back door, get out of someone's kitchen, and look for a bird to strangle," says Sheavly. "It's about pollution, but people are the source of it, in terms of their actions and how they function." No one doubts that people are the source of all the plastic in the ocean. The question is, which people? The answer is more complicated than Sheavly's six-pack-ring scenario suggests.

According to Steve Fleischli, "some of the pollution originates at the beach. Most does not." Data from the International Coastal Cleanup at first seems to contradict Fleischli's analysis. Measured by number rather than weight or volume, 81 percent of the "debris items" that ICC volunteers collected in 2006 came from two somewhat vaguely described sources, "smoking related activities" and "shoreline and recreational activities." But the ICC collects most of its data on recreational beaches. Travel away from the sun-worshipping throngs and the results begin to change.

In 2001, the Southern California Coastal Water Research Project sampled forty-three randomly selected beaches in the L.A. area and found that "debris density on the remote rocky shoreline was greater

than on high-use sandy beaches for most debris items." There was a
fairly simple explanation for the discrepancy: municipalities periodically
cleaned up "high-use sandy beaches" with beach-sweeping machines.
The study's data set reads like the inventory of a really big convenience
store: straws, 84,990; cigarette lighters, 5,810; toys, 2,159. But the single
most abundant item—nurdles—had nothing to do with recreation, or
smoking, or litterbugs refusing to give a hoot. Plastics manufacturers
ship many kinds of virgin resin to extruders and molders in the form of
little pellets. These are nurdles. If you're curious about what nurdles look
like, disembowel a Beanie Baby. The toys are stuffed with them. Even
when compared by weight, on beaches in the L.A. area, nurdles were
fourteen times more abundant than cigarette butts, the fourth most
common item found.

Although quoted in the press far less often than Moore or Ebbesmeyer, the man who deserves the most credit for discovering and mapping the so-called Garbage Patch is an oceanographer named Robert
Day. In the mideighties, Day and a team of other scientists began trawling for plastic in the North Pacific, inventing the sampling and sorting
methods that Charlie Moore would later adopt. Those surveys showed
that most varieties of plastic pollution turn up in unlikely places, including the Gulf of Alaska. The largest concentrations were found north of
Hawaii, in the Subtropical Convergence Zone, which Day, like most
other scientists, has never taken to calling the Garbage Patch.

But things grew thornier still: Sponsors of the International Coastal
Cleanup include a number of corporations that make and sell the sorts
of products the volunteers most commonly pick up, corporations like
Coca-Cola, Dow Chemical, and ITW-Hi Cone, manufacturer of the
notorious six-pack ring. Every year, along with funding and volunteers,
these sponsors contribute inspirational homilies about saving the planet.
"Working together we can keep our coasts clean," ran Coca-Cola's contribution to the ICC's 2006 report, titled "A World of Difference." Marine debris, declared Dow Chemical, is a "problem that we, the citizens
of the world, have the power to stop." Is it? Over the past two decades,
the ICC has shown no sign of stopping it.

Shavelson asked me if I remembered the old "crying Indian" ad, an anti-littering public service announcement that ran on TV in the seventies. Of course I did, as would any sentient TV-watching child of the seventies. If you were born too late, or if your memory's fuzzy, or if you were alive and sentient in the seventies but didn't watch TV, you can now view the crying Indian ad as often as you like thanks to the magic of the Internet. The preservation of pop-cultural ephemera is perhaps one of the most underappreciated blessings the Internet has bestowed on humanity; reviewing the crying Indian ad, I find myself experiencing something akin to those time-traveling intimations of immortality ignited in Proust's mind by a petit madeleine.[11] Said Indian was played by an actor who went by the stage name—ironic for a crying Indian—of Iron Eyes Cody. Even off camera Iron Eyes Cody tried to pass as an Indian of Cherokee-Cree descent. In truth, like many of my Greenwich Village in-laws, he was descended from Sicilians. Irony Eyes Cody he should be called, for his real name was Espera Oscar De Corti. What, in the ad, makes him cry, is a bag of trash tossed onto a highway shoulder from the window of a passing car.

First broadcast on Earth Day in 1971, the ad appeared to be the heartfelt if heavy-handed work of environmentalists. It wasn't. It was part of the Keep America Beautiful campaign. If like many Americans you thought that Keep America Beautiful was an environmental group, you'd be mistaken. It was created by beverage and packaging executives in 1953. By organizing volunteer cleanups and running public service announcements, the group has over the past half century managed to present pollution as an aesthetic problem for which litterbugs, not industries, are to blame. Meanwhile, the group's sponsors—the American Chemistry Council among them—continue to lobby against regulatory actions.

In Bob Shavelson's opinion, GoAK was comparable to Keep America Beautiful. It was an Astroturf group, a Trojan Waterkeeper. Politicians and corporations "love beach cleanups," he told me, "because of the metrics." By metrics he meant measurable results. Results that lend themselves to spectacular photo ops. Results that can be reckoned in tons, rather than in parts per billion. Show a scientifically illiterate layperson the chemical formula of bisphenol A, or try explaining to them the phenomenon of lipophilic bioaccumulation, or endocrine disruption, and their attention will drift off on currents of boredom or doubt. They'll seek refuge in pictures of Jessica Alba or go dipnetting in Bird Creek. Show them photographs like those Pallister had taken at Gore Point, before and during the cleanup, and wallets begin to open. "I have a friend who used to be a Democratic congresswoman," Shavelson said. "She says, 'You know, there's nothing that resonates more with senators than that stuff.'" It's no wonder, says Shavelson, that Ted Stevens, Alaska's famously pro-development Republican senator, cosponsored the 2006 Marine Debris, Research, Prevention, and Reduction Act.

Nor is it any wonder that the commercial fishing industry supported it. Speaking on the condition of anonymity for fear of losing funding, the director of another Alaskan environmental group would later tell me, and I would later confirm, that the largest Alaskan beneficiary of the Marine Debris Program was the Marine Conservation Alliance Foundation (MCAF), the group from which Pallister had just won a $50,000 grant. Since the MCAF had been given the power to disburse an outsize portion of the federal funds allocated to Alaska for marine debris programs, this unnamed environmentalist was, like Pallister, beholden to their largesse. Hence, her request for anonymity. Like Keep America Beautiful, and Alaskans for Litter Prevention and Recycling, the MCAF is no more a conservation group than Iron Eyes Cody was an Indian or GoAK a chapter of the Waterkeeper Alliance. It's a tax-exempt nonprofit created in 2005 by the commercial fishing lobby.

"A lot of what they lobby for favors the big guys," this environmentalist told me. By big guys she meant the owners of the factory ships and processing plants, as opposed to "the little guys," the "cowboy" fishermen like Larry Calvin who own a boat or two. To the MCAF, she said, "beach cleanups are a PR issue."

A few days later, I flew down to Juneau to meet with representatives from the MCAF. Their spokesman, Bob King—an erstwhile radio news anchor who'd crossed over to the journalistic dark side of public relations—readily acknowledged his employers' connections to commercial fishing. The connection was logical and legitimate, King argued. The fishing lobby could more effectively than mistrusted outsiders teach fishermen how to prevent gear loss. Then he took me to the Alaskan Brewing Company, where a flack named Amy Woods plied me with brochures and beer. The beer, I will admit at the risk of product endorsement, was quite tasty. The brochures, less so.

Teaming up with the MCAF, the Alaskan Brewing Company had committed "1% of all proceeds from Alaskan IPA to provide grants that support the cleanup and preservation of the Pacific Ocean and its coastlines." When the word *proceeds* replaces the word *profits*, you know you've entered the realm of smoke and mirrors known as PR. How, exactly, would the Alaskan Brewing Company support the cleanup of the Pacific Ocean and its coastlines? Perhaps by endorsing a bottle bill exacting a deposit on Alaskan IPA? Or perhaps by endorsing a ban on disposables? Or perhaps by opposing more insidious polluters—those that generated greenhouse gases, for instance? Or those that spilled oil or mining tailings into watersheds? Nope. The Alaskan Brewing Company would sponsor cleanups and decorate its packaging with public service announcements.

About half the debris befouling Alaskan coasts is derelict fishing gear, and although most of that gear appears to be foreign in origin, much of it isn't. The MCAF with some truth claims that Alaska's fishermen are, compared with their foreign competitors, exemplary stewards of the waters from which they earn their living; they work with scientists to prevent overfishing, adhere far more faithfully than their foreign competitors to anti-littering laws such as Marpol Annex V, and in fact there is some evidence to suggest that marine debris prevention efforts have had more success in Alaska than elsewhere, slowing the rate of accumulation if not reversing it. Regrettably, under the rough conditions Alaskan fishermen encounter in the Gulf of Alaska and the Bering Sea, some amount of accidental gear loss is inevitable, but it would be unfair, the MCAF contends, to hold law-abiding Alaskan fishermen accountable for debris that is accidentally lost or foreign in origin.

Such tangled threads of liability help explain why the problem of plastic pollution has proved so intractable. If you find PCBs in the Hudson River, detective work can eventually lead you back to the General Electric plant, the "point source," whence they came. Most plastic pollution, by contrast, comes from what environmental regulators call "nonpoint sources," meaning that the polluters are too numerous and too mobile to identify. Which is another reason corporations and pro-development politicians like beach cleanups, Shavelson told me: no one gets sued or fined for fouling the ocean with debris.

Thirty years ago in *Coming into the Country*, John McPhee wrote memorably about what he called "the Alaskan paradox," the irreconcilable cohabitation in Alaska of "the Sierra Club syndrome" and "the Dallas scenario"—of the impulse to sanctify the wilderness and to exploit it. Those two impulses are still at war in Alaska, but the allegiances have shifted and the plotlines have blurred. On the one hand, after the *Exxon Valdez* fulfilled the dire prophecies of environmental Cassandras, threatening Alaska's fisheries and tourism industry as well as the ecosystem of Prince William Sound, fewer Alaskans now regard the Sierra Club as "a netherworld force" than when McPhee visited the state in the early seventies. On the other hand, in Alaska, as everywhere else in America, it's no longer quite so easy to tell who's on the Sierra Club's team and who's playing for Dallas, which is why it pissed off Shavelson so much that Pallister and GoAK had obfuscatorily appropriated the Waterkeeper brand.

He had no problem with environmental groups accepting private donations; he himself relied on donations from local businesses as well as from individuals. But he insisted that there was a difference between donations made with no strings attached and those that were, in effect, a stealth form of "greenwashing." Take those garbage bags emblazoned with the sunflower logo of BP. Just a year before I went to Gore Point, in 2006, BP had spilled 200,000 gallons of crude oil onto the tundra of Alaska's North Slope. In response, the Cook Inletkeeper had sent an engineer to Washington to testify on behalf of HR 5782, an act known as PIPES, for Pipeline Inspection, Protection, Enforcement, and Safety. The act passed. Three years later, in 2009, the U.S. Justice Department would eventually hold BP negligently liable for those 200,000 gallons of spilled oil.[12] In the meantime, here was GoAK, parading BP's logo around on the evening news.

As for why Pallister hadn't been made director of the Prince William Soundkeeper, Shavelson had an alternative explanation: "Let's just say Pallister has a hard time working with other people." I had an idea what he was talking about. To keep the *Johnita II* from revolving around its anchor, Doug Leiser had run a thick line from the yacht's stern to a tree onshore. One morning while we were out at Gore Point, convinced that the anchor was slipping and the line going slack, threatening to entangle the prop, Pallister started yelling and cursing, trying to summon help. Then he'd blasted the *Johnita II's* horn until, from the other side of the isthmus, Leiser and the boys came running. By the time they emerged from the surfgrass, Pallister was sawing through the line with a knife, cursing under his breath about the waste of a brand-new three-hundred-dollar rope. While the boys hoisted and relowered the anchor, he stormed about the outer deck hollering orders and curses. "When Chris gets into one of his whirlwinds, I keep my distance," Leiser had said.

Homer, unlike Juneau and Anchorage, is a geographically bipolar town. Most of the 5,332 year-round residents live up on the forested hills overlooking Cook Inlet. The summertime transients—itinerant fishermen, backpackers, tourists—tend to favor the town's other pole, Homer Spit, a sandbar about a hundred feet wide extending four and a half miles into the seemingly pristine waters of Kachemak Bay, beyond which rise the snowcapped peaks and ice fields of the Kenai Mountains. Thanks to those mountains, the weather in leeward Homer is unusually moderate for the region. Out on Gore Point, Raynor and the boys were stormbound. Here, just fifty miles away as the chartered helicopter flies, the skies were mostly clear.

Driving the length of the spit, you pass boatyards, then a fishing hole stocked with salmon for lazy anglers. On a gravel beach at the shoulder of the two-lane blacktop, visitors in tents and motor homes have set up camp. You come to a commercial strip of gift shops and restaurants and art galleries designed to resemble the quaint shacks of an old-timey fishing village. Then you reach the grounds of the Homer Jackpot Halibut Derby, where during the summerlong halibut season, for a fee, men in splattered aprons will take your day's catch and—with an expert flourish of knives—clean and butcher it. Finally, near the end of the spit, by the docks, you arrive at a three-story house clad in cedar shingles. Fishing floats are strung like bunting along the railing of the front porch. That morning, an aluminum fishing boat was parked in the muddy yard, up on blocks for maintenance and repairs. This is the home of Brad Faulkner, the "jerk" and "idiot" who'd given Raynor and the boys "crap about cleaning beaches." Bob Shavelson, a bit eagerly, had produced Faulkner's phone number.

Inside, the house was unexpectedly nice, airy, with tall ceilings and lots of skylights and lots of art, and Faulkner was unexpectedly thoughtful if also, when it came to the topic of GoAK, expectedly foulmouthed. On his living room wall, beside a Balinese batik silk-screen print of multicolored fish, there branched and curled big dendritic fronds of cold-water coral. Though he used to be a fisherman, Faulkner is now the president of Alaska Custom Seafoods, a dealer in halibut and cod, which he buys from independent fishermen and sells at a profit to processing plants and retailers. The coral branches on his walls were gifts from his suppliers, who'd snared it as bycatch on their trawling lines. Business in recent years had been good, allowing Faulkner to take lots of time off, which is how he liked it. When he wasn't out on his boat, he liked to read. His sizable home library contained many first editions, including one of *Farthest North*, Fridtjof Nansen's narrative of his Arctic expedition aboard the *Fram*. During the course of our conversation, Faulkner kept jumping up to retrieve books he thought I should read.

Pallister's description of Faulkner had led me to expect a closeminded ecophobe, but Faulkner said he had nothing against legitimate environmentalists. He applauded the work the Cook Inletkeeper did. "Somebody has to be watchdogging the oil companies," he said. He also applauded the watchdogs of the timber industry; it was scandalous, he said, the way the National Forest Service had encouraged the building of pulp mills in the Tongass National Forest down by Sitka, many of which were now Superfund sites. That man-made emissions, especially those pouring out of coal plants, were accelerating the pace of climate change was to him "a no-brainer." Although he'd made his living from Alaska's fisheries, he felt that the owners of big factory ships had too much influence setting fisheries policy. He was opposed to bottom trawling of the sort that ravaged cold-water coral. "You don't have to trawl the bottom," he said. "The fish they're trawling for, you can catch at midlevel depths. It costs a little more, but you can do it."

So, no, he wasn't opposed to environmentalists. He simply felt that what GoAK had done on Gore Point was take "this remarkable collection of flotsam and jetsam from the whole freakin' Pacific," and turn it into something resembling "the fucking city dump." The only reason anyone ever stopped there, he said, was to go beachcombing. "Other places collect, but no place collects like Gore Point," he said. "Between here and Cordova most of the coastline is clean, far cleaner than it used to be."

My experiences as a GoAK volunteer lent credence to this claim. How Rockwell Kent lasted six months on Fox Island I do not know. Nor did I know, by my third day on Gore Point, how GoAK's crew had lasted two weeks, with no other company besides one another. It was astonishing, really, that they hadn't already gone all *Lord of the Flies.* Astonishing that Ted Raynor or his pit bull Bryn hadn't been offered up as placatory sacrifice to the local population of grizzly bears in repayment for the Styrofoam playthings GoAK had pirated away into garbage bags. On my last full day at Gore Point, in a cold rain that alternated between drizzle and downpour, the boys in Zodiacs, the other volunteers and I aboard Ted Raynor's boat, the *Cape Chacon,* Bryn once again serving as figurehead, had motored around Port Dick, searching for debris.

Judging from what I'd seen that day, Faulkner was right. Except at Gore Point, there was little trash and little treasure to be found along that stretch of the Kenai Peninsula's outer coast. There were, however, dozens of superior anchorages and camping spots, Faulkner said. Try camping on Gore Point's windward shore and you'd get "hit in the head by a rubber duckie!" He guffawed loudly at the thought. Pallister, to his mind, was "some dingbat," and GoAK was a "phony baloney organization" making misleading use of the Keeper name. He could see why you might want to remove the derelict nets that entangle wildlife, but why excavate all that harmless flotsam?

I explained to him the plastic-poisoning hypothesis.

"The fuel they burned in their boats going out there, back and forth, is probably worse for the environment than that stuff breaking down," he said—a notion, I had to concede, that had occurred to me too.

Furthermore, Faulkner continued, even if there were sound ecological reasons to clean up Gore Point, these "misguided do-gooders" were in his opinion doomed to fail. No way was a helicopter airlift going to work. "That shit's going to be there for years, until somebody comes along and spends a bunch more money to fix it. It just looks like a scam."

You might be inclined to see in the controversy over Gore Point the sort of tempest that invariably brews in provincial teapots, especially in geographically isolated provincial teapots like Homer. Rightly or wrongly, I was inclined to see in it something more: a parable of environmentalism in the information age. All the time, all over the world, in the coal towns of Appalachia and Hunan, in the fishing villages of New England and Indonesia, near landfills in Virginia and the Philippines, near incinerators in Hawaii or the Bronx, on farms in the valleys of the Mississippi and the Amazon, similar debates about our vexed relationship to the natural world are playing out. The more I thought about it, the more it seemed to me that the fundamental unstated question at the heart of the arguments I heard in Homer was this one: How do you measure the value of a place?

In America, this question has always pressed on our minds with unusual urgency. Our contradictory thoughts about nature in general and this so-called New World in particular eventually gave rise to a kind of pantheistic religion, the early scriptures of which were written by New England Transcendentalists. Like all religions, American nature worship has since its inception undergone a series of schisms and reformations and inquisitions, prophecies of the Transcendentalists ossifying into sentimentalities and platitudes, only to give way to new prophecies (the Book of Muir, the Book of Teddy, the Book of Leopold, the Book of Carson, the Book of Hardin, the Books of Brower, Berry, Wilson, Dillard, Lopez, McKibben, Pollan), spawning denominations as numerous and as internecine as those of the Protestant Church: Preservationism, Conservationism, Agrarianism, Wise Use, Environmentalism, Ecology, Deep Ecology, Organic Agriculture, Biodynamic Agriculture, Biophilia, Locavorism, Green Christianity, Green Consumerism. Now, in the information age, Nature goes by as many names as Allah or Yahweh, and assumes as many forms as the Bodhisattva.

What should we value in a place? Beauty, Emerson said, for in the "immortal" beauty of nature a sensitive poet might perceive the discreet but harmonizing designs of "the mind of God." And the organ with which both beauty and Divinity could best be perceived was, for Emerson, the eye. "The eye," he wrote "is the best of artists." Herein, I would suggest, lies one of the seeds of our present confusion. Beauty, like beautification, can be deceiving.[13]

If beauty is the paramount value of a place, what sort of aesthetic adjustor gets to make the assessment, especially now, post-Darwin, when it's hard to credit the notion that in nature we can glimpse "the mind of God"? If Brad Faulkner finds Gore Point more beautiful with its midden heap than without, and Pallister more beautiful without than with, who resolves the dispute? Perhaps then we should measure the value of a place by its usefulness. Usefulness to whom? A miner? A beachcomber? An ecotourist? An albatross? A copepod? A chronically restless artist, such as Rockwell Kent, seeking contact with primal energies? Our generation? Generations of the future?

Such distinctions might seem academic, but how we imagine a place determines how we value it, and how we value a place determines how we allocate our tax dollars or charitable donations—what actions we choose to take, which places we choose to save and what it means to save them. Is beautification tantamount to salvation? Sometimes, perhaps, but not always. And in the information age, which is also the age of images, and the age of public relations, and therefore the age of make-believe—when beautification can be deceiving, when what appears to be a crying Indian turns out to be a con artist for hire—our relationship to the natural world is, it seems to me, more vexing than ever. Our eyes alone are not enough.

THE RESURRECTION OF CHRIS PALLISTER

Four weeks after my voyage to Gore Point, for the first time in twenty years, I'm in San Francisco, my hometown, vacationing with Beth and Bruno, when Chris Pallister's call finally comes: the airlift is on, and if I want to witness it, I'd better get back to Alaska, pronto. Wincing at the price, and at the environmental impact of all this jet travel, I go ahead and book yet another round-trip ticket to Anchorage.[14]

Ryan Pallister meets me outside baggage claim, at the wheel of his father's red Pathfinder. The next day we three drive to Homer, stopping for lunch at a Dairy Queen. The elder Pallister boys are driving down with a friend. When we reach Homer, late in the afternoon, the NOAA forecast is ominous—thirty-five-knot winds, eleven-foot seas. There's a gale warning in effect, and a small-craft advisory in effect. If he goes ahead with the airlift, Pallister could end up paying $4,000 a day for the amphibious barge to lay up at anchor, waiting for this storm to blow through.

If the NOAA forecast weren't enough, Otto Kilcher, uncle of the one-named pop-folk crooner Jewel and the owner of the amphibious barge, calls to say that his water pump is acting up. He can fix it, but it might take a few hours, and he wants to know if his mechanic should keep working through the night. Are they leaving in the morning or aren't they? Pallister decides to give it a day. "*Jeee*sus Christ!" he says, when he gets off the phone with Kilcher. "Things look like they've un-raveled now!" He and his boys are staying with an acquaintance in Homer. I retire for the night to the Driftwood Inn, wondering whether I've just blown several hundred dollars on a last-minute flight to Alaska.

Out at Gore Point, aboard the *Cape Chacon*, Ted Raynor makes his daily entry in his logbook: "Otto Kilcher and the crew are supposed to be here late tonight. But there's a gale warning out for tonight and to-morrow so who knows if they'll make it. They and the helicopters are supposed to get the garbage loaded tomorrow. Fat chance."

The following morning the forecast hasn't improved, but the barge's water pump has. Otto Kilcher's meter is now running, and a local TV news team is already on its way down from Anchorage. This might be GoAK's only shot. If Pallister waits another day, the forecast could im-prove, but it could worsen. He decides to chance it. The elder Pallister boys will fly out ahead in a floatplane to help Leiser and Raynor ready the Super Sacks. Pallister and Ryan, his youngest, will come out tomor-row night on the barge, weather permitting. There's an extra spot on the floatplane, mine if I want it. If the barge is delayed by bad weather, I could end up stuck at Gore Point and miss my return flight home, at considerable expense. I've never ridden in a floatplane. I, too, decide to chance it.

The Pallister boys, invariably polite, let me have the good seat, up front beside the pilot, a tall, lanky fellow named José de Creeft, whom we meet beside his dock on the lake that serves as Homer's unofficial airport. De Creeft wraps our luggage in garbage bags and stuffs them into the plane's hollow pontoons. We buckle ourselves in, clap miked headsets over our ears, and go taxiing across the water. Through the blurry discus of the prop I watch a pair of swans and a gray cygnet paddle in a hurry out of our path. Then the motor accelerates into a deafening whine, then we are aloft, the silver waters of Kachemak Bay flashing far below. De Creeft pilots us through a foggy notch in the snowcapped Kenai Mountains. Here and there through the fog you can see waterfalls pouring from ledges, feeding into the Red River far below. "Been raining," De Creeft says through his headset mike. "Lot of water coming out those rivers." We bounce around a little. "Tends to be pretty squirrelly in here for the wind," De Creeft says.

With disorienting speed we reach the fjords of the outer coast. Soon Gore Point's leeward lagoon comes into view. Toylike at first, the *Cape Chacon* and the *Johnita II* grow life-size as we make our descent. De Creeft touches down, skis up to the *Cape Chacon*, unloads our bags, and minutes later is taking off again, anxious to make Homer ahead of the anticipated storm. The plane's drone fades into the distance and the wake of silence it leaves behind is immense.

With no time to waste, we work late loading 1,800-some-odd garbage bags of flotsam and jetsam one by one into Super Sacks. My postoperative convalescence now complete, I help out, too. Rain has puddled in the folds of plastic, and when we lift a bag it pours off, drenching us. When we finally call it quits, we are all wet and exhausted and eager to learn what tomorrow will bring.

"It is now 11:04 P.M.," Raynor will write in his logbook at bedtime. "Bryn and I are aboard for what I hope is the last night at Gore Point."

Waking early, I am relieved to see, out a port window, in the dark, a pair of lights like eyes above the water. Dawn breaks to reveal Otto Kilcher's amphibious bow-loading barge, a one-hundred-foot steel box called the *Constructor*, anchored on the far side of the lagoon. It is shaped something like a very long black shoe, the white tower of the wheelhouse resembling a spat. To the east the sky begins to turn all peachy, then the upper slopes of the western mountains go from black to green. As the sun rises, the brightening line of green slides slowly to sea level, the darkness withdrawing like a sheet. There are few clouds overhead, and only a mild breeze blowing, which goes to show how much you can trust NOAA forecasts out here on the unpredictable coast.

As we're tugging on our Sitka sneakers and preparing to board the

Zodiacs, Ted Raynor, like a high school football coach before a big game, gives us an unfortunate pep talk: "Mother Nature's saying, 'Please, please finish what you started. Please give me the big O.' We've got her stroked and stoked and now we've just got to finish her off."

A visitor who happened by Gore Point this morning might well wonder if a reenactment of the *Valdez* oil spill is under way. Here again is the barge, here again the garbage bags of debris, here again the remediation contractors assembled on the pebble beach. All that's missing is Exxon's oil. Slumping into overstuffed Super Sacks as if they were Barcaloungers, dressed in jeans and Polarfleece, Raynor and his crew gaze west, beyond the yacht and barge, to the Kenai Mountains, above which, any moment now, they expect the helicopter to appear. It's foggy on the other side of those mountains. You can barely see the clouds, rising over the crests like foam over a brim. "God's smiling," Raynor remarks of the favorable weather. "God's saying, 'Thank you. Thank you for cleaning up Gore Point.'"

A half hour later, he isn't so sure what God is saying. The helicopter was supposed to have arrived by ten. It's getting on ten thirty. Has something gone wrong? Is Homer weathered in?

"Wait a minute," Raynor says. "I think I see it."

"Where?" asks Leiser, who's already thrown on a hard hat, in anticipation.

"There, to the left of that notch. See it?"

Everyone squints.

"Sorry," says Raynor. "False alarm. Just a cloud."

The Pallister boys rise from their Super Sacks, walk down to the surf, and begin amusing themselves with strands of bull kelp, whipping the slick green ropes toward the water as if casting lines. Then they drop the kelp and skip a few rocks. Then they sit down on the beach and study the horizon. I do likewise.

At last, from the direction opposite from the one expected, the unmistakable throb of a rotor can be heard, growing louder. Almost in unison, we rise, turn around, and shade our eyes with our hands, waiting for the helicopter to burst from behind the treetops. But then the noise fades. The treetops toss around in the wind. We continue to stare. "They must be doing a flyover of East Beach," Leiser says. "Probably the

TV crew wants an aerial shot." The treetops keep tossing. At this distance the rotor sounds like a neighbor's lawn mower. Now, thundering, it appears, swooping magnificently past, dark blue, alive with gleams, flying low enough that it's easy to read the words MARITIME HELICOPTER written in white along its underbelly. Here in the wilderness it seems angelic. The pilot banks over the lagoon, over the *Johnita II*, over the *Constructor*, where Chris Pallister stands on the deck looking up.

To my surprise and Pallister's relief, the airlift went almost perfectly according to plan. Out of the treetops the orange hook came snaking down on its 125-foot cable. One of the elder Pallister boys, wearing a hard hat, leaped forth and snatched it. The other lashed on a bundle of Super Sacks and fishing floats. Meanwhile, Raynor stood by filming with his camcorder or snapping photographs. I stood by scribbling notes. Doug Leiser stood by barking something into his handheld CB, something only the helicopter pilot could make out. Then we all stepped back in anticipation, ready to dive into the devil's club should the Super Sacks, like a berserk wrecking ball, swing our way.

Two hundred feet above us the helicopter pilot, Drew Rose, began slowly, artfully, to ascend. The cable drew taut. The loop of rope threaded through the handles of the Super Sacks drew taut. The plastic of the Super Sacks crackled and stretched. The rainwater that had the previous night puddled into their folds gushed off. Then came the oddly beautiful moment of levitation, when this bulging, cumbersome bouquet of derelict flotsam, still dripping rain, spinning and swaying a little, took improbably to the air, snapping a few spruce branches as it went, and vanished into the blue asterisk of sky. In the quiet that followed the helicopter's departure, the Pallister boys began readying the next load. Meanwhile, across the isthmus, aboard the *Constructor*, Pallister and his youngest son, Ryan, were receiving Drew Rose's delivery.

And so the day passed. For six hours, the helicopter flitted back and forth among the treetops like a bee among blossoming stalks, its busy collecting interrupted by surprisingly few unforeseen events. Once, late in the afternoon, as the helicopter passed over the isthmus, a portion of a dock fender, made of galvanized rubber, broke loose and plummeted out of the sky, landing among the chocolate lilies near the forest's edge

with an anticlimactic plop. And once, on the *Constructor*'s deck, after
Pallister had unburdened it of Super Sacks, the swinging, thirty-pound
hook clocked him in the forehead, drawing a trickle of blood. But the
injury wasn't serious, and by the time the airlift ended, he'd recovered
his wits enough to sit down on a driftwood log before the television
camera and give Channel 2's blond anchorwoman a valedictory interview in which he thanked GoAK's donors without mentioning them by
name and hailed the tireless and triumphant labors of his crew.

He looked strangely melancholy for a man who after many long
months of arduous preparation, enduring many unforeseen events, had
just regained a paradise. He seemed even smaller than usual. His smile
still looked like a scowl. He still had that pop-eyed squint. His monk-
ish hair was molded into the bowl shape of his hard hat. Watching him,
I thought of what the television audience wouldn't see—the lonely condominium to which Pallister would soon return, the estranged wife
who, several months later, would file for divorce, the toxins adsorbed
onto all that flotsam, toxins that continued to circulate through our watersheds and through the ocean's currents despite GoAK's labors.

I thought, too, of the day, back in July, on my first visit to Gore Point,
when, while Pallister stayed behind to finish his repairs to the *Johnita II*,
the rest of us had motored around the Gore Peninsula in search of debris. Northwest of Gore Point's leeward lagoon, we'd landed in Takoma
Cove. High tide there reaches all the way to the surfgrass, green droopy
bursts of which grow thick along the forest's edge. Unlike that of Gore
Point, Takoma Cove's littoral zone teemed with flora and fauna. On the
small dark-sand beach were outcroppings of rock, lumpy and pocked as
meteors. In their concavities tide pools had formed, and over every ten-
able surface grew a slippery green pelt of sea sac (*Halosaccion glandi-
forme*). Below the waterline, the bulbs of the sea sac swelled into fingers
that seemed to be feeling around as they swayed in the surf. Above the
waterline, they lay limp—golden-green deflated balloons—atop con-
stellations of limpets and barnacles. Just past the tideline, in a clump of
surfgrass, one of the Leiser boys, Bryan, discovered yet another Floatee,
a duck. Unlike those specimens entombed in the Gore Point midden
heap, this one seemed to have only recently arrived, tossed up into the
surfgrass, perhaps by the last storm. Had it been at sea all that time, circling between Alaska and Kamchatka? Had it spent a season on Attu

Island, near the end of the Aleutian chain? Or had it been here on the Kenai Peninsula all along, washing in and washing out? There was no telling.

Bryan Leiser already had in his possession a complete set of toys, and, knowing how badly I wanted a duck of my own, a duck found in the wild, he gave me this specimen. Truth be told, when I set out on my quest I hadn't expected to find any one of the four varieties of bath toy I was chasing, let alone a duck. I studied Bryan Leiser's gift for answers, gazing into its blank, banana-white eyes, wondering where it had gone and what it had seen—a silly thing to wonder, since it hadn't seen anything, since it was nothing more than polyethylene, but I wondered nevertheless, because fancy is often more powerful than fact. There were other questions I wished the dumb thing could answer: Was it coated in pollutants? Had it taken dominion over the slovenly littoral wilderness of Takoma Cove the way the anecdotal jar in the Wallace Stevens poem takes dominion over the woody wilderness of Tennessee?

Since the afternoon I'd spent on Kruzof Island, I'd reread Stevens's "Anecdote of the Jar." It was more ambiguous than I remembered. When the speaker places the jar in Tennessee, the wilderness is no longer wild, he tells us. Why not? After all, this isn't a jar of dioxin or DDT. It is, the speaker makes emphatically clear, empty, like "a port in air." The jar does not exterminate the birds and bushes of Tennessee.

Literary historians have speculated that Stevens had in mind a particular brand of jar, now obsolete, then common, with which a Tennessean could lay up fruits of the harvest. The brand name? *Dominion*, a word now fraught with contentious environmental connotations thanks to the King James translation of Genesis 1, verse 26: "And God said, Let us make man in our image, after our likeness: and let them have *dominion* over the fish of the sea, and over the fowl of the air, and over the cattle, and over all the earth, and over every creeping thing that creepeth upon the earth." The most mysterious thing about the poem to my mind is this personification of the jar. You'd think that the speaker, in placing it there, had taken dominion. How can a mere jar, even one embossed with the brand name *Dominion*, take dominion?

Here's what I think. It only takes dominion in the mind of the speaker, by changing the look of the scenery—not the ecology of the scenery, or the chemistry of the scenery, mind you, but the look of it.

What the jar exterminates is the appearance of wilderness, the *idea* of wilderness. The wilderness is "slovenly," whereas the jar is anything but. It's gray, bare, round, inanimate, man-made. It is, above all, a symbol of human order imposed on nature's entropy. It creates a kind of round center that the previously centerless wilderness seems to surround, but only when viewed by the godlike speaker does this center hold. Perceived by a bird, the jar would be, what? An inedible bit of kiln-fired clay? A nice place to build a nest?

Stevens's orderly poem, organized into shapely quatrains, with its irregular but nevertheless discernible iambic meter and its internal rhymes, is itself a jar of sorts, and so for that matter is a wilderness park. Much as Stevens's jar takes "dominion everywhere," the State of Alaska had taken dominion over the Kachemak Bay Wilderness. We might conclude, given the contrast Stevens draws between the sterility of the gray jar and the vitality of the slovenly wilderness, that the poem favors the latter. This is hard to say. The poem itself isn't wild, after all. And in other poems about the natural world, "The Snow Man" for instance, Stevens's speakers gaze into natural landscapes but fail to detect the mind of God, or the evidence of design, or of harmony, and they see the beauty of such places as a human creation, a projection of the imagination onto a kind of blankness.

A hairline crack ran around the duck Bryan Leiser gave me, and I could see the shadow of seawater sloshing around inside its translucent body. I shook as much seawater out as I could without beheading the thing completely, and zipped it into the pocket of my raincoat. Back in New York, I stashed it in my freezer, subjecting it to Arctic temperatures, performing an experiment. The beaver I'd found I sent to an environmental toxicologist, Lorena Rios, at the University of the Pacific, who'd agreed to analyze it, pro bono, with her mass spectrometer, a machine that, by measuring the mass of molecules, thereby distinguishing among them, can see chemicals invisible to the human eye. Months later Rios would send me the results: her tests would accurately identify the polymer—polyethylene. They would also detect twelve different polychlorinated biphenyls (PCBs) that my beaver had adsorbed on its journey to Gore Point[15].

Did Pallister feel a sense of satisfaction, the blond anchorwoman wanted to know.

"It's satisfying," Pallister said, squinting into the camera on its tripod, his rubber boots planted on the gravel beach, a microphone clipped to his fleece collar. Visible behind him was the *Constructor*, thousands of white bags piled high above the bulwarks, and beyond them the dark lagoon, and beyond that the snowcapped peaks of the Kenai Mountains—a dramatic backdrop that would no doubt do wonders for GoAK's fund-raising efforts. "Satisfying, but it sure was a grind," Pallister said. "Right now I'm just tired."

THE THIRD CHASE

In passing, it is worth noting that the morality of an act cannot be determined from a photograph. One does not know whether a man killing an elephant or setting fire to the grassland is harming others until one knows the total system in which his act appears. "One picture is worth a thousand words," said an ancient Chinese; but it may take 10,000 words to validate it. It is as tempting to ecologists as it is to reformers in general to try to persuade others by way of the photographic shortcut. But the essence of an argument cannot be photographed: it must be presented rationally—in words.

—Garrett Hardin, *The Tragedy of the Commons*

Mark, how when sailors in a dead calm bathe in the open sea—mark how closely they hug their ship and only coast along her sides.

—Herman Melville, *Moby-Dick*

SOUTH POINT

The southernmost edge of Hawaii is also the southernmost edge of the United States of America and feels like the southernmost edge of the world. Two months after the airlift at Gore Point I traveled there, staying in Naʻālehu, whose name means "the volcanic ashes." Naʻālehu is the southernmost town on the Big Island, and for frugal reasons I took a room at the Shirakawa Motel, the southernmost motel and quite possibly the cheapest one on the entire Hawaiian chain. Just up the road from the southernmost motel is a bar advertised as the southernmost bar, a restaurant advertised as the southernmost restaurant, and a bakery advertised as the southernmost bakery. Geographical extremity seems to me of dubious relevance to baked goods, but the tourists come by the busload to sample a morsel of southernmost bread.

The windward side of the Big Island is not what most of us imagine when, finding ourselves stuck in traffic or a bad job, we dream our Hawaiian dreams. Downtown Hilo, once the capital of a booming sugar trade, reminded me of cities in the American Rust Belt, only with palm trees and rain, lots of rain—a sunbaked, rain-soaked, tropical Sandusky. To get to Naʻālehu from Hilo, you drive over the Kilauea volcano, which is active but tame, the lava bubbling forth in a steady simmer but never, in recent history, shooting forth a shower of fire and brimstone. You pass through dilapidated towns made derelict by the collapse of Hawaii's sugar industry, past bungalows with corrugated metal roofs, past a barn-size movie theater with a sea turtle painted on its corrugated metal roof and a CLOSED sign hanging in its dark entrance even on Saturday nights (or at least on the particular Saturday night I saw it), past walls of lava rock stacked in the manner of fieldstones in rural New England.[16]

Then you reach the Shirakawa Motel, advertised by a rusty metal sign protruding out of the foliage. The Shirakawa Motel is a dilapidated compound of buildings also roofed in metal and encircled by a jungle of banana trees, and palm trees, and monkey pod trees, and ti plants, which have big droopy leaves and resemble the sort of tropical flora one encounters in the waiting rooms of dentists. The sky here is smoky gray and the rain falls in torrential outbursts lasting, at least on the afternoon I checked in, roughly seven minutes. It is falling now, loud as drumsticks on the canopy of leaves. And now it has stopped. Within moments, butterflies emerge to suck nectar from the flowering bushes, the flowers of which resemble little red starbursts.

My room is monastically spare—no television, no telephone, no clock, only a tattered orange paperback copy of *The Teaching of Buddha*, left out, like the Gideon Bible, as if to entice converts, on the round table between the two saggy queen-size beds. On the cover of *The Teaching of Buddha* is a setting sun. The binding has cracked and pages have come loose. Someone has stuck some of the loose pages randomly back in. The frontmost page is page 166, which begins midparagraph: "Was it a man or a woman? Was it someone of noble birth, or was it a peasant? What was the bow made of? Was it made of fiber, or of gut?" In my present bewildered state of mind, these seem like good questions.

From outside my little room come the sounds of insects—cicadas, perhaps, or crickets, or some tropical variety of noisy insect I've never heard of. A dog is barking. The striated world visible through the slanted jalousies above the queen-size beds is bright and green. Someone is playing Jawaiian reggae. Chickens run wild in southern Hawaii. This is the sort of habitat in which they evolved, after all. We think of chickens as animals indigenous to a Nebraskan farmyard, but they emerged from tropical forests like those encroaching all around the Shirakawa Motel. Every so often I can hear a rooster crow. They crow at dusk and dawn both. The flora and fauna here are as lush as the economy and the architecture are moribund. If the residents left, it wouldn't be long before the motels and movie theaters vanished into the foliage. Yes, this feels very much like the edge of the world. It feels much farther from the tourist-infested boulevards of Waikiki than it is—the boulevards of Waikiki, where, a few days ago, I saw white Christmas lights twinkle in the palm trees and street vendors peddle ukuleles made in China and the eternal flames of gaslit tiki torches burn on the facades of the big hotels.

One morning I drive farther south still, to the stretch of coastline known as South Point. The lush jungle gives way to feral sprays of sugarcane. Then to ranchland. The sun comes out. The road is cracked. Horses and cattle graze on windswept pastures over which loom rows of derelict windmills—an alternative energy plan gone bust. I drive on, until the pavement ends, at which point I park my rental car and continue on foot down a red dirt road that branches and splits among green hills carved by storm surges and rain. Every now again an SUV or pickup will rattle past. At the terminus of the land blue breakers obliterate themselves prettily against gnarled tumors of black igneous rock. I keep walking, searching, combing, stumbling over the crumbled lava rock in my canvas boat shoes, canvas boat shoes now dusted with red dirt. I follow the shoreline, the sun beating down. I wish I'd brought more water.

A week ago, if you had stood on these lava rocks amid the crashing waves and stared out to sea, you would have seen a catamaran under sail. And if you had aimed a telescope at that catamaran you might well have seen, dressed in Adventure pants, unzipped at the knee into Adventure shorts, baring my pale ankles, me, cross-legged atop the catamaran's cabin, on watch, scanning the horizon for obstacles and debris. And if you'd aimed your telescope at the cockpit, a foot or two below me, you would have seen, through the tinted windshield beaded with spray, at the helm, in his big captain's chair, Captain Charles Moore considering a dashboard aflicker with soundings and readings and bearings.

CAPTAIN MOORE

It had taken months of nagging and cajoling and begging to persuade Charles Moore that I was worthy to sail under his command. He'd had bad experiences with other landlubber journalists. A film crew for *Vice* magazine, shooting an online video about the Garbage Patch, had been particularly, almost mutinously, annoying, worried more about getting good shots than about standing watch or hauling in sheets or reefing sails or washing dishes. When I first reached Moore, by cell phone, from my classroom, back before I'd given up schoolteaching for seafaring, he'd said, gruffly, "There are no passengers aboard the *Alguita*; only crew."

Fine by me, I'd told him, thinking of Melville's Ishmael, who whenever he goes to sea always goes as a sailor, before the mast, on the forecastle deck, never as a passenger. It was early spring, that afternoon, and on my chalkboard was a stanza of Dickinson's "There's a certain slant of light," between the lines of which I'd marked out the stresses of Dickinson's irregular hymn meter, and in the pockets of my corduroy blazer were chalk nubbins and red pencils, and outside my classroom windows, on Manhattan's East 16th Street, the fruitless pear trees were in bloom, and on the sidewalk below, schoolchildren, just dismissed, were purchasing colorful balls of Italian ice from a cart, and in my mind the tropics of the North Pacific were a vague, blue dream.

Moore was planning a short expedition that would take place the following November, and maybe there'd be room for me among the crew. "Do you know how to sail?" he asked. "Because our number-one priority is safety. It can be dangerous out there." November is still hurricane season in the tropics of the North Pacific, he said.

I did have some sailing experience, I assured him. The summer I was thirteen, back when I was still considering a career in marine biology, my parents had enrolled me in sailing school. Weekday mornings for several weeks, I'd traveled by bus from San Francisco, across the Golden Gate Bridge, to the Sausalito Cruising Club, where behind the protection of a breakwater, I and a handful of other pubescent greenhorns had learned the difference between leeward and windward, starboard and port, luff and leech, to tack into the wind and run before it. On land, at age thirteen, I was a lummox, chubby, bad at sports, asthmatic, bespectacled, overly fond of a particular pair of striped wristbands that always stank of sweat. Out on the water, at the helm of a Laser, a fiberglass sloop, I was half dolphin, half bird. That, at the time I first contacted Moore, was the sum total of my sailing experience.

The synopsis I gave Moore of my training didn't seem to leave a favorable impression, and so in August, the week before the airlift at Gore Point, I flew to Long Beach, California, hoping to close the deal.

"If nothing else, you'll need to know how to tie a clove hitch and a bowline knot," he said as we stood on the aft deck of the *Alguita*, Moore's custom-made oceanographic research vessel, tied up at his private dock, on an inlet lined with palm trees and patio umbrellas and pleasure boats. Across the street from the dock was the boxy, two-story, many-chambered yellow stucco house, the second-story windows like port-holes, to which his parents had brought him home from the hospital as a newborn six decades before. Around it Moore grew—organically, of course—a small, thriving jungle of tropical plants.

In a few weeks he would be sailing to Hilo, where he planned to spend most of the winter, returning to Long Beach in late January. On the passage to Hilo and back he would go trawling yet again for plastic in the Eastern Garbage Patch. This would be the fourth time he'd collected samples there in ten years. Why repeat the same experiment? By comparing old data with new, he hoped to determine the rate at which pelagic plastic was accumulating.

Aboard the *Alguita* that sunny afternoon, you could feel the air of frenzied preparation. A mechanic named Tomas, busy hooking the boat's hydraulic system to an array of solar panels, kept popping out of an open hatch to snatch one of the shiny wrenches scattered about the black rubber mats of the aft deck, then popping back below. When Tomas was up, Moore conversed with him in Spanish—fluently, it sounded to my ignorant ear.[17] The solar panels were a recent acquisition, paid for with a grant from none other than BP. "We're going to be able to run a three-horsepower motor," Moore said, "and the gantry crane here, and the six hundred meters of trawl cable here, and the anchor—all with solar power." Made to order by a Tasmanian shipwright, Moore's fifty-foot aluminum-hulled catamaran is a handsome vessel, handsome as the *Opus* is homely, and its new solar panels would make it state-of-the-art. Since Moore launched it into Hobart Harbor more than two decades ago, the *Alguita* has logged some 100,000 blue-water miles, most of them on research cruises. When he isn't conducting his own research, Moore charters the boat to other scientists. Earlier that summer he'd taken a group of wildlife ornithologists bird-watching in the Sea of Cortez.

During the time I'd travel with him, I'd never get to see Charlie, as everyone who knows him calls him, in the captain's outfit that he sometimes wears—light-blue shirt, little name tag pinned to the breast pocket, the striped epaulets, everything but the white cap with the gold anchor above the black visor. On the day I met him he—potbellied, with the permanent tan and sinewy limbs of a man who spends a lot of time both in and on the water, dressed in a grease-stained khaki shirt

with holes at both elbows, the top four buttons undone to expose a griz-
zled wedge of chest fur—didn't look much like a sea captain. He looked
like the organic-farming, utopian beach bum that, when not at the *Alguita's* helm, he is. Almost every day, when he isn't sailing, he surfs. That
morning off Huntington Pier, he'd "shredded it up pretty good." He has
a bushy head of sun-bleached brown hair shot through with filaments of
gray, but otherwise bears a striking resemblance to the balding actor
Robert Duvall. Moore could be Duvall's younger, shorter, tanner, hair-
ier brother. His heavy-lidded eyes and wrinkly neck give him the aspect
of a sun-drugged tortoise. He speaks in a gravelly, almost inflectionless
yet hortatory drone, as if reading a prepared speech badly from a tele-
prompter. When he laughs he opens his mouth just a little, once again
bringing to mind a tortoise, and makes a coughing sound, as if trying to
clear his throat: *heh*.

Giving me a tour of the *Alguita*, he delivered a kind of extemporane-
ous sermon that ranged widely, from the chemistry of polybrominated
diphenyl ethers to the social critic Thorstein Veblen to Rell Sunn, the
deceased Hawaiian high priestess of surfing. Moore sounded at times
brilliant, a font of facts and expertise ("our research indicates that 2.3
billion pieces of plastic go down the L.A. Basin in three days"), and at
times like a half-cocked conspiracy theorist ("in our economy a series of
short-lived and sickly generations is more profitable than a series of
long-lived and healthy ones").

For a utopian organic farmer without an advanced degree in oceanography or anything else, Moore has been lead author on an impressive
number of scientific papers. He has successfully lobbied for stronger
regulation of Southern California's plastics industry. Think of Los Angeles, and if you don't live there you probably think of Hollywood, but
Los Angeles is the American epicenter of virgin plastics processing. In
2006, the World Federation of Scientists (WFS) invited Moore to Erice,
Sicily, to deliver a presentation on plastic contaminants, which the
WFS, thanks in part to Moore, has now added as a subcategory to its
growing list of planetary emergencies. Also on the list: Missile Prolifer-
ation, Cultural Pollution, and Defense Against Cosmic Objects.

Moore was recently inducted into the Explorers Club, the elite association of scientific adventurers whose ranks have included Sir Edmund
Hillary, Roald Amundsen, Neil Armstrong, and Jacques Piccard, the

deep-sea explorer who, in 1960, with copilot Don Walsh, set the unbroken world depth record—35,800 feet—for a descent in a manned submersible. In 2006, with his wife, Samala Cannon, known as Sam, Moore attended the club's annual dinner, infamous for its adventurous menus. At the Waldorf-Astoria in New York, Sam and Charlie had been offered such hors d'oeuvres as toasted crickets and kangaroo testicles, then watched as the club's president made a grand entrance on a hired camel.

Back before he became a scientific adventurer, when he was still a humble furniture repairman, one of his regular clients was the Hanjin shipping line, which hired him to restore furniture damaged in transit. He'd restored, for instance, a grand piano that Gregory Peck had shipped to Los Angeles from Marseille. The first time Sam saw her future partner, in the early seventies, he was sitting in the back of a pickup truck outside a food co-op, his hair even bushier than it is now, a joint in one hand, a beer in the other. He has since renounced all mind-altering substances in favor of organic fruit. This renunciation, and Moore's heterogeneous résumé, suggest that his has been a wayward path, that he was personally acquainted with the experience of drift.

Around the same time he started the Algalita Marine Research Foundation, Moore started another nonprofit whose mission is to teach the wayward, drifting youths of Long Beach to grow and tend organic community gardens. He handed me a business card. Long Beach Organic, it said. Slogan: "For gardens, trees, & kids!" He is in effect a kind of nonprofit entrepreneur. At first the organic nonprofit was the more successful of his two ventures, but in recent years, thanks to growing public awareness of the Garbage Patch, the Algalita Marine Research Foundation has surged past organic farming.

Summoning me to follow, he headed up the gangway, across the street, to his organic garden. "Don't bump your head on the cucumber," he said as we entered this frugivorous, bee-loud, homemade glade, a kind of personal Eden equipped with a solar-powered hot tub and a chlorine-free swimming pool purified by ozone. With compost below and fruit-heavy branches overhead, the air smelled of both ripeness and rot. A giant avocado tree grew up the back of Moore's house like a leafy chimney. That avocado tree was almost as old as he was, Moore said. In the corner, a surfboard leaned against a fence, and a pair of wet suits,

one bright blue the other bright red, hung from a clothesline. They
looked like melted superheroes. A wooden bridge of the sort you see in
tea gardens spanned the chlorine-free pool. Without explanation Moore
excused himself, leaving me loitering by his back door, near the entrance
to a shed in which could be seen a photo of Moore on his surfboard,
shredding it up pretty good.

He emerged from his kitchen with a glass of passion-fruit juice
freshly decanted from the blender. "Here," he said, in the manner of a
medicine man, "drink that." There were bits of pulp and seed and who
knew what else—gnats, probably—floating around in it, and I hesitated
a moment before taking a sip. It was the best juice I'd ever tasted. Then
Moore reached up, plucked a dwarf banana from a branch, and peeled it
for me as if I were a child. "Try that." It was the best banana I'd ever
tasted, tangy and vegetal as well as sweet.

I'd like to share Moore's faith in the arc of progress, but even there,
in the never-never land of the man's garden, I had a hard time imagining the bright future he saw, in which we Americans would trade conspicuous consumption for cradle-to-cradle manufacturing practices,
disposable plastics for zero-waste policies and closed ecological loops. I
had a hard time because such a future seemed to me inimical to the
American gospel of perpetual economic growth, and because my up-
bringing in Northern California had taught me to distrust utopian
strangers who talked about "economic paradigms," no matter how tasty
their bananas. Nevertheless, when Moore offered me a goodie bag—
paper, not plastic—of fresh fruit, I accepted it. Off I went, laden with
grapes and figs and papayas and bananas and an understanding that
Moore and I would meet again the following November, in Hawaii.

THE BELLY OF THE ALBATROSS

My first morning on the Big Island, I reported to the laboratory run by
Karla McDermid, marine biologist at the University of Hawaii–Hilo.
Students in her marine debris seminar stood around long metal tables
analyzing the latest haul of plastic that Moore had collected from the
Eastern Garbage Patch. The lab had the feel of a workshop. One team
of students was investigating the origins of flotsam and jetsam that had

yet to photodegrade—ropes, bottles, sticks of deodorant. "It's from India!" a student at a computer shouted. Another team was sieving through sand retrieved from Kamilo Beach. "That's the world's dirtiest beach, according to Charlie Moore," McDermid told me. "So dirty that the sand is like plastic sand." I asked her where Kamilo Beach was. "Down near South Point."

McDermid—in a button-down blouse and sandals and glasses and pink glass earrings—turned to address the busy room. "If you need beakers, you guys, beakers for floating things, or petri dishes, I'll be back, you just yell at me what you need." She led me outside, where at the edge of a parking lot an aspiring wildlife veterinarian, Sarah Ward, had volunteered to perform the savory task of pawing through albatross "upchuck balls." Wearing purple rubber gloves, Ward crouched beside a bucket from which there rose the nauseating stench of death and in which could be seen a partially digested bolus coated in what looked like engine oil. In a way it was engine oil—the oil of an albatross engine. Albatrosses live fifty years or more, and spend 95 percent of those years at sea, and for 90 percent of their time at sea, they're flying—gliding, really. Their wings locked fast, rarely flapping, riding the same winds favored by mariners of the age of sail, they can go thousands of miles without touching down. They even fly asleep. By the estimate of Carl Safina, the naturalist and author of *Eye of the Albatross*, an albatross can clock around four million air miles before it dies. To fuel these epic flights, an albatross's stomach distills albatross prey—squid, fish roe, herring—into an oil almost as caloric as diesel. In the not too distant future, two or three generations hence, when the planet's reserves of fossil fuel begin to run out, perhaps some entrepreneur will start farming albatrosses for their oil.

This sort of avian prospecting would be no innovation. It would be a return to the technology of the nineteenth century, when hunters slaughtered albatrosses for their oil, as well as for their feathers, desirable as millinery plumage. In the breeding grounds of the Japanese island of Torishima, feather hunters killed an estimated five million birds between 1887 and 1902, when the island's volcano erupted, exterminating the hunters; some of the birds, however, survived. On the Northwestern Hawaiian Archipelago, as on other island chains, marauding feather hunters would pillage the rookeries where docile adults refused to quit their nests, pluck a quarter pound of feathers from every bird, and leave behind a heap of carcasses. In many ways, most ways, the lot of the albatross has improved greatly since the nineteenth century. In other ways, not. So far Sarah Ward had found oily tangles of monofilament fishing line and scads of oily squid beaks, which, disembodied, look a great deal like the beaks of birds. While I watched, she found something else, an oily little plastic flower that had perhaps once decorated a child's hair barrette.

Back indoors, down the hall, in another lab, five students bent before Leica dissecting microscopes, searched petri dishes for minuscule grains of plastic. The samples had previously been sieved, and each student was working with particles of a different magnitude—one with 0.70-micron particles, another with 0.99-micron particles. With tweezers they poked around, sorting particles by color and type, separating nurdles from monofilament from foam from film. Once sorted, the particles would be set aside to dry. Once dry, they'd be weighed on a scale so sensitive that even a breeze blowing over its surface could tip the balance. The goal of these forensics: to determine as irrefutably as possible the plastic-to-plankton ratio in the Eastern Garbage Patch.

While I was marveling at the painstaking process, the door swung open and in walked a surfer chick. She had long, apparently unwashed blond hair pulled back into a messy Rastafarian knot. Circumscribing her head was a knitted band in a black, yellow, red, and green Rastafarian motif. Tucked into her knotted hair above the Rastafarian headband was a pair of sunglasses. She was wearing plaid Bermuda shorts and a black tank top, and carrying a skateboard under her arm. Her name was Amy Young. She'd crewed on the *Alguita* before, and had a word of warning for me: bring Dramamine. "I've been out on a lot of boats," she said, in a surprisingly high-pitched, surprisingly childish voice, "and *Alguita's* different even for a small boat, because it's wider. That boat got me." I had brought Dramamine and, at Moore's behest, one of those flesh-colored transdermal motion-sickness patches that you're supposed to stick behind your ear. I'd also brought some of those ginger candies that had saved me aboard the *Opus*.

Back in McDermid's main lab, in a short-sleeved blue-and-white shirt ornamented in some sort of vaguely Balinesian pattern, Charlie Moore was delivering, for the benefit of a local TV news crew, a spiel similar to the one he'd given me in Long Beach: "The ocean obeys the basic law of ecology, which is that in an ecosystem everything is used. But humanity is breaking that basic law." Or: "We say we're throwing stuff away, but there is no away. The ocean is away." He described seeing, on "a night dive" a thousand miles off the coast of California, caught in the beam of his waterproof flashlight, a plastic grocery bag. "I'm still upset," he said. "As a member of the human race I find it embarrassing."

One of McDermid's students found, among the flotsam she was sorting, a stick of men's deodorant denuded of its label and, camera still rolling, McDermid handed it to Charlie. "Is that a hint?" he asked, uncapping it and swabbing his armpits, then said, "I think I smell better."

It was a good performance, humorous yet full of the factoids and sound bites that make news producers happy. It was also a performance he'd given many times before, most recently the previous morning, on the *Today* show. There was one datum that was news to me: sailing from Long Beach to Hilo, detouring yet again into the Eastern Garbage Patch, he'd collected samples containing forty-six times as much plastic as plankton, an eightfold increase since 1999. In 1999, plastic had outnumbered plankton by a factor of six.

Moore, environmental Barnum that he is, presented to the camera a bag of plastic sand collected on Kamilo Beach. It looked like the colorful gravel you used to see at the bottoms of decorative fish tanks. By comparing Moore to Barnum, I don't mean to suggest that he's a charlatan; I mean to suggest only that he plays his role exceedingly well. Now, in the information age, which is readily transforming into the disinformation age, if you wish to combat, as Moore does, the armies of corporate publicists waging endless ad campaigns, including greenwashing campaigns of the sort devised by Keep America Beautiful, then you have to be media savvy, and like a guerrilla warrior on the information battlefield you have to exploit your adversaries' weaknesses. The problem is that in the universal propaganda arms race that has issued from the disinformation age, no one trusts anyone, and doubtful citizens are free to luxuriate in distractions, while overconfident citizens are allowed to pick and choose whichever factoids and sound bites confirm their self-serving predispositions. Make-believe, I was beginning to think, is inescapable.

Holding up the bag of plastic sand, Moore explained into the reporter's microphone the mission of the research cruise on which he and I were about to embark. We were going to determine whence this bag of plastic sand came. From fishermen Moore had heard that a "feeder cell," a sort of miniature garbage patch, had formed fifty miles south of Hawaii in an eddy of convergent currents, an eddy that was in turn generated by the winds that spin between the crests of the Big Island's two active volcanoes, Mauna Loa and Kilauea. From NOAA oceanographers he had procured satellite images showing the convergence zones of the North Pacific. Mostly the convergent currents lie to the north, east, and west of Hawaii, at the heart of the Subtropical Gyre. But on Moore's colorful satellite-generated maps, fifty miles south of South Point, there was a dot of purple, the color that, as a caption to the satellite images explained, represented a zone of "greatest convergence." It was into this zone that we would sail.

Before leaving McDermid's lab, Moore made it known that the *Alguita*'s crew was still one deckhand short. Amy Young, surfer chick, was quick to sign on. "I'm totally stoked," she said. I swear, that's exactly what she said. Filling out our crew would be two other students in McDermid's marine debris course, Jeff Ernst, a puckish, towheaded kid with a wispy, towheaded soul patch, and Cory Hungate, a buff surfer dude who liked to go around topless, showing off his pneumatic pectorals between which lay, suspended from a hemp lanyard, a pendant fish hook. Carved from bone, the fish hook is an artisanal symbol of Polynesian culture when worn by a native Hawaiian. When worn by a *haole* like Hungate, it is perhaps a symbol of either dudeliness or douchiness. Then there was Moore's wife, Sam Cannon, and co-captain Laurent Pool. A marine biologist with the Audubon Society and a seasoned sailor, Pool had the look of a Jazz Age bohemian. He kept his balding head shaved, sported a pair of little wire-rim spectacles, and, when not on the windy deck of a sailboat, protected his bare scalp with a feathered fedora. Then there was me.

INTO THE CONVERGENCE ZONE

We sailed out of Radio Bay at dusk, intent on making South Point by morning. Conditions at our departure: sea state 3 on the Beaufort scale, wind a little under ten knots, skies cloudy but opalescent, casting a rime of shine on the *Alguita*'s brightwork. Even tablespoons of rain puddled on the cabin roof caught the light, and the tinted Plexiglas portholes in the cabin's roof, when the angle was right, turned into celestial mirrors. To shore you could see the green island and the white honeycombs of vacation condominiums. The trade winds were blowing, and conditions promised to worsen when we emerged from the lee of the Big Island, but our main worry, said Moore, was the volcanic smog, or "vog." "Vog Keeps Students Indoors; Mountain View Gets 'Worst Ever Seen' Emissions," ran the front-page headline in that morning's *Hawaii Tribune-Herald.* Vog smells sulfurous, and for good reason; it consists mostly of sulfur dioxide. The trade winds would speed us along, but they could also blow the vog into our path, or whip up rough conditions. "We can't just beat our way through a trade wind," Moore said just before we slipped the lines. "We have to finesse our way through."

Moore divided the crew into pairs and put a watch schedule into effect. Watch lasted four hours. My watch partner was Moore himself, not, I think, because he particularly enjoyed my company or wished to give me extra journalistic access but because he distrusted my seamanship. Moore asks every "prospective research team member" to fill out a questionnaire. My score on this questionnaire, it should be stated, was pathetically low. Did I hold any seaman's documents? No. Did I have CPR and/or first-aid certification? No. Was I a certified scuba diver, a licensed ham radio operator, a scientist? No, no, no. Yes, I had a valid U.S. passport. Yes, I would be "willing to take an active part in the actual operation of the ship." Yes, I could swim, though I'd rather not do so if there were sharks about. Yes, I had taken "a basic boating course."

In Hilo, on the eve of our embarkation, as Moore, Sam, and I were walking back to the *Alguita* from dinner at a nearby restaurant, I pulled from my pocket a little length of rope I'd been carrying around in my pocket and executed my bowline. "Good," Moore said. "Now let's see the clove hitch." My clove hitch was shakier. I'd only really mastered it during a layover in Minneapolis en route to Hilo. We happened to be standing by a koi pond, beside a highway, and between us and the koi pond was a railing around which, a bit doubtfully, I executed a clove hitch. Moore gave the bight a tug. "Good enough." The Big Island was loud that night with the singing of coqui frogs, an invasive species that had reached Hawaii, the thinking goes, by stowing away on landscaping plants shipped from Puerto Rico. "A lot of stars for Hilo," Moore had said, considering the night sky. He swung his hand up as if saluting an angel or an albatross and squinted one eye, sighting down the length of his arm. "We're at 19 degrees north, so that's definitely the North Star." He slid his arm over a little. "And that's Cassiopeia." *Kokee, kokee,* a frog sang from a nearby bush. "*Kokee,*" Moore sang back.

We took the graveyard shift, Moore and I, standing watch between ten and two, by day and by night. In fact, except when he was stealing a few hours of sleep, Moore was effectively on duty, very much the captain of his ship. I spent that first watch up on the stippled, supposedly slip-proof fiberglass roof of the *Alguita's* cabin. Most of the crew was out there—not yet weary or seasick, full of anticipation, thrilled by watery beginnings, wearing transdermal patches behind our ears. The clear night sky teemed with stars, some of them shooting. You could see the Milky Way. The tinted Plexiglas hatch above the helm swung open, and out, like a gopher, popped Charlie Moore. He was standing on his captain's chair. Across his brown ball cap swam a sea turtle and the words NEW ENGLAND AQUARIUM. He had a pair of binoculars, which he trained to starboard. Excitedly, he pointed out to the rest of us the simmering lavas of Kilauea. Even without binoculars you could make out the orange tendrils and ropes and balls and droplets of molten rock splashing and leaping forth from the black horizon of the land.

That first night at sea, when my watch ended, I did not sleep well. Each of the *Alguita's* two cabins is set down in one of the two aluminum hulls. Captain's quarters are in the port hull, crew's quarters in the starboard hull. I was sharing the crew's quarters with Jeff Ernst and Amy Young. Their berths were up on a kind of loft, right under the tinted Plexiglas hatches. Between their pillows oscillated a little plastic fan. My berth was way down in the humid depths, below the waterline. It was hot down there, but I couldn't complain; consigned to bunks on the port side of the main cabin with only a curtain for privacy, co-captain Laurent Pool and surfer dude Cory Hungate had drawn the worst lot.

I woke, just before seven, to the sound of music, and stumbled, sleepy, rank with sweat, in a caffeine-deprived stupor to the aft deck. Ernst was strumming an acoustic guitar, Young accompanying him on bongos. The manta trawl, already overboard, was still unspooling from the white steel A-frame crane that spanned the stern. It was easy to see how the manta trawl got its name. The resemblance was obvious: from the side of the trawl's boxy aluminum maw flared hollow aluminum wings that, together with the long tail of netting, brought the giant ray to mind. The netting terminated in a cylinder called a cod end. Considering me through a pair of silvered sunglasses, her hair balled up inside a knit cap, Young stopped drumming and said, in her girlish voice, "You going to get some sun on those pasty legs? You've got a library tan, friend."

Laurent Pool, shirtless, bald head shining, about to take his morning piss off the stern, stopped to consider whether he might end up tainting the sample we were presently collecting and availed himself of the head instead—an unnecessary detour, it seemed to me; wasn't the ocean full of the eliminations of whales and seals and copepods? Having made a hundred miles during the night, we were right on schedule, right off South Point. You could see up on the green foothills of the Big Island a row of white windmills. If an explorer such as Captain Cook were to have arrived here and discovered this radiant, towering display, I wonder what he would have made of it. I suppose he would have taken the windmills for armaments, or monuments.

Moore was out on the aft deck, too, snapping digital photos of the tropical sunrise, dressed in a pair of baggy black shorts, and flip-flops, and a big baggy tank top that hung from his midriff like a tent. On it was a cartoon finch. His face was whiskery, and his expression seemed by turns melancholic, stoical, or bemused. The wind was up, now twenty to twenty-five knots. Motors off, we were under sail. The waves sloshed pleasantly against the hull. "Quite a lot of potential wind power they've got up there," Moore said of the windmills on South Point, shadowing his eyes. "Most of those are derelicts."

Stickers from other institutions and organizations dotted the windows of the *Alguita*'s main cabin: Scripps, Woods Hole, Waterkeeper, American Chemical Society (not to be confused with the American Chemistry Council), a website called Pursuebalance.org. Leis made from real flowers hung around a brass wall clock, beneath the sound system. In the galley, above the sink, strung along a piece of kitchen string, were those little colorful plastic tabs that clip shut bags of sliced bread.

Moore tried to reuse them instead of throwing them out. Tucked on a few small shelves, the *Alguita* possessed a cozy library, with such volumes as *Pacific Coast Pelagic Invertebrates, Stars and Planets, Royce's Sailing Illustrated,* and even a tattered pulp paperback copy of *Moby-Dick.* During the course of our research cruise, as we sailed through oceans, no one went for so much as a dip in this library, at least not while I was looking, though I did notice Laurent Pool reading a book he'd brought with him: *Ishmael,* a didactic ecological novel named for its main character—not a greenhorn embarking on a whaling trip, nor the outcast son of Abraham, but a telepathic gorilla, who thinks lots of profound ecological thoughts. In Pool's opinion, *Ishmael* was primo stuff. The rest of us were otherwise preoccupied, by trawling, and by music, and by the scenery.

In a way, the *Alguita* was a kind of floating eco-friendly utopian experiment, powered mostly by the sun and the wind, and by the youthful enthusiasm of the volunteer crew, and perhaps most of all by the propulsive force of Moore's charisma. Our food on this trip was local and organic, purchased at the Hilo farmers' market. To exterminate the fruit flies that circulated over the mangoes and bananas piled atop a cabinet, Moore likes to suck them out of the air with a solar-powered vacuum cleaner. No pesticides here. All biodegradable trash went in one bin, nonbiodegradable trash, of which we generated little, went in another, to be recycled upon our return.

After forty-five minutes of sampling, Moore fired up the solar-powered hydraulic winch and reeled the manta trawl in. Once we'd wrestled the trawl aboard, our captain unscrewed the cod end, dumped its contents into a bowl of saltwater resting on the winch table, and delicately scraped the gauze clean with a toothbrush. In the bowl, zooplankton—sea bugs, some species of little jelly or salp—swam through a sprinkling of plastic. Moore fixed the sample in formalin and transferred it to a labeled specimen jar. Then we sent the trawl overboard again.

This was a somewhat awkward process, the trawl swaying around under the arc of the A-frame crane, its gauzy tail blowing around like a wet wind sock, Amy Young, in Bermuda shorts and tank top, poised to starboard; Cory Hungate, topless, to port; both gripping lines that held the trawl in place, waiting for Moore's signal. When the signal came, Pool gave the thing a shove, and Moore, at the winch, paid out the line. The trawl splashed into our wake, where it floundered around as if threatening to sink. Then the line tightened and the trawl, righting itself, seemed to swim after us, bobbing up and down through the troughs and crests, gulping great mouthfuls of ocean. In his recent survey of the gyre, Moore had sieved some sixty thousand cubic meters and had found, on average, one plastic particle per meter, which might not seem like much, roughly equivalent to finding a single particle of plastic in your bathtub. But when you consider how many cubic meters of water there are in the upper layers of the North Pacific Subtropical Convergence Zone—billions, though the exact number changes with the seasons—a particle per meter amounts to a lot of plastic.

At 8:10 in the morning, the second lowering complete, we stopped for breakfast. Complaining, in a good-humored way, that it's hard to find deckhands willing and able to cook, Moore headed to the galley and whipped up a batch of organic banana pancakes. When my daytime watch began at ten, I resumed my position on the cabin roof and scanned the turquoise horizon for debris. In my peripheral vision there flashed what I, turning, at first took for a flock of small birds. Then I saw the silvery flock dive below the waves. *Flying fish!* I said to myself. *Like the ones suspended over the plastic waves in the dolphin diorama at the American Museum of Natural History!*

By noon on that first day, the Hawaiian Islands, by some measures the most isolated islands on earth, had sunk below the northern horizon. We were off the grid, out of contact, out of sight, beating our way south, in oceans three miles deep. By dinnertime, we were fifty miles south of South Point, conducting our fourth sample of the day, letting the sails luff. The seas had moderated a little. According to Moore, we were now at sea state 5 on the Beaufort scale. Developed in 1805 by Sir Francis Beaufort, an Irish admiral, the Beaufort scale includes descriptions of sea conditions that read like found poetry. This is how the Beaufort scale, revised several times since its author's death, describes seas in light air, level 1: "scaly ripples, no foam crests." This is seas in a strong breeze, level 6: "Larger waves 8–13 ft, whitecaps common, more spray." In near gale, level 7, the "sea heaps up, waves 13–20 ft, white foam streaks off breakers." Level 12, hurricane, the scale's maximum, Sir Beaufort memorably described as "that which no canvas could withstand."

"One of the things we're trying to discover is if there's any correspondence between sea state and the size of the plastic," Moore told me.

Considering his lack of credentials, you might be inclined, like plastics executives, to dismiss Moore's investigations as wishful, amateuristic grandstanding. After witnessing the meticulous sorting and analyzing in Karla McDermid's lab, I was inclined to give Moore the benefit of the doubt. But I still had doubts.

In the scientific community, Moore's work remains somewhat controversial. Even marine biologists who share his alarm have misgivings about the sensationalism with which the Garbage Patch is sometimes described in the press—by Moore and others. A 2006 Greenpeace report attempted to rechristen the Garbage Patch "the trash vortex," bringing to mind a maelstrom that had sucked entire landfills into its swirling maw, and one magazine article went so far as to call the patch "evil," comparing it to both a "trash tsunami" and a "twenty-first century Leviathan." Other articles call it "plastic island" or the "island of plastic in the North Pacific." Moore himself objects to such descriptions. He even objects to the term "Garbage Patch." Despite Moore's efforts to suggest different metaphors—"a swirling sewer," "a superhighway of trash" connecting two "trash cemeteries"—"Garbage Patch" appears to have stuck.

Since the plastic debris in the North Pacific Convergence Zone is spread out unevenly across millions of miles of ocean, and since most of it is fragmentary, flowing through the water column like dust through air, the Garbage Patch bears little resemblance to a floating junkyard, it turns out—or to an island. And yet, on one thing scientists, plastics executives, and environmentalists agree: the amount of plastic in the ocean is increasing, more so in convergence zones than elsewhere.

Far more than his institutionalized counterparts ensconced in universities and government agencies, Moore resembles the pioneers of oceanography, among the last of the natural sciences to professionalize for the simple reason that wealthy, swashbuckling, yacht-owning amateurs were often the only ones who could afford ship time at sea. The

first large-scale government-sponsored professional oceanographic research "cruise"—the charmingly vacational term modern oceanographers have bestowed on their seasickening fieldwork—was the
three-and-a-half-year British circumnavigation completed in 1876 by
the HMS *Challenger* (for which an ill-fated space shuttle would later
be named). The "scientifics," as the naturalists on the *Challenger* were
called, didn't think of themselves as belonging to a single discipline.
The term *oceanography*, a German neologism derived from *geography* and
hydrography and *cartography*, had yet to gain widespread currency in
English. The scientifics of the *Challenger* referred to their interdisciplinary pelagic fieldwork as "the science of abysmal research."

Before 1876 a naturalist of modest means who wished to conduct
abysmal research had two choices: he (always a he, until the middle of
the twentieth century) could either study coastal waters, as did the Venerable Bede, the medieval monk who first recognized that freshwater
spreads over saltwater. Or else, like Joseph Banks on the *Endeavor* or
Darwin on the *Beagle*, he could avail himself of voyages of opportunity,
hitching rides with the Royal Navy, collecting whatever evidence he
chanced upon. Even well into the twentieth century, oceanography like
the lawless ocean itself remained open to uncredentialed amateurs—
amateurs like Jacques Piccard, he of the bathyscaphe, who had no formal training in marine science; or like William Beebe, who trained as
an ornithologist but became famous as an explorer of the deep. Henry
Stommel, who perhaps did more than any other oceanographer to solve
the riddles of ocean circulation, never completed an advanced degree.
An entrance exam Stommel took his freshman year at Yale indicated
that he had little aptitude in science, and he initially considered going
into the law or the ministry, ending up at the Woods Hole Oceanographic Institution largely by accident, while conscientiously objecting
to World War II.

Perhaps the swashbuckling amateur whom Charlie Moore most resembles is Prince Albert I of Monaco. Seeking relief from luxurious
boredom, the young prince in 1873 bought a schooner, the *Hirondelle*,
and became a full-time yachtie. After twelve years of recreational sailing, he began conducting a series of drift experiments, setting bottles,
beer kegs, and floats adrift—1,675 in all, of which 227 were recovered, or
14 percent, a better return than Ebbesmeyer has ever gotten from a con-

tainer spill. From their trajectories the prince was able to discern the clockwise rotation of the North Atlantic Subtropical Gyre. He later turned his resources and his mind to other oceanic mysteries. Of dolphins he made living sampling devices, harpooning them, necropsying them, and studying the contents of their stomachs.[18] He collected abyssal species of squid vomited by a dying sperm whale—"precious regurgitations," he called them. When Moore's researchers examine the upchuck balls of albatross chicks, they are following Prince Albert's lead.

The amateur oceanography practiced by Prince Albert and Captain Charlie do differ in one fundamental respect, however: Albert, a huntsman as well as an amateur scientist, went about harpooning cetaceans with guiltless delight. He gave no thought to humanity's impact on the sea. In the nineteenth century, next to no one did. The oceans were indomitable, unfathomable, dangerous, and divine. The Victorian ocean was still the watery chaos of J. M. W. Turner, or the teeming hunting grounds of *Moby-Dick*, not yet the fragile wonder world of Carson and Cousteau.

THE INFINITE SERIES OF THE SEA

Time now seemed to dilate. We weren't on Greenwich mean time, or Hawaiian-Aleutian time. We were in a time zone of our own, *Alguita* time, our rhythms determined neither by the hours our busy contemporaries kept on land nor even by the circadian revolutions of the earth but by the schedule of watches and meals and naps and trawls. One imagines, before setting sail, that seafaring promises excitement or romance, but on calm tropical seas, the hours pass through one's mind like cubic meters of water through a manta trawl, leaving a sprinkling of impressions snared in memory's gauze.

At night, while the rest of the crew slept, Moore and I kept watch mostly in silence, he in his captain's chair, I standing by, studying the shimmering instruments and every twenty minutes making entries in the logbook: speed over ground, set of the current, direction of the wind, state of the sea. Moore kept the hatch above the helm open. Through it we could watch the stars. They seemed to swing back and forth across the black abyss of space, though really it was the *Alguita* that swung, its mast an upside-down pendulum. Sometimes Moore let me take a turn standing on his chair, poking my head above decks. With Moore below and the rest of the crew asleep and the stars above, I felt more alone than I've ever felt.

"There you stand, lost in the infinite series of the sea, with nothing ruffled but the waves," Melville writes in the "Mast-Head" chapter of *Moby-Dick*. "The drowsy trade winds blow; everything resolves you into languor. For the most part in this tropic whaling life, a sublime un-eventfulness invests you; you hear no news; read no gazettes; extras with startling accounts of commonplaces never delude you into unnecessary excitements; you hear of no domestic afflictions; bankrupt securities; fall of stocks; are never troubled with the thought of what you shall have for dinner."

At midnight, my head poking out of the *Alguita*'s cockpit, the only white noise I heard was the hiss of the water along the hull, the wind giving the canvas the occasional shake; the only colorful noise, the halyards chiming against the aluminum mast, the creaks and clinks that accompanied the boat's motion, keeping time.

In the heart of that convergent eddy, that purple dot on Moore's satellite image, there was no flotsam or jetsam to be seen. Was the map wrong? Had the eddy shifted south? Or east? Or west? I could tell that Moore was disappointed. He kept consulting his satellite-generated charts, wishing aloud that they showed more detail. Meanwhile, conditions had begun to deteriorate. We'd gone almost as far south as we'd planned to go. "Getting a little sundowner," Moore said one day at sundown. I asked him what a sundowner was. "Winds picking up as the sun goes down. It's going to be a little hairy tonight, coming about."

We were already at sea state 6 on the Beaufort scale. Before morning it seemed likely we'd get a taste of sea state 7 or even sea state 8. In 1866, Mark Twain rounded South Point aboard the schooner *Emmeline* (a name he would later give to the suicidal poetess Emmeline Grangerford in *Huckleberry Finn*). The *Emmeline* had run into "contrary winds." So, almost a century and a half later, did we—forty-knot winds to be exact. When my watch began at ten, the boat was yawing and pitching in twelve-foot seas, spray flying. At midnight, Moore called all hands on deck, and the crew appeared, as Twain put it, "plunging about the cabin with the rolling of the ship."

I wish that I could give an accurate account of the maneuver we performed, but circumstances permitted no note taking, nor did they permit voice recording (the wind was too loud, the rain too short-circuiting), and my memories are a sleep-deprived blur. This, the following morning, scribbling in my notebook, is what I remembered: I remembered gathering with my crewmates around the winches on the aft deck. I remembered Charlie Moore consuming packets of Emergen-C, a vitamin supplement that sailors swear by. I remembered hauling on the sheets, hand over hand, while Moore, hauling too, shouted at us to haul harder. I felt like a weakling—understandably, because I am, a weakling with a bad back. As the mainsail came down, crackling and snapping like a bonfire, a white maniac against the black sky, Moore shouted, "Somebody's going to have to climb up onto the boom and lie down on the sail!"

Jeff Ernst volunteered for the job, and the rest of us volunteered to accompany him. Wearing emergency harnesses equipped with rescue beacons that would automatically start flashing should we fall overboard, we made our way forward (froward, as we seafarers say), following the rails, grasping at the wire stays, stumbling to the main mast as the spray came bursting over the bow. There, Ernst threw himself up onto the boom and smothered the lowering sail as if making love to it, or wrestling a crocodile. Meanwhile, Laurent Pool furled the angry canvas until, Ernst having slithered off, it was lashed tight inside a kind of green sleeve.

Taking the helm, Moore hove to, and the *Alguita* came about into the prevailing swell. I stumbled to the head, heaved my dinner into a bucket, then crashed into my berth, and slept through the remainder of the storm.

Bob Marley was singing over the sound system when I awoke the next morning, telling me not to worry about a thing, because every little thing was going to be all right. Evidence suggested that it would. In the galley, beside the doorstep of the crew's quarters, Sam was making coffee. The sun was shining. Moore was at the helm. Laurent Pool was pissing off the stern. We were still almost fifty miles out. No land in sight. But the waves were calm. Sea state 3 on the Beaufort scale.

We sent the trawl over again, for a fifth time. Forty-five minutes later, when we were reeling it in, a gust plucked Moore's ball cap from his head, the one with the sea turtle on it, and set it adrift. Mournfully, he watched it float away. This wasn't the first time the wind had cast something overboard. The day before, a gust had plucked the nylon-upholstered mattress from a chaise longue. A great commotion had ensued. While Moore brought the boat about, Jeff Ernst, perhaps the most devoted of our captain's youthful votaries, had plunged headlong into the water and, crawl-stroking through the waves without a life vest or an emergency harness or anything besides his swim trunks, rescued the rectangle of nylon-upholstered foam from decades of drift and his captain from embarrassment. The last thing Moore wanted me to witness was him contributing to the Garbage Patch. His cap, at least, was biodegradable, but not by now—as the currents and winds swept it out of reach, making it appear and disappear, lifting it atop a crest, then sending it down into a trough—retrievable.

The fifth lowering came up totally empty—no plastic, no plankton. "Nothing. Nada," said Moore, failing to hide his disappointment. Was the ocean here that clean and that lifeless? Had the trawl ridden too high? Rinsing out the codpiece, Moore discovered the answer: "a hole big enough to poke your finger through." He poked his finger through it.

While Moore repaired the hole with needle and thread, Laurent Pool, shirtless, hatless, his bald head shining, stood at the stern, fishing. Before long he landed a rainbow runner, long as my arm, beautiful, silver with bright yellow fins, and a blue stripe equatorially dividing the dorsal from the ventral. As it slapped about on the aft deck, Pool bludgeoned it with a wooden cudgel, then gutted it clean. At midday, when I was up on the cabin roof, on watch, Moore popped open the skylight above the helm and handed me a plate of fresh rainbow runner sashimi salad—raw fish, cubed, drizzled with sesame oil and seasonings. I sat there leaning back on the fiberglass of the *Alguita*, gazing out between the rubber toes of my canvas boat shoes at the blue Pacific, degusting the best, freshest sashimi I'd ever had.

On our last day of sampling, sails raised again, beating north toward the Big Island, we had better luck, or worse luck depending on your point of view: lots of granulated plastic in our samples. Moore, elbows propped on winches, face inches above the sampling bowl, studied the contents like a clairvoyant inspecting tea leaves: "About ten pieces in there, plus some fishing line. A velella baby. Actually, scratch that, I can see a good twenty pieces in there even without looking under a microscope." He fished around with a teaspoon, poked around with tweezers, plucked out a black ball the size of a jujube. "Hard to tell whether that's a nurdle."

Our third trawl of the day brought up more plastic still. "Yeah, this has got all the stuff we look for," said Moore. "Some big stuff. Nurdles." He plucked an unmistakable nurdle out with his tweezers. "Slightly less than a gyre sample, but this is definitely the worst place we've found around Hilo so far. I think we can definitely say that the debris is statistically associated with the plankton. More plankton, more debris. Certainly looks that way." In the next trawl, we collected even more plastic.

Success brought about a change in mood, futility giving way to some moderate sense of accomplishment. Amy Young, who'd recently acquired a waterproof camera that she was notably proud of, went around snapping photos. She asked me to give her my best "adventure pose."

The Big Island came into view: first the clouds that ring the volcanic peaks, then the dark green-black line of land on the horizon slowly growing into cliffs, and then at the edge of the cliffs a row of figures— fishermen with poles, dropping their lines hundreds of feet off the sheer rock into the ocean. As we approached, Moore luffed the sails and announced a swim call. We were to go snorkeling, looking for "hand samples," semibuoyant flotsam and jetsam suspended in the water column below the surface. On the Greenpeace website I'd seen footage of Moore doing this, swimming around in the Eastern Garbage Patch with a dive mask on, pointing out the derelict objects around him—a toothbrush, a briefcase, a plastic shopping bag. I decided that it would be humiliating and unprofessional of me to sit the swim call out.

After three days in close quarters a camaraderie had developed among the members of the *Alguita*'s crew. I'd ended up confessing my fear of sharky waters. Young, who wants to be a scuba scientist, doing her fieldwork underwater, admitted that she, too, was afraid to swim out here, in blue water. Swimming on reefs, she didn't mind; in the shallows of reefs the sharks tend to be of the smaller, fish-eating variety, and you can see them coming. It's in blue water, near rocky coasts, that the seal-eaters, who out of hunger or confusion become man-eaters, tend to remove the knobby limbs of scientists enjoying a swim call.

When he goes swimming or snorkeling, Moore wears, in addition to flippers and a mask, red trunks and a shiny blue long-sleeved shirt that together give him the aspect of an aging superhero—Snorkel Man, Captain Plastic. He plunged in flippers first. Most of the crew, Ernst, Hungate, and Pool, quickly followed, outfitted with snorkels and flippers and little green nets of the sort with which people remove dead guppies from fish tanks. Sam was at the helm. Young and I stood on deck, panicking, she in her bikini, me pasty and pale, flippered and masked, in my red trunks. Young, overcoming her fears, made the plunge. Left with no other choice, I followed.

Then I was in the deep ocean, underwater, hair swirling. Bubbles bubbled before my breath-fogged mask. I'd read somewhere that to prevent goggle fogging you're supposed to spit on the plastic before jumping in. Too late for that, and if I opened my mask now, the water would rush in, and I might well lose a contact lens. Better to keep the mask on and peer through the fog. Everything hued to blue except my pasty limbs and the little green net I clutched in my right hand and the colorful swimwear and luminescent limbs of my shipmates. In no time I was sucking ocean through my snorkel and commenced to hyperventilate. Hyperventilation tends to interfere with snorkeling. Hyperventilate with a snorkel on and you tend to snorkel down big mouthfuls of the Pacific. So research has found. Upon surfacing, upon blasting the water from the snorkel, peeping over the waves lapping at my mask, recovering my bearings, I splashed madly in the sort of way that, I'd read, tends to attract sharks, whose sensory organs are well-attuned to the panic of seals—for the *Alguita*'s swim ladder. [19]

I hauled myself back out, flippers flopping, onto the aft deck, water dripping from my red trunks onto the rubber mats. There, safe, taking slow deep breaths, making sure that my wet trunks weren't clinging to my anatomy in an embarrassing way, plucking the cloth away from my anatomy in the way men do, I felt more ridiculous and more ashamed than ever. And so, having caught my breath, having fetched my little green net from where I'd dropped it while clambering aboard, having spat onto my mask, I gave it another go.

The water met my flippers. My face met the water. Here I was again, overboard. Mask on, no longer foggy, snorkel in my mouth, no longer full of water, I decided to take a look below, and what I saw made me hyperventilate again—no sharks, but nothing else either, no plastic bags, no semibuoyant sneakers, no fish, just blue water shading away toward dark, unfathomable depths. Swimming in the tropics feels, Moore had said, "like you're swimming in the sky." An apt comparison that in him inspires a kind of reverence and in me, gazing down, a kind of swimsuit-pissing vertigo. Who the hell but a suicide wants to go swimming in the sky? Again I splashed hysterically over to the *Alguita*'s swim ladder and scampered—you might think that it would be impossible to scamper in flippers, but I can attest otherwise—back aboard, and stayed there.

Moore and the rest flippered merrily about with their little green nets, but they, too, came up empty-handed. Moore again seemed disappointed that we hadn't found the miniature garbage patch he'd expected to find—no suitcases, no toothbrushes, no grains of plastic visible to the naked eye. "Next year we'll come at it from the south," he said, on board, toweling off lustily. There was still some spirit in him. "Maybe the countercurrent here acts as a barrier." Truth be told, I was disappointed, too; having hunted for plastic on the high seas, I'd found it, but the sprinkling of polymers that had turned up in our sampling jars fell far short of the maelstrom of garbage, the Sargasso of the imagination—refrigerator doors, basketballs, sneakers—I'd been dreaming of. "People always ask me for pictures of the Garbage Patch," Moore had told me. "You can't take pictures of it." You have to watch for it, and trawl for it.

After toweling off, we raised the mainsail, and the staysail, and the spinnaker, and—"Tupelo Honey" playing on the sound system, Young playing her bongos, Ernst his guitar, Hungate playing his long blond hair, I standing by the mast, striking adventure poses, adventure shorts and adventure shirt fluttering in the breeze, Moore standing on his captain's chair, head, now hatless, poking from the hatch—we ran before the wind. It was dusk again when we motored into Radio Bay.

THE PASSION OF THE ALBATROSS

Looking down at the Pacific from fifteen thousand feet, on a flight from Hilo to Honolulu, I kept remembering what it felt like to be out there, down there, swimming in deep water, gazing doubtfully and fearfully into the bottomless ambiguities, my vision obscured by the foggy snorkel mask, fishing in vain with my dainty green net. How to reconcile the vertiginous emptiness I'd glimpsed underwater with what I'd seen in Karla McDermid's lab and Pallister's paradise? Or with that photo Ebbesmeyer had shown me back in Sitka—of 252 plastic tidbits extracted from a single albatross cadaver. That photograph was taken, I'd since learned, by a photographer named Susan Middleton and appears in *Archipelago,* a collection of zoological portraits that Middleton and another photographer, David Liittschwager, made on a tour of the Northwestern Hawaiian Archipelago, the remote volcanic chain to which Laysan Island belongs.

Both Liittschwager and Middleton trained with the fashion photographer Richard Avedon, and his influence on their work is easy to see. Instead of depicting animals in their habitats, or in stylized versions of their habitats, the customary approach of wildlife artists since Audubon, Liittschwager and Middleton place their subjects before a white or black backdrop, light them with strobes that vivify the colors, and shoot what Liittschwager describes as "close-up, formal portraits"—a technique borrowed, he admits, from commercial advertising. Their stated aim in *Archipelago* was to "create an intimate connection between the subject and the viewer," but to my eye what their portraits mostly do is abstract and aestheticize the tropical sea creatures and seabirds they portray. Shot like advertisements, the portraits look like advertisements, and the connection they create resembles the one that advertising creates—more seductive than intimate—between product and consumer. The protected flora and fauna of the Northwestern Hawaiian Archipelago aren't for sale, of course; what is for sale is yet another vision—colorful, desirable, pleasing to the eye, often cute, since the photographers make a point of choosing juvenile specimens—of the natural world.

If one had any doubt about the thoughts with which Middleton and

Liittschwager burden the beasts they portray, one only need read the text that accompanies their images. On Seal Kittery, a tiny volcanic atoll, "astonished" by the sight of an endangered native plant, Middleton experienced the ecological equivalent of an epiphany. "I recognized the same sensation I experienced when I saw the Mona Lisa for the first time," she writes. Her images of the epiphanic plant, *Solanum nelsonii*, are postcards brought back from nature's Louvre. "On Laysan," where human beings are not the dominant species, "my soul felt nourished," Middleton writes.

It's not that I don't admire these photographs. I admire them the way I admire portraits of the saints—with the aesthetic appreciation of an apostate. I'm grateful for the glimpse that Liittschwager and Middleton give me of a lipspot moray eel swirling like a flamenco dancer through a crepe cloud of pink sea lettuce. But admiration is different from idolatry, and such photographs create no more intimacy between me and their subjects than the *Mona Lisa* creates between me and Lisa Gherardini, the sixteenth-century Florentine noblewoman who is the ostensible subject of Leonardo's painting. Far more than by Middleton's captions, I'm convinced by Susan Sontag, who in her famous book-length essay *On Photography* writes, "Photography implies that we know about the world if we accept it as the camera records it. But this is the opposite of understanding, an approach which starts from *not* accepting the world as it looks."

Two photographs in *Archipelago* are on their way to becoming environmental icons. The first shows the necropsied cadaver of a six-month-old Laysan albatross named Shed Bird from whose downy breast spills a slimy casserole of bottle caps and cigarette lighters and unidentifiable plastic shards. The second is the one Ebbesmeyer showed me in Sitka, of the contents of that slimy casserole, arranged, carefully, artfully, onto the photographers' white backdrop—a "mosaic of death," the caption in their book calls it.

National Geographic published the two pictures of Shed Bird in 2005, and Greenpeace has since used them in an ad campaign captioned with the slogan "How to starve to death on a full stomach." Shed Bird is now a poster child, and it's easy to see why. The images are not merely powerful, or shocking; they're persuasively accusatory. It's as though the photographers had sailed off into the mists of our collective oblivious-ness and returned with forensic evidence. Look, dear consumer, these two pictures seem to say; look at what you've done, look where what you throw away ends up. Or as Charlie Moore likes to say, "There is no 'away.' The ocean is away." Shed Bird is away.

Nearly everyone I've asked about the impact of plastics pollution—oceanographers, environmentalists, policy makers, plastics executives—invokes these images of ornithological death-by-plastic. "You've seen the pictures of the seabirds?" Pallister had asked me while we were tow-ing the *Opus* down to Resurrection Bay. Several months before, after taking a tour of the Northwestern Hawaiian Archipelago, First Lady Laura Bush had been asked the same question by a reporter for the *Honolulu Star-Bulletin*: "You've seen the picture of what came out of one little bird?"

"That's one bird from Susan Middleton's book," Laura Bush replied.

When I spoke to Benjamin Grumbles, an assistant administrator in the Environmental Protection Agency's Office of Water, he mentioned the First Lady's concern for those "ocean birds dying or having died be-cause they were eating some toys or other types of plastics."

"We've all seen the pictures of the impacts on sea life," said Sharon Kneiss, vice president of the products division of the American Chem-istry Council. "You can't help but be moved with concern. We're citi-zens of the world, too. We don't want to see that happen."

The story goes that the First Lady took up the cause of marine con-servation after seeing Liittschwager and Middleton's photographs, and then persuaded her husband to take it up as well. In 2006, President Bush designated the entire Northwestern Hawaiian Archipelago and the surrounding 137,797 square miles of ocean a marine national monu-ment, the highest level of environmental protection that the federal gov-ernment can bestow. Next to no commercial activities are permitted within the monument's bounds, not even tourism. You can only visit the place with government permission, and even then, to prevent bringing invasive stowaways with you, you must wear brand-new clothing that has been frozen for at least forty-eight hours. "For seabirds and sea life, this unique region will be a sanctuary for them to grow and to thrive," Bush said at the signing ceremony. Waxing ecological about "the

destructive effects of abandoned nets and other debris," he called for
"robust efforts to prevent this kind of debris from polluting our—
polluting this sanctuary, this monument."

Most environmental groups applauded the designation. Nevertheless, having watched the Bush administration roll out cynically titled
policies like the Clear Skies Initiative, which in fact weakened emission
controls, or the Healthy Forests Initiative, which gave the logging industry more access to national forests, Jon Coifman, a spokesman for
the Natural Resources Defense Council, was wary of the administration's motives. "In the rare instances when they've done the right thing,
it's because the political cost has been nil," Coifman told me. "There was
no one fat and happy on the other side of it who had an obvious interest
at stake." Designating the monument was, in Coifman's opinion, such
an instance. After all, the territory is largely uninhabited, and the only
people whose economic interests will suffer from the designation are a
handful of unlucky fishermen. Still, even if the designation were politically expedient, who could complain? Said Coifman: "Whatever motive
brings somebody to the table to do something good on the environment,
big or small, terrific, welcome aboard, we need all the help we can get."
This would be the largest marine park in the world, after all, home to
some seven thousand species, some of them, like the monk seal, seriously endangered, and the Laysan albatross, seriously threatened—
though less threatened than when feathered hats were in fashion. As for
the former president's idyll of thriving and growing seabirds, and Greenpeace's ad campaign (what unlikely allies!), there was, I learned a few
days after disembarking from the *Alguita*, one small problem.

I went to Honolulu to meet with scientists who possessed firsthand
knowledge of the Garbage Patch, hoping to learn whether or not they
concurred with Moore's findings and shared his alarm. Beth Flint's re-
action was fairly typical. A biologist with the U.S. Fish and Wildlife
Service, Flint has been for many years the person in charge of protecting the threatened seabirds of the Northwestern Hawaiian Archipelago.
"I suspect that you've seen some of those photos of the dead ones with
all of the plastic in their guts," she told me one morning in the conference room of a federal office building. "We find that when we take people up to the Northwestern Hawaiian Islands, the debris is the most
compelling thing they see—sometimes to the exclusion of everything

else. They can't tear their eyes away from it or think about any of the other issues because it's so disturbing to them."

To her it was disturbing too, but ambiguous. Wildlife biologists don't know for certain that plastic killed the albatross. The pathology is unclear. Did sharp shards perforate the intestines of fledglings? Sometimes. Did it obstruct the digestive tract or make a bird "starve to death on a full stomach," as the Greenpeace campaign put it? Possibly. On the other hand, Flint speculated, albatrosses eat squid, and chitinous squid beaks are also impossible to digest. Furthermore, Flint, who clearly respects the animals she studies, is tired of reading that albatrosses mistake plastic for food. "That's always the assumption: those stupid birds, they can't tell a ham sandwich from a plastic bottle cap." In fact, adult albatrosses seek out floating plastics because nutritious treats like barnacles tend to grow on them.

Despite these caveats, Flint still believed that all that plastic in the albatross diet was "clearly not good for them." Were the toxins in and on plastics getting into the food chain? According to Flint, studies have confirmed that long-lived seabirds like albatrosses have "high contaminant burdens," and they also have "plastic in their guts in pretty prodigious amounts," but "it's still all sort of circumstantial," Flint said, and the greater dangers may be the invisible ones.

The greatest known threat to the albatross is commercial fishing. Adult birds will swoop down for a morsel and end up as "bycatch" on a longliner's fishing hook. Or in the trawl fishery, said Flint, they'll "collide with the third wire, the thing holding the trawl, and break their wings and die." And if even one of its parents dies, a hungry fledgling back in the rookery will likely starve. Then there's the golden crownbeard, an invasive plant that "displaces everything in its path," and "becomes this impenetrable thicket so that birds can't even get into their nests, and if they do manage to raise a chick, the chick is surrounded by thick, thick vegetation that cuts off the breeze." Chicks overheat, dehydrate, and die, and "having a big gutful of plastic" probably makes dehydration all the more likely. So does global warming. Finally there's the lead paint flaking off the derelict military compound at Midway, which it would take around $6 million to clean up—considerably less than we're spending on marine debris. Listening to Flint catalog the plague of perils that face the Laysan albatross even there in that sanctuary where seabirds are supposed to thrive and grow, I was overcome by a sense of gloomy futility. Why bother with beach cleanups, or anti-littering campaigns, or plastic-bag bans, or bottle deposits, or any of the "robust efforts against marine debris" that we could possibly make? At least the future of the Laysan albatross wasn't as dim as that of the Laysan duck, of which there are now only around 465 left; of Laysan albatrosses, by contrast, there remain 2.4 million.

Confusingly, despite her litany of caveats, Flint's praise for Moore was unequivocal: "I think that he's done a tremendously valuable service to humanity by pursuing this when none of the big oceanographic or academic institutions or government institutions did," Flint said. She predicted that other researchers would soon "get on his bandwagon." Her prediction seems to be coming true. In the last few years several studies of plastic poisoning have appeared in reputable journals.

The hardest question to answer about the Garbage Patch, it turns out, isn't whether plastic debris threatens animals and ecosystems, but what if anything can be done about it. "We haven't been able to hatch up any good ideas," Flint admitted. Albatross fledglings don't forage on land, she explained. In fact they don't forage at all. Their parents do, flying far and wide across the Pacific, swooping down to skim morsels off the surface, which they bring back home and regurgitate into a hungry fledgling's mouth. That's where all the detritus in that Greenpeace ad came from. Even if we were to clean every beach in the world, it wouldn't keep albatrosses from stuffing their offspring full of plastic. "You'd have to clean the entire ocean," Flint said.

Back on the *Alguita*, I'd described for Charlie Moore the tonnage of debris I'd witnessed on Gore Point. "That's not unusual," he'd said. "I have pictures of Japan, where that's the case. I've got pictures of Hawaii where that's the case. Any windward side of an island's going to have situations like that. The question is, How much can we take? We're burying ourselves in this stuff." Moore sympathized with Pallister's motives, and believed that GoAK's efforts might help "raise awareness." But he also agreed with Bob Shavelson that cleanups alone serve little purpose. If Pallister thought he was saving Gore Point from plastic pollution he was fooling himself. "It's just going to come back."

Evidence I'd collected on the day of the airlift lent credence to this prediction. Before I boarded the helicopter and began my long journey home, I'd hurried across the isthmus for one last solitary walk along the windward shore. At the strand line I'd found the day's deposit: toothbrush, Clorox bottle, surfboard fin, capless deodorant stick, sixteen water bottles, all Asian in origin. This, in Moore's opinion, is why the 2006 Marine Debris Research, Prevention, and Reduction Act is likewise doomed to fail. "It's all been focused on cleanups," he said of the action the federal government had taken. "They think if they take tonnage out of the water, the problem will go away."

Moore is right in this respect: current federal policy does treat symptoms more than causes. In the Northwestern Hawaiian Islands, whose shores are washed by the southern edge of the Garbage Patch, federal agencies are staging one of the biggest marine-debris projects in history. Since 1996, using computer models, satellite data, and aerial surveys, NOAA, the Coast Guard, and the U.S. Fish and Wildlife Service have located and removed more than five hundred metric tons of derelict fishing gear in hopes of saving the endangered Hawaiian monk seal from entanglement. Administrators at NOAA's Marine Debris Program point to the project as an example of success, but the results, vindicating Moore, have been mixed at best. NOAA incinerates the debris it collects at a power plant on the outskirts of Honolulu, converting it into electricity. But unless they're supplemented by a metropolitan supply of garbage, such incinerators operate at a loss, and in places that cannot support them, like maritime Alaska, the only option is to bury debris in landfills, and in many coastal communities, landfill space is already running short. Furthermore, although wildlife biologists are now finding fewer monk seals entangled in debris, they are also finding fewer monk seals, period.[20]

In some respects, however, Moore's indictment of the Marine Debris Program is not entirely just: NOAA isn't focusing only on cleanups. The agency is also investing in educational programs to teach litter prevention to both beachgoers and fisherfolk. And NOAA scientists known as "net nerds" are submitting derelict gear to the sort of rigorous scrutiny to which archaeologists submit arrowheads and potsherds. Although it's not a job I'd want, the expertise of net nerds, and the Siddharthan reserves of patience and attention they surely must possess, does command a kind of admiration. From the colors of the fibers and the size and style of the weave, net nerds can distinguish old nets from new ones, American nets from Indonesian ones. Toward what end all this taxonomizing? To hold the guilty parties accountable and persuade them to change their ways.

Then, too, the NOAA scientists I spoke to in Honolulu do not believe that cleanups alone will make marine debris go away. In removing debris from the water column, they merely hope to spare as many animals as possible—sea turtles, seabirds, whales, monk seals—from the tortures, often fatal, inflicted by a ghost net or plastic bag. "If animals do get entangled in gear and they aren't able to get out of it, it's a very slow, rather painful death," a freckle-faced, black-haired NOAA cetologist of mixed Scottish-Japanese descent told me. Her name was Naomi McIntosh. Consider what happens to a juvenile humpback with a loop of netting snared around its fin: As the whale grows, the nylon line will carve through flesh and bone until the animal dies—slowly, painfully—from its infected wounds. "With animals that are entangled, you can see the weight loss, their bones coming through. When we see animals that have been entangled for a long time there are parasites," McIntosh said.

Despite her grisly, heartfelt descriptions of the horrors inflicted by marine debris, I found myself, as usual, entangled in doubts. "If I'm a taxpayer in Kansas," I asked her, "even if I'm concerned about the environment, shouldn't I be more concerned about investing in alternative energy, or reducing CO_2 emissions from power plants, than about the suffering of whales? I mean they're whales, and they're cute, and I love them, but they've got a lot of competition right now for the tax dollar."

Perhaps I'd chosen my words poorly, insensitively. To a cetologist, a humpback is not merely cute, after all. To a cetologist, a humpback inspires wonder, and empathy. McIntosh had already described for me, with childish delight, the curiosity of whales, how they like to investigate and play with any floating thing they find. Now, listening to my question, she was visibly distraught. Fighting back tears, she managed a reply: "Sure," she said, pausing to compose herself, wiping her eyes with the backs of her hands. "A friend of mine that I spoke to has been in federal government for a long time, and he's looking at the number of years he has left, and he said to me, 'You know, Naomi, there are certain issucs that are going to right themselves on their own, and there are certain issues that, if we don't do something now, we can't change that.' And to me this is one of those issues because if we lose that particular species or population, we can't bring it back." And then, unchaperoned by a press officer, she permitted herself an indiscretion, an outburst of misanthropic wrath: "Or maybe eventually we'll just kill ourselves off, and the earth will return to its natural state."

As nearly everyone I spoke to about marine debris agrees, the best way to get trash out of our waterways is, of course, to keep it from entering them in the first place. But experts disagree about what that will take. The argument, like so many in American politics, pits individual freedom against the common good. Both Seba Sheavly, the marine-debris consultant, and the American Chemistry Council, one of Sheavly's clients, prescribe voluntary measures—voluntary, industry-sponsored research into the environmental impact of plastics in the ocean, voluntary public awareness campaigns in which industry and local governments work together to discourage littering, encourage recycling, and teach more consumers the lesson that volunteers in coastal cleanups learn: that manufacturers don't throw their products into the ocean; consumers do. "Don't you tell me I can't have a plastic bag," Sheavly says, alluding to plastic-bag bans like the one San Francisco enacted in 2007. "I know how to dispose of it responsibly." But proponents of bag bans insist that there is no way to use a plastic bag responsibly. "If you go to Subway, and they give you the plastic bag, how long do you use the plastic bag?" says Lorena Rios, the chemist at the University of the Pacific who subjected my plastic beaver to mass spectrometry. "One minute. And how long will the polymers in that bag last? Hundreds of years."

"The time for voluntary measures has long since passed," Steve Fleischli, president of Waterkeeper Alliance, told me. Fleischli would have us tax the most pervasive and noxious plastic pollutants—shopping bags, plastic-foam containers, cigarette butts, plastic utensils—and put the proceeds toward cleanup and prevention measures. "We already use a portion of the gasoline tax to pay for oil spills," Fleischli says. Such levies shouldn't be seen as criminalizing the makers and sellers of plastic disposables, he argues; they merely force those businesses to "internal-

ize" previously hidden costs, what economists call "externalities." This
market-based approach to environmental regulation, known as extended producer responsibility, is increasingly popular with environmental groups. By sticking others with the ecological cleaning bill, the
thinking goes, businesses have been able to keep the price of disposable
plastics artificially low. And as Pallister learned at Gore Point, the
cleaning bill may be greater than we can afford.

In Charlie Moore's opinion, the solution may require more radical
sacrifices, sacrifices that the citizens, governments, and corporations of
the world are reluctant to make. Eventually we will have to abandon
planned obsolescence, he believes, and instead manufacture products
that are durable, easily recyclable, or both. In short, we'll have to abandon the American way of life.

Such eco-utopianism sounds to my ears far-fetched, but there are
other smaller, more practical actions we could, in theory, take. In 1999,
the Natural Resources Defense Council successfully sued the U.S. Environmental Protection Agency for permitting municipalities to pollute
watersheds around Los Angeles. As a result of the lawsuit, Los Angeles
County had to comply with stricter total maximum daily loads, or
TMDLs, the local pollution limits that the EPA places on a region's
waterways under the Clean Water Act. The new TMDLs, the first in
the country to treat floatables as a pollutant, will require the county to
reduce the amount of solid waste escaping its rivers and creeks from 4.5
million pounds a year to zero by 2016. To meet that target, cities will
have to invest in "full-capture systems," filters that strain out everything
larger than five millimeters in diameter. In theory, every region in the
country could follow suit, but already cash-strapped governments in
Southern California are complaining that these "zero-trash TMDLs"
are too costly and ambitious to implement. Moore, meanwhile, has collected data showing that even full-capture systems would allow tens of
thousands of plastic particles to escape the Los Angeles River every day.

Forty years ago, *Science* published an essay called "The Tragedy of the
Commons" in which the ecologist Garrett Hardin challenged what
might be called the American Comedy of Progress—the cherished notion that with time, technology, entrepreneurialism, and, if need be, ac-

tivism, all problems can be solved. In America, even prophets of environmental doom subscribe to the Comedy of Progress. Thus, at the end of *An Inconvenient Truth*, Al Gore follows his alarming forecast with an uplifting recipe for salvation. In a similarly comedic vein, politicians promise to save the planet and revive the economy by investing in "green-collar jobs." Hardin's truth is more inconvenient than Gore's, so inconvenient it amounts to an American heresy.

In a finite world of diminishing resources, Hardin's reasoning went, the freedom of individuals will not lead hopefully to progress but fatalistically to destruction. Here's why: Picture a pasture—a commons—on which all herdsmen are free to graze their flocks. "Such an arrangement may work reasonably satisfactorily for centuries because tribal wars, poaching, and disease keep the numbers of both man and beast well below the carrying capacity of the land," Hardin wrote. But if an era of peace and prosperity comes about and the flocks multiply, "the inherent logic of the commons remorselessly generates tragedy." Each herdsman, acting rationally in his own self-interest, will keep adding to his herd. He alone will profit from an additional animal, whereas the environmental impact will be shared by all. What's one more cow? Overgrazed, grass eventually gives way to dust. "Each man is locked into a system that compels him to increase his herd without limit, "Hardin concluded. "In a world that is limited, freedom in a commons brings ruin to all."

In twenty-first-century America, there are still resources that we share, air and water being the best examples, and so the tragedy of the commons still obtains. It explains the depletion of fisheries and aquifers. It explains the pollution of skies and seas. Technology may forestall the tragedy—by increasing crop yields or fuel efficiency, for instance—but so long as the human population continues to grow it cannot in the long run avert it. Only a decline in population or consumption can do that. And here Hardin leaves us with a profound dilemma, forcing us to choose between environmental health and economic growth. If Hardin is right, we cannot have both.

No one—not Charles Moore, not Chris Pallister, not Greenpeace—will tell you that plastic pollution is the greatest environmental threat our oceans face. Depending on whom you ask, that honor goes to global warming, or ocean acidification, or overfishing, or agricultural runoff. In a way, plastic's greatest threat may be symbolic, which is not to say that it is empty or cosmetic. Most pollutants are invisible. Saturated with CO_2, our oceans have begun to acidify, our scientists tell us—you can't discern pH levels with the human eye. But unlike many pollutants (mercury, for instance, or CO_2), there is no natural source of plastic and therefore no doubt about how to apportion blame. We're to blame. Where plastics travel, other, invisible pollutants—pesticides and fertilizers from lawns and farms, petrochemicals from roads, sewage tainted with pharmaceuticals—usually follow. There is no such thing as natural plastic, and because it is so visible, it provides a meaningful—and alarming—bellwether of our impact on the earth. Then, too, as numerous conservationists have told me, compared with other environmental problems this one should be easy to solve. And yet we show no sign of solving it.

PLASTIC BEACH

Down at South Point, after my sojourn in Honolulu, after dropping off my suitcase at the Shirakawa Motel and driving south in my rental car, after reaching the end of the road and continuing on foot, I begin to see what I've come looking for—the colorful confetti of plastic debris. In almost every cove formed by black lava rock there are drifts of it, piled up in crescents many feet deep. There's fishing gear, shattered fishing floats, tangles of nets, but if you stop and stoop down and look closely you can pick out more commonplace objects—detergent bottles, Nestlé lids, golf balls. The coves sort the plastic. In one there's an abundance of jar lids. In the next an abundance of nets. In another, there are dozens of plastic popsicle sticks. Do fishermen like popsicles? Or did these formerly belong to children on a seaside holiday? I stumble upon the top of a garbage can on which I read the word RUBBERMAID, and then upon one of the kind of flip-flops I'd seen in Curtis Ebbesmeyer's basement, with the lightning bolt zigzag running through it. I don't find any Floatees, but I do find a green plastic soldier aiming his rifle at a pebble of pumice, and then a few feet away the wreckage of a plastic jeep that by the look of it has been terrorized by an albatross. On and on I walk, under the tropical sun, and on and on stretches the lava rock littered with wrack. In one cove, I can see the colorful confetti of plastic floating in on the blue waves. Charlie Moore was right to wonder where all this plastic was coming from.

How beautiful the Pacific looks from here! No catamarans today. Just the blue waves. The spray of the surf. The black rocks. The riddles written in debris. The sun is high. I wish I'd brought a hat. I wish I had some water. It's nine miles to Kamilo Beach—farther, I decide, than I can hike in this heat.[21]

The next morning, after a breakfast of granola bars and Gatorade purchased at the gas station up the road from the Shirakawa Motel, a gas station on the side of which someone has painted a pretty, pastoral Hawaiian mural—white church, tropical flowers, guy playing a ukulele, another guy (Mark Twain?) wearing a mustache and a tuxedo, another guy hacking at sugarcane with a machete—I station myself in the motel courtyard, where a peace pole, a little obelisk inscribed with pacifistic messages in many languages, tilts amid an island of flowers from which butterflies sip.

A silver pickup truck arrives. At the wheel is Bill Gilmartin, a retired, ponytailed biologist formerly employed by the National Marine Fisheries Service. On the door of his truck is a green sign that reads HAWAII WILDLIFE FUND. His truck isn't your ordinary truck. Gilmartin paid mechanics to turn his pickup truck into a dump truck. The bed, hoisted by some sort of hydraulic mechanism, can tip. You wouldn't know it looking at it. I climb into the passenger seat, and we zoom south, past the lava rock walls, toward South Point, bound for Kamilo Beach. "I started out in debris collection on the Northwestern Hawaiian Islands in 1982," Gilmartin says. "You get into this business in your gut and your heart and it's hard to back out. I got into this business because of the first monk seal pup I rescued."

Here again are the windmills and the cracked road leading to the end of the world or at least to the edge of the United States. Asphalt gives way to dirt. We stop at a gate and I clamber out to open it, noticing beside it, lashed to a post, a yellow sign commanding me not to approach monk seals. We ramble along the red dirt road, slowly, jouncing, and soon enough we've traveled farther, I can tell, than I made it the day before on foot. "This coast is so bad," Gilmartin says, "debris comes up faster than you can keep it clean. It gets continuous onshore winds and it's in the way of the easterly currents." Perhaps Moore should have pointed the *Alguita* west instead of south, I find myself wondering. By Gilmartin's calculations, twenty tons of trash wash up along the eastern coast of South Point every year. "Even if everything stopped today, it would keep washing in." Unlike Pallister, Gilmartin doesn't expect to win. He just can't think of a better way to spend the weekends of his retirement than this—hauling debris in his truck.

The clouds this morning are gray and low. Out the window, through the shrubbery, I catch glimpses of ocean. We come to a big pile of bright blue bags, left here by a cleanup group akin to GoAK. Gilmartin and I hurl these into his truck. Then we come to a big mess of derelict net, which Gilmartin hooks to his truck's winch, mounted just behind the cab. The engine whines, the cable hauls the heavy net in. At last the tires of Gilmartin's truck crackle up to the edge of Kamilo Beach, which is everything Moore had led me to expect: a beach of plastic, finely ground. There are twigs of driftwood, and igneous pebbles, and a few errant coconuts, but they're far outnumbered by the shrapnel of debris. This is where the flotsam I saw yesterday ends up. Washing in and out, the surf and the sun grind it down and the currents deposit it here, and here, until a flood tide or a storm surge sweeps it out to sea, it remains; Gilmartin can't bother with the plastic sand. There's simply too much of it.

It's almost beautiful, all those unnatural colors and shapes in such a natural landscape, beautiful because incongruous. It occurs to me now, as it has before, that this is what I have been pursuing these past months, this is what I found so spellbindingly enigmatic about the image of those plastic ducks at sea—incongruity. We have built for ourselves out of this New World a giant diorama, a synthetic habitat, but travel beyond the edges or look with the eyes of a serious beachcomber and the illusion begins to crumble like flotsam into sand. Incongruities emerge, and not just visual ones.

For instance: In 1878, nine years after its invention, a sales brochure promoted celluloid as the salvation of the world. "As petroleum came to the relief of the whale," the copy ran, so "has celluloid given the elephant, the tortoise, and the coral insect a respite in their native haunts; and it will no longer be necessary to ransack the earth in pursuit of substances which are constantly growing scarcer." Ninety years later, in the public mind, plastic had gone from miracle substance to toxic blight. In 1968, at the dawn of the modern environmental movement, the editor of *Modern Plastics* argued that his industry had been unfairly vilified. Plastic was not the primary cause of environmental destruction, he wrote, only its most visible symptom. The real problem was "our civilization, our exploding population, our life-style, our technology." That 1878 sales brochure and that 1968 editorial were both partly, paradoxically right. Petroleum did save the whale, or at least some whales; plastics did save the elephant, not to mention the forest. Modern medicine would not exist without them. Personal computing would not exist without them. Safe, fuel-efficient cars would not exist without them. Besides, they consume fewer resources to manufacture and transport than most alternative materials do. Even environmentalists have more important things to worry about now. In the information age, plastics have won. With the wave of a magical iPod and a purified swig from a Nalgene jar, we have banished all thoughts of drift nets and six-pack rings, and what lingering anxieties remain, we leave at the curbside with the recycling.

Never mind that only 5 percent of plastics actually end up getting recycled. Never mind that the plastics industry stamps those little triangles of chasing arrows into plastics for which no viable recycling method exists. Never mind that plastics consume about 400 million tons of oil and gas every year and that oil and gas will in the not too distant future run out. Never mind that so-called green plastics made of biochemicals release greenhouse gases when they break down. What's most nefarious about plastic, however, is the way it invites fantasy, the way it pretends to deny the laws of matter, as if something—anything—could be made from nothing; the way it is intended to be thrown away but chemically engineered to last. By offering the false promise of disposability, of consumption without cost, it has helped create a culture of wasteful make-believe, an economy of forgetting.

At Plastic Beach, I crouch down, scoop up a handful of multicolored sand, and sift it through my fingers. This then is the destiny of those toy animals that beachcombers fail to recover: baked brittle by the sun, they will eventually disintegrate into shards. Those shards will disintegrate into splinters, the splinters into particles, the particles into dust, the dust into molecules, which will circulate through the environment for centuries. The very features that make plastic a perfect material for bath-tub toys—so buoyant! so pliant! so smooth! so colorful! so hygienic!—also make it a superlative pollutant of the seas. No one knows exactly how long a synthetic polymer will persist at sea. Five hundred years is a reasonable guess. Globally, we are currently producing 300 million tons of plastic every year, and no known organism can digest a single mole-cule of the stuff, though plenty of organisms try.

Just a few weeks before my voyage in the *Opus*, Sylvia Earle, for-merly NOAA's chief scientist, delivered a speech on marine debris at the World Bank in Washington, D.C. "Trash is clogging the arteries of the planet," Earle said. "We're beginning to wake up to the fact that the planet is not infinitely resilient." For ages humanity saw in the ocean a sublime grandeur suggestive of eternity. No longer. Looking at the debris at Gore Point or South Point, we see that the ocean, vast as it is, is perhaps smaller and more vulnerable than we'd thought; that we have, perhaps, taken dominion over the watery wilderness, too. Now it is the sublime grandeur of our civilization but also of our waste that inspires awe.

THE FOURTH CHASE

Chuckedy-chuckedy-chuck goes the rubber duck machine.

—Eric Carle

RED MAGIC

Entering the Toys & Games Fair at the Hong Kong Convention and Exhibition Center felt a bit like falling down the rabbit hole into Wonderland. Alice, however, was nowhere to be seen, and upon inspection the white rabbit turned out to be a battery-powered plush toy with an MP3 player tucked inside its foamy bowels. And as they watched the synthetic fauna flash and beep and dance, there were no expressions of wonderment on the faces of my fellow travelers to Wonderland, unless you count as wonder the naughty twinkle in the eyes of the silver-haired Chinese man peeping through his spectacles at a dozen little plastic dogs industriously pantomiming the procreative act. Humping Dog, the toy was called—I HUMP UNTIL DISCONNECTED, the tagline ran.

At a booth up in Expo Hall 2, a crowd had gathered around an upholstered table on which a salesman was demonstrating a remote-controlled dune buggy that could leap from a ramp, land with a bounce, and zip back around for another go. With every jump, the crowd murmured approvingly. I'd never seen so many adults among so many toys. I passed a British sales rep piloting a remote-controlled helicopter that actually flew—no wires or anything—circling his head like a performing dragonfly. Down the aisle, another crowd had formed in front of the booth of Creative Kid, a "Camera-Interactive Entertainment System," where a woman dressed in black was playing a purportedly educational video game called Bubble Music. Gesticulating with a plastic baton, a look of grim concentration on her face, she sent an on-screen avatar, some sort of jellyfish, chasing after cartoon bubbles that chimed out the melody of "London Bridge" as they popped. The woman's performance was strangely mesmerizing. If you didn't know the screen was there you might have thought she was swatting at invisible flies, or practicing tai chi.

This trade show, the toy industry's largest in Asia, was the first stop on what I intended to be a kind of economic safari, a sightseeing expedition into China's industrial wilderness. It wasn't really China I'd traveled halfway around the planet to see but the means of production, which were to me, a child of consumerism, unimaginable. We are not meant to know where our possessions come from, we American consumers, or from what ingredients and by what mysterious processes they were spun and by whom. And so long as our possessions pose no risk to us or to our loved ones, I don't think we really want to know. Such knowledge would be overwhelming. Willfully suspending disbelief, we prefer instead to pretend that our possessions were begotten, not made, and the marketers of consumer goods are happy to assist in the illusion.

Consider for a moment the aesthetics of packaging, the wrappings of enchantment, the clamshells and plastic blisters that serve as both miniature shop windows and seemingly sterile cocoons. Consider the little transparent sticker placed like a hermetic kiss across the cardboard flap. *This box has never been opened*, it is there to tell us, or even, more superstitiously, *The contents herein have never been touched*. Slit that seal with a fingernail and something changes. Magic escapes. Unused, untouched, the contents of that box are nonetheless no longer brand-new. The difference between the new and the brand-new is like the difference between youthfulness and chastity. Think of the components individually wrapped inside their little plastic sachets, the power cords crimped into perfect coils. Think of the nested loaves of Styrofoam—Styrofoam, which is quite possibly the cleanest, whitest, lightest, chastest substance chemists have ever confected (until, that it is, it goes adrift on polluted seas). It is functional, no doubt, preventing breakage while minimizing shipping costs. But it is also symbolic. The sound that snug Styrofoam makes as you coax it from the cardboard box is a Pavlovian signal: the squeak of the new.

And yet mystery has always acted like a pheromone on the human imagination. Browsing through the colorful circulars that spilled from the Sunday newspaper like candies from a piñata, or noticing yet again the ubiquitous phrase MADE IN CHINA embossed on one of Bruno's toys, I'd found myself having vaguely mystical thoughts about the places and lives with which the chain of production invisibly entangled me. Having spent my life at the receiving end of that chain, curious and eager to learn what the business end was like, I set out to follow it back to its source.

I'd read that 70 percent of the toys we Americans buy—about $22 billion worth—are wrought by low-wage factory workers in Guangdong Province, and that 70 percent of those toys are blow-molded or injection-blow-molded or extruded out of plastic resin. I'd read disturbing reports about the Chinese toy industry, and now when, at bedtime, I read to Bruno Eric Carle's *Ten Little Rubber Ducks*—which was itself printed and bound in China—and came to the scene of the woman in the brick-red dress painting brick-red beaks with her little paintbrush, I couldn't help but think of Huangwu No. 2 Toy Factory, where, according to the nonprofit group China Labour Watch, in order to earn the legal minimum wage of $3.45 for an eight-hour day, a piece-rate worker in the spray department "would have to paint 8,920 small toy pieces a day, or 1,115 per hour, or one every 3.23 seconds."

Did workers make the Floatees under similar conditions? Way back at the outset of my journey, before boarding the ferry to Sitka, I'd called The First Years Inc., which had recently been bought out. The current management seemed to know less about the Floatees than I did, or pretended to. There was no way they could tell me which factory produced that yellow duck of mine, they said. I assumed that my chances of finding the factory were slim, my chances of gaining access to it, nil.

At Curtis Ebbesmeyer's suggestion, I contacted T. Berry Brazelton, founder of the Child Development Unit at Children's Hospital Boston, and a professor emeritus of clinical pediatrics at Harvard Medical School. On the back of the 1992 packaging of the Floatees—alongside the claims that these toys were "dishwasher safe" and conformed to "ASTM Standard Consumer Safety Specification on Toy Safety F9693," whatever that was—there had appeared the following endorsement: "Our products are inspired and pretested by parents, for parents. They are designed in consultation with Dr. T. Berry Brazelton and staff members of the Child Development Unit, Children's Hospital in Boston. The First Years is a benefactor of the Child Development Unit." Understandably perhaps, Ebbesmeyer had concluded that Brazelton, a celebrity among pediatricians, had designed the Floatees—understandably but mistakenly.

Brazelton had serious misgivings about his relationship with The

First Years. Pediatric colleagues had criticized him for accepting donations from a corporate benefactor. In his own defense, he insisted that in lending his name to The First Years' products, all he hoped to do was encourage the development of safe, educational toys that even the poor could afford. He played no part in designing the products, he said. Once a toy was in development, he and his staff would "sit around and discuss how a child might play with it." And if the toy seemed worthy, then, and only then, would Brazelton bestow the imprimatur of the Child Development Unit. If I wanted to know more about the Floatees, Brazelton suggested that I contact his acquaintance Ron Sidman, the former CEO of The First Years Inc.

To my surprise, Sidman agreed to meet with me. In September of 2007, while receiving a crash course in oceanography at the Woods Hole Oceanographic Institution, at the height of the Great Chinese Toy Scare, when every month seemed to bring news of yet another scandalous recall, I traveled from Woods Hole to Sidman's Cape Cod home. Started by Sidman's parents in the fifties, The First Years develops and markets safety products and playthings for infants and toddlers—rattles, teethers, plastic utensils, disposable sippy cups, bath toys. After chasing the Floatees, after recovering a pair of them from the Alaskan wilderness, I hoped to determine their origins, which I feared would be even more shadowy than their fate.

The plastic animals marketed under the brand name Floatees are no longer in production, and Ron Sidman is no longer in the toy business. In 2004, he and his shareholders sold The First Years for $136.8 million to RC2 Toys, the same Illinois company, it so happens, that in the summer of 2007, while I was messing around on boats with Chris Pallister, had recalled more than a million Thomas & Friends toy trains after routine tests detected "excessive levels of lead" in the paint. The main lesson to be learned from that and other recalls, in Sidman's opinion, was not that the system was broken but that it had worked. "In the old days, these products would have been out there with no one knowing about it," he said. Now the threat of "criminal and civil penalties" and the fear of bad publicity compel toy companies to test their products vigilantly and report any defects they find. "This is the toughest business in the world from a regulatory standpoint," Sidman said. "When it comes to babies, it's so emotional. If there's any publicity about a safety hazard—even if it's not a real hazard—there's an overreaction to it."

But wasn't that what was so frightening to consumers—that toymakers had failed to meet product-safety standards despite those tough regulations? If the companies themselves couldn't police their Chinese manufacturers, I asked Sidman, then who could? The Chinese government?

"I don't see that that's where the responsibility lies," he said, sitting in a leather armchair, stroking the enormous head of his slobbery black Labrador. Behind him a picture window looked out over Cape Cod's deciduous hills.

Sidman thought the responsibility lay with the manufacturers themselves. He wasn't sure which Chinese manufacturer had made those plastic animals lost at sea, but he gave me the name of his old trading partner in Hong Kong, Henry Tong, vice president of the Wong Hau Plastic Works & Trading Company Ltd. If anyone still knew who'd made those toys it would be Henry, Sidman thought.

Yes, sure, he knew, Tong told me when I called his office in Hong Kong. They were made at the Po Sing plastics factory in Dongguan. I asked if it might be possible to arrange a tour. Yes, probably, Tong said. The general manager there was an old friend of his, he said. He'd just need a few weeks' notice to set everything up.

Dongguan, I learned, is an industrial town in the Pearl River Delta Economic Zone, an alluvial maze of factories and shipping routes radiating outward into Guangdong Province from Hong Kong. The Pearl River Delta is mainly what people have in mind when they talk about China's "economic miracle." Newspapers often refer to it as "the workshop of the world," a phrase, first applied to England in the nineteenth century, that has in the twenty-first century drifted east. The iPod is manufactured in the Pearl River Delta, and so is Chicken Dance Elmo. So are most of the cheap, ubiquitous goods labeled MADE IN CHINA that we Americans buy.

Although, reading the newspaper, I tended to imagine the Pearl River Delta as a polluted wasteland where workers toiled miserably away in dark Satanic mills, not all my Pearl River dreams were bad ones. In *The Oracle Bones*, his exquisitely reported book about life in China at the turn of the millennium, Peter Hessler quotes a song sung in commemoration of the country's thirty-year experiment with capitalism, an experiment that began when Deng Xiaoping established the Pearl River Delta Economic Zone. "In the spring of 1979," one verse of this song goes,

> An old man drew a circle
> on the southern coast of China
> And city after city rose up like fairy tales
> And mountains and mountains of gold
> gathered like a miracle. . . .

For much of the past two decades, the Delta's economy has been the fastest-growing in the world. The millions of itinerant laborers from China's rural interior who moved there seeking work on the assembly lines ended up participating in what is often called the largest migration in human history. In Shenzhen, known as the Overnight City because of the speed with which it sprouted up, mushroomlike, out of the rice paddies and fishing towns that preceded it, the annual growth rate in some years surpassed 30 percent. Despite recent competition from Beijing and Shanghai, the Pearl River Delta remains China's most productive region. Home to just 3 percent of the country's population, it nonetheless accounts for more than 25 percent of China's foreign trade.

The late Chinese sociologist Fei Xiaotong once described the Pearl River Delta as "a store at the front and a factory in the back." Hong Kong is the store; Guangdong Province is the factory. Journeying from the one to the other, I imagined, would be like traveling upstream toward the headwaters of my material universe. The Toys & Games Fair, where buyers from Western toy companies come to find Chinese suppliers, seemed like a good place to start.

At another booth an olive-skinned man—Mediterranean, perhaps, or Middle Eastern—sat across the table from a Chinese woman, catalogs and laptops open before them, colorful plastic creatures gazing down from modular walls. "Now I can give you the recording, right?" the man was saying. "Yes," the woman said, "you give me the sound, I put it in the toy."

Another booth, two men in suits, both with laptop cases slung over their shoulders. "Dude, check this out," one said to the other. On a shelf were a row of plush animals, below each animal, a button. The first man pressed a button below a cat; the cat mewed. He kept pressing the button: *mew, mew, mew, mew*. His colleague joined in on a cow. They stood there mewing and mooing.

Children of the twenty-first century evidently do not like silence. Close your eyes and you could hear, along with the barnyard ruckus of onomatopoeia, a fantasia of whirring gears, the disturbingly convincing simulated gunfire of disturbingly realistic plastic assault weapons, the clicking feet of motorized puppets, the soothing voices of promotional videos ("by keeping a close eye on market trends and by acting simultaneously, Eastcolight launches products in a timely fashion"), and fading in and out as you passed through the Educational Toys and Games Section, a synthesized measure of Mozart or Chopin.

Wandering through this animatronic funhouse, I kept thinking of two-year old Bruno, of how terrified—or even terrorized—he would have felt if he'd accompanied me here. For him this festival of commerce would have been a giant shop of horrors. The imaginary child implied by the toys on exhibit in Hong Kong was impossible to reconcile with my actual child. I didn't think I'd like to meet the imaginary child they implied. That child was mad with contradictions. He was a machine-gun-toting, Chopin-playing psychopath with a sugar high and a short attention span.

I sought refuge in the exhibit of The Toy Company, whose simple name reflected its simple wares. The Toy Company had created a lilliputian world out of wood. How tranquil this wooden world was, how unlike the other exhibits, how quiet. I gazed down giantlike at a farm where a little wooden farmer drove his little wooden tractor past a little wooden cow over a meadow of green felt. Ah, the pastoral. While I was admiring her samples, Xenia Kuzelka, the German-born managing director of The Toy Company, mistook me for a buyer. That's what most of my fellow visitors were: buyers from toy companies. I explained apologetically that I was in fact a writer conducting amateurish research into childhood. Did she, I asked while I had her ear, ever think about children? What they are? What they want? What goes on inside their little inscrutable brains?

"Yes, of course," she said, glancing awkwardly around, as if for a hidden camera. "I am myself a mother, so I think about that." She told me wood was better than plastic because it's warmer to the touch—an opinion shared by none other than Roland Barthes, who wrote, "Current toys are made of a graceless material, the product of chemistry, not of nature. Many are now moulded from complicated mixtures; the plastic of which they are made has an appearance at once gross and hygienic, it destroys all the pleasure, the sweetness, the humanity of touch." I resisted the temptation to quote Roland Barthes to Kuzelka, now directing my attention to a wooden jumbo jet, explaining that this was the sort of toy that a whole troupe of children could play with. She removed its roof and let me peek inside at all the little wooden people in their little wooden seats. "Everyone says the future is in electronic toys," she said, "but I prefer toys that encourage children to interact."

Among her wares I found something I'd hoped to find, a wooden duck. I hoped to find it because I'd arranged to return from the Far East via container ship, and upon approaching the site of the 1992 container spill, I hoped to commemorate the moment by chucking a duck overboard, but knowing what I knew about plastic pollution I felt that I could not, in good conscience, contribute to the Garbage Patch. I told Kuzelka to name her price. She told me that unfortunately these toys were for display only. It was against trade show policy to make sales. Opening my wallet, I offered to pay twice the wooden duck's retail value. No. Thrice? No. She was a woman of principle or at least of prudence. Kuzelka's toys were beautiful in a quaint sort of way, but I wondered whether her anonymous, anachronistic figurines stood much of a chance against Bob the Builder or Elmo or a PVC duck molded in the likeness of Ernie's, or any of those other celebrities of the nursery.

Down in the basement of the convention center there was a low-ceilinged room called Expo Drive Hall, and upon entering I sensed that all was not well. Upstairs in the premium exhibition spaces, the exhibitors hailed mainly from Europe, America, and Hong Kong, but down here they were all from the Chinese mainland. Up on the mezzanine, while enjoying fried rice or an espresso at one of several concession stands, you could gaze through the building's glass shell at the picturesque junks and antique ferryboats out on Victoria Harbor and at the neon-bedecked towers of Kowloon, Hong Kong's peninsular borough, rising up behind them. Down in windowless Expo Drive Hall, the only food was popcorn peddled from a pushcart whose glass case gave off a lurid, buttery glow.

Upstairs the English-speaking sales reps trusted their toys and their spectacular displays and their promotional videos and their fancy adjectives—"creative," "educational," "interactive"—to reel the buyers in, but down here the sales reps far outnumbered the buyers, and as you passed their displays of gizmos and trinkets, they would glide out of the dark like buskers. "Here, here!" "Take, take!" And into your hand they would thrust a business card and a promotional brochure containing such language poetry as "Our main products include gift, rainbow ring, Tinsel Pom Poms, Eva warhead gun, water bomb balloons, hand-knitted of beads, beauty set etc."

Most of the toys down here were cheap knockoffs of the ones upstairs, and the names of many of the companies sounded like cheap knockoffs too: Baoda Baby Necessities Manufacture Factory, Believe-fly Toys, Combuy Toys. It was hard not to admire the unvarnished directness of that last moniker. This, after all, was the subtextual refrain that could be heard throughout the convention center, no matter how good or bad the salesman's English. "No more PVC," they said. "For smart kids only," they said. "Warm to the touch," they said. But what they really meant was "Come buy! Come buy!"

Although the desperation was louder in Expo Drive Hall, if you eavesdropped upstairs, you could hear it there, too:

"We depend on Christmas, and it was an awful Christmas."

"Before I put more money out, I need more coming in, see what I'm saying?"

"Oil's up, dollar's down. It's a fucking mess."

Depressed by cheap oil, cheap Chinese labor, and the bargaining power of retailers like Toys "R" Us and Wal-Mart, toy prices in the United States had declined by 30 percent since 1996. According to an industry veteran interviewed by Eric Clark, author of *The Real Toy Story:*

Inside the Ruthless Battle for America's Youngest Consumers, the toy business by the end of the twentieth century had become "the game of trying to put the least amount of plastic in a toy so you can keep the price low, of trying to get first in line when Hollywood comes out with a blockbuster, of squeezing the last half cent out of something." Luckily for toymakers, even as margins kept slimming, volume kept ballooning, inflated by the helium of American desire. In 2006, Americans spent $31 billion on toys and video games, almost as much as the rest of the world's countries combined.

Then came 2007, the year of Thomas the Lead Paint Delivery System and the Polly Pocket Magnetic Intestinal Obstruction and the Date Rape Arts-and-Crafts Beads. The year that an American candidate for president, the eventual Democratic nominee, the eventual president of the United States, campaigned on the promise "to stop the import of all toys from China"—80 percent of the toys Americans buy, roughly 60 percent of the toys the Chinese export. The year the stock of the world's largest toy company, Mattel, lost 19 percent of its value in a single month. The helium in the toy balloon seemed to be leaking away, replaced by some less buoyant vapor—the exhaust fumes of malaise, the off-gas of dread.

At an information desk on the mezzanine, a video about product safety looped endlessly, but for whose benefit I couldn't tell. No one else stopped to watch. "Only toys that do not fit inside the gauge are not considered small parts," the British-accented voice-over said, while on the screen a disembodied hand unsuccessfully attempted to insert a plastic strawberry into a steel cylinder. On the first day of the toy fair there had been a special seminar on "Risk Management and Brand Development in International Trade of Toys Industry." Tomorrow there would be yet another seminar, this one titled "Latest Product Safety Directives of Toys Industry & Good Practice in Achieving Safety Standard." New monitoring and safety protocols had already driven production costs up 10 percent, a spokesman for the Hong Kong Toys Council reported. Small margins were growing smaller still.

TOYS TO TRUST/MADE BY HONG KONG, billboards outside the convention center declared—not, the publicists meant to imply, MADE IN CHINA. But their prepositional sleight of hand couldn't change the fact that comparatively little is made in Hong Kong anymore besides money.

Ambiguity is now Hong Kong's major asset; translation, its major industry. Hong Kong translates Chinese labor into Western goods, Asian exports into American imports. It is a semipermeable membrane as well as a semiautonomous region. In 2007, more than sixty thousand factories in the Pearl River Delta belonged to Hong Kong interests. Those factories are the primary source of both the city's prodigious wealth and its equally prodigious smog, a sulfurous whiff of which, up in Expo Hall 7, had penetrated the air-conditioning.

A carpeted room on the convention center's third floor. Gray stackable chairs arrayed into three rows. At the front of the room rise three projection screens and in their midst a lectern, yet to be approached. I have rented a translation device, a black box with headphones, and through this battery-powered medium I will soon receive divinations from an oracle by the name of Richard Wong, founder, CEO, and chief designer of Red Magic Holdings Ltd., who any moment now is to deliver a PowerPoint presentation on the future of toys. The room fills. Ushers fetch more chairs.

I am expecting our speaker to resemble one of the Chinese businessmen seated around me, sporting a necktie and a Bluetooth earpiece and an ID card on a lanyard and an air of fastidious professionalism. Instead, five minutes late, a baby-faced character in ripped jeans and sunglasses swaggers to the front of the room. He has a blazer over his T-shirt, the shaved head of a Tibetan monk, and around his neck a thin gold chain. In Cantonese he asks an assistant to cue his first slide. Behind him, in unison, the three projection screens flash fire-engine red, and above the words "Red Magic" (the "i" dotted with a star) appear a pair of white bunny ears, the right one rakishly lopped as if waving "Hello."

The translation device proves unnecessary. Wong prefers broken English. "Okay, what is kidult?" he asks, slouching over the lectern. "Anyone know?"

"Is it a combination of a kid and an adult?" an American in the second row suggests.

"Right," Wong replies, visibly annoyed. He'd meant the question to be rhetorical. "Okay, probably he is just like me." By his own admission,

Wong has the body of a thirty-five-year-old man and the mind of a five-year-old boy. "I like video games, toys, model, comics book, everything. This is kidult," he says. To "mix the imagination world and the real world—this is kidult." Wong is hungover, he confesses, which explains the sunglasses. To give a PowerPoint presentation at a trade show while wearing sunglasses and recovering from a hangover, this also is kidult. By day, Wong is a CEO, but at night he likes to imagine he's Batman. This is kidult. Growing up in Hong Kong, Wong was forever pining after toys. "For example, when I was ten years old," he says, "I saw a toy. It's a robot, but my mom she never buy it for me. At that moment the toy was 150 Hong Kong dollars. Now it's 5,300, forty times as much. I still buy it. Why is it forty times expensive? Because of the kidult market." A kidult is not to be confused with an *otaku*, a Japanese term Wong recently learned. "*Otaku*, it's like a freak," Wong says. "They imagine they are the character of the comics book all the time. The kidult is different. At least they are working at the daytime."

What Wong's company mostly does by day is design and market collectible figurines—little plastic creatures that look like the mutant love children of Hello Kitty and Pikachu. Flipping through PowerPoint slides, he gives us a quick tour of this menagerie. Red Magic isn't really in the toy business, he explains; it's in the character business, the trend business. "Red Magic is happiness," he says, sounding a bit like Chairman Mao. "Red Magic is style."

Instead of merchandising licensed characters from movies and television shows, as most toymakers do, Wong decided to invent characters of his own. Red Magic's dolls are like merchandise for movies and television shows and comic books that do not exist. Each one has a name and a psychological profile. There is, for instance, Hiro, a thumb-size brown fellow with the big head and stumpy limbs of a fetus, who "pretends to be brave but is really fragile" and "hates losing face." Po, a little plastic teddy bear with nipples, "loves fish biscuits, cooking," and a pink plastic rabbit named Bo. Deri, a thumb-size gray fellow who also resembles a fetus, "savors the destruction process and loves watching others suffer."

Red Magic claims to have sold more than twenty million of these toys in twenty different countries, mostly to boys and men between the ages of fifteen and thirty-five, by deploying a marketing technique that

McDonald's pioneered thirty years ago, when, in an effort to boost declining sales, it introduced the Happy Meal, luring children and therefore their parents to restaurants with the promise of cheap toys. It could be argued that the Happy Meal, and similar gimmicks, went some way toward saving the fast-food business. Red Magic uses the same marketing technique on teenagers and adults. And it works. For a while in Hong Kong, when you bought eight bottles of San Miguel beer, you got one of fifteen different limited-edition Red Magic collectible dolls, but you couldn't choose which one. This "gambling element" kept customers "drinking and drinking and drinking and paying," Wong explained. "When I go to see the customer—they buy and buy and buy, and they can't get the one they want—I feel very happy inside."

For several years, toy-industry insiders have fretted over a trend known by the acronym KGOY, kids getting older younger. At ever earlier ages, market research shows, children are putting away their childish things in favor of adolescent and adult varieties of entertainment—cell phones, movies, social-networking websites. The kidult, Wong believes, could be the industry's deliverance.

JOSHUA THE MOUSE

What is childhood? Ever since I learned I was to become a father, this question has been on my mind. Developmental psychologists like T. Berry Brazelton will tell you that infancy and toddlerhood and childhood and adolescence are neurologically determined states of mind—developmental stages through which all of us progress. Sociologists and historians, meanwhile, tell us that childhood is an idea, distinct from biological immaturity, the meaning of which changes over time. In his seminal 1962 study of the subject, the French historian Philippe Ariès argued that childhood as we know it is a modern invention, largely a by-product of schooling. In the Middle Ages, when almost no one went to school, children were treated as miniature adults. At work and at play, there was little age-based segregation. "Everything was permitted in their presence," according to one of Ariès's sources, even "coarse language, scabrous actions, and situations; they had heard everything and seen everything." Power not age determined whether a person was treated as a child. Until the eighteenth century, the European idea of childhood "was bound up with the idea of dependence: the words 'sons,' 'varlets,' and 'boys,' were also words in the vocabulary of feudal subordination. One could leave childhood only by leaving the state of dependence, or at least the lower degrees of dependence." Our notion of childhood as a sheltered period of innocence begins to emerge with the modern education system, Ariès argues. As the period of economic dependence lengthened among the educated classes, so too did childhood. These days education and the puerility it entails often lasts well into one's twenties, or longer.

Twenty years after Ariès published his book, the media critic Neil Postman announced in *The Disappearance of Childhood* that modern childhood as Ariès described it had gone extinct, killed off by the mass media, which gave all children, educated or otherwise, premature access to the violent, sexually illicit world of adults. Children still existed, of course, but they'd become, in Postman's word, "adultified." I was ten years old when Postman published his book, and in many respects my biography aligns with his unflattering generational portrait. In Postman's opinion the rising divorce rate indicated a "precipitous falling off in the commitment of adults to the nurturing of children." My parents divorced just as the American divorce rate reached its historical peak. After my mother moved out for good, my brother and I came home from school to an empty house where we spent hours watching the sorts of television shows Postman complains about (*Three's Company, The Dukes of Hazzard*). Reading Postman's diagnosis, I begin to wonder if he's right. Maybe my childhood went missing.

But then I think of Joshua the Mouse. One day at the school where I used to teach I stopped to admire a bulletin board decorated with construction paper mice that a class of first graders had made. Above one mouse there appeared the following caption: "My mouse's name is Joshua. He is 20 years old. He is afraid of everything." I love this caption. I love how those first two humdrum sentences do nothing to prepare us for the emotional revelation of the third. And then there's the age: twenty years old. What occult significance could that number possess for Joshua's creator? When you are six, even eight-year-olds look colossal. A twenty-year-old must be unfathomable as a god. Contemplating poor, omniphobic, twenty-year-old Joshua, I was convinced that children might impersonate adults, but they would never become them. I doubt that childhood has ever been the safe, sunlit harbor adults in moments of forgetfulness dream about. I suspect that it will always be a wilderness. When you are a child, almost nothing makes sense—not the expressions on your father's face or the intonations in your mother's voice, not the cruelties and affections of your classmates, certainly not prime-time television or the evening news.

"For as this appalling ocean surrounds the verdant land," Ishmael philosophizes midway through his whale hunt, "so in the soul of man there lies one insular Tahiti, full of peace and joy, but encompassed by all the horrors of the half known life. God keep thee! Push not off from that isle, thou canst never return!" We canst never return, but oh how we try, how we try.

Postman does not only argue that television produced "adultified children"; paradoxically, it also produced "childified adults"—"kidults," Richard Wong would say. As evidence, Postman points to the absence on television of characters who possess an "adult's appetite for serious music" or "book-learning" or "even the faintest signs of a contemplative habit of mind." One wonders what he makes of the popular culture of centuries past—the pornographic peepshow boxes, the slapstick vaudeville acts, the violent and salacious Punch and Judy shows, the bear-baitings and cockfights, the dime novels and penny dreadfuls. The great difference to me seems not one of quality but of quantity: entertainment has become so cheap and ubiquitous that it is inescapable. Even the material world has become a "Sargasso of the imagination." Despite my "book-learning," I still feel bewildered, even in adulthood. The world still makes little sense, no matter how much I study it. Life is still half known. I have brought with me on every leg of my accidental odyssey a portable library. I have read about the science of hydrography and the chemistry of plastics and the history of childhood. I have learned the Beaufort Scale of Winds and the chemical structure of polyethylene and the medieval ages of man. And the more I read, the more lost I feel.

Listening to Richard Wong's prophecies, I find myself thinking that those who complain about the commercialization of childhood have it wrong. The real problem isn't that childhood has been commercialized but that our economy has been infantilized.

UP THE PEARL RIVER

My third morning in Hong Kong, Ron Sidman's old trading partner, Henry Tong, meets me in the lobby of my hotel. Richard Wong and Red Magic may represent the future of the toy business, but Tong and the Wong Hau Plastic Works & Trading Company are far more typical of its recent past. America's toy industry emerged at the end of the Civil War, when factory owners, faced with excess production capacity, started milling merchandise of little inherent worth—tin trinkets stamped from scraps, paper dolls, windup bears. Like their counterparts today on the Chinese mainland, few early American toymakers bothered to innovate, preferring instead to make cheap knockoffs of handcrafted European classics. After World War II, when the marketplace was suddenly awash in surplus plastic resin and molding machines, the toy industry pioneered globalization. It was then, in the 1950s, that Mattel outsourced production of the first Barbie doll to a factory in Japan, and it was then that Henry Tong's grandfather started the family toy business that Tong and his brothers operate today.

A paunchy, bespectacled, double-chinned man in khakis, black sneakers, and a checked Oxford, pen poking from his breast pocket, Tong, thirty-nine, has a courteous, self-conscious manner—avoiding eye contact, punctuating his sentences with a little close-lipped smile, asking me politely whether my hotel has been to my liking. Like most of Hong Kong's business class, he would seem at home in corporate America, whose offices he occasionally visits. He speaks English with an accent but well. At the University of Michigan in Ann Arbor, where he earned a bachelor's degree, he enrolled in the university's ambitious yearlong survey of Western literature and thought. His favorite book he read that year, he tells me as we speed north through the suburban outskirts of Hong Kong in the backseat of a minivan chauffeured by a company driver, was the *Odyssey*. He preferred the *Odyssey* to the *Iliad* because it has a happier ending, and because it has monsters. The only book of Dante's trilogy that he liked was *Inferno*. "Paradise and Purgatory were boring," he said. "But Hell is a fun place." He doesn't read much anymore, he admits. Instead he likes to watch American television shows, especially *Lost*.

Tong's driver drops us off at the train station at Lo Wu, where we are to pass through Chinese customs separately, reuniting on the other side. The Pearl River Delta is so densely developed that some demographers regard it as a single, uninterrupted megalopolis, a Delaware-size economic organism with two nuclei: Hong Kong in the south and, in the north, Guangzhou, the city formerly known to Westerners as Canton. But the political unification remains less complete than the economic one. Unlike residents of China, residents of Hong Kong still enjoy most of the civil liberties guaranteed by British common law (freedom of speech, freedom of religion), and there is a plan to start holding local elections there by 2017. An American needs no visa to visit Hong Kong, as he does to visit China, and the half million or so locals who regularly commute across the Chinese border have to pass through one of six checkpoints, an arrangement reminiscent of that found along the border between Mexico and California.

Here, as there, the border security serves mainly to control the tide of illegal immigrants seeking better wages, though in the Pearl River Delta that tide flows south instead of north. Here, as there, many of those immigrants speak a foreign language—Mandarin or a provincial dialect, not Cantonese—and if they make it across the border, they, too, can expect to be treated as second-class citizens in their new home. There is one striking difference between the two borders, however: to enter China by car you need a special permit, and such permits are hard to come by, even for a successful businessman like Henry Tong. We will cross the border by train, he explains. Another driver awaits us on the other side.

As we join the crowds streaming to and from the quays, Tong gestures toward an old man and woman hurriedly packing manila-wrapped parcels into a little wheeling cart. The cart's wire basket is lined in plaid nylon. "Smugglers," Tong says. After that, I notice them everywhere, those little plaid carts, those manila parcels. I ask Tong why the border guards let them through. He shrugs. Probably there are simply too many of them.

This isn't what I'd expected the Chinese border to be like. I'd expected vigilance. I'd read stories of foreign journalists shadowed by government officials, or deported for traveling on a tourist's visa, as I was. A request for a journalist's visa takes weeks or months to process, and the request itself is an invitation to surveillance, I'd read. Pretend you're a tourist, acquaintances of mine had advised—sound advice for most journalistic visitors to China. My personality, however, presented a problem: Interrogated by authority figures, I am fabricationally challenged. Interrogated by authority figures, confessions come burbling unbidden to my lips.

Hoping to elude suspicion, I'm wearing a sport coat and khakis rather than my usual adventure wear, doing my best to pass as an American businessman on holiday. I've also left my implicating voice recorder back in my hotel. As we approach customs, Tong and I part ways, he heading to the queue for Hong Kong citizens, I into the queue for foreigners. The line shortens. I ready myself to be cross-examined about the nature of my visit. Instead, when my turn comes, the customs officer glances once at my face, stamps my passport with the bored, silent efficiency of a grocery store cashier, and sends me on my way, to meet up with Henry Tong.

It might be different elsewhere in China, but here in the Pearl River Delta the entropy of the marketplace has overwhelmed much of the bureaucratic order, for better and for worse. The crime rates in some of the Delta's boomtowns, among the highest in China, are almost downright American, and so, almost, are the freedoms.[22]

If anything, it's piratical capitalists not bureaucratic Communists that one has to worry most about in the Pearl River Delta.[23] The previous June, when a business correspondent for the *New York Times* paid a surprise visit to the RC2 Industrial Park in Dongguan that had produced those leaded toy trains, a factory boss held him hostage for several hours, refusing to surrender him even after government officials arrived to negotiate his release. And a few months after my trip across the border with Henry Tong, a Guangzhou newspaper called the *Southern Metropolis Daily* would uncover a child-labor ring in the factories of Dongguan, factories not far from the one to which Tong and I, once again in the backseat of a chauffeured minivan, were now bound. Taking custody of more than a hundred children, most of them between the ages of thirteen and fifteen, Chinese police would speculate to the press that there might be hundreds or even thousands more yet to be found. Corrupt employment agencies enticed or abducted the children from impoverished towns in Sichuan Province and auctioned them off to employers who paid them wages well below the already penurious legal minimum. According to the Chinese government, some were forced to work three hundred hours a month.

On the highway to Dongguan, billboards appear, advertising varieties of molding machines and plastic—PP, PET, PVC. It occurs to me that I have never seen plastic of any variety advertised before. The day is unseasonably warm. A gray haze hovers at the horizon, but overhead the sky is clear. I've been trying to learn a little Cantonese and ask Henry Tong to help me with my pronunciation. We start with "thank you," the single most important word for linguistically challenged travelers to pick up. My guidebook spells the word *doh je*, but when I utter these syllables Tong snickers. He says the word, smiling widely on the long *a* of *je*. I do my best to imitate him, smiling widely on the long *a*, but my efforts evidently remain comical.

We pass vacant lots where rag-and-bone pickers are bundling recyclables into bales. Then comes a vast field of idle backhoes, hundreds of them, their jointed shovels pointing upward in yellow, blue, and green triangles, a strangely beautiful forest in the midst of this industrial sprawl. A little farther we pass an outlet selling office supplies. The outlet is several stories tall, festooned with red plastic flags, and every square foot of its facade has been papered in poster-size photographs of brand-name office equipment—Canon printers, Panasonic faxes, Xerox machines. Three decades ago, this freeway didn't exist, and the Pearl River Delta was still mostly farmland. Peasants still cultivate what little undeveloped land remains, growing banana trees among warehouses, cabbages under highway interchanges. We pass silos and smokestacks, soot-streaked factories and soot-streaked housing projects stretching densely in all directions.

The street level scene in Dongguan is livelier. Outside an open garage, a mechanic is repairing upside-down bicycles. Outside an eatery, a woman is stir-frying something in a wok poised over a fire in a rusty barrel. An old man pedals past on a kind of oversize tricycle loaded with empty straw baskets. There is a public park at the center of town, and all the apartment buildings around it look spanking new. Trimmed hedges line the sidewalk. Even the laundry hanging in the windows looks cheerful. Finally, on a side street, we turn into the gated entrance of a three-story complex of buildings encircled by a high wall along the top of which embedded shards of glass sparkle menacingly, and prettily, in the late-morning sun. On the gate, painted in red, a pair of Chinese characters spell the factory's name: Po Sing, or "Treasure Star."

A gatekeeper rolls the gate open, and we enter a courtyard where bicycles crowd beneath a sheet of corrugated tin elevated atop four poles. The factory reminds me of a postwar American elementary school in some district deprived of tax revenues. Through an open ground-floor window, I spy a Ping-Pong table and a television—a rec room for the workers, currently unoccupied. From the main entrance emerges factory boss Tony Chan, grandson of Po Sing's founders. Trained as an engineer at the City University of New York, Chan is a middle-aged, square-jawed man whose sartorial tastes tend, purposefully it seems, toward the managerially forgettable—chinos, button-down, etc. His black hair is so carefully cropped and combed that it looks detachable. He leads us up a dingy flight of stairs to the factory's main office, a tiled room, noticeably lacking in secretaries, furnished with wooden benches, a great big desk, and a display case with sliding glass doors. The contents of the display case appear to have been curated by a toddler. Jumbled about on the glass shelves are plastic flowers, plastic water pistols, plastic knockoffs of the Floatees (the red beaver replaced by a red fish), plastic rings of the sort that babies like to stack onto cones. From the bottom shelf, Chan extracts, proudly, several unopened packages of Floatees, including one identical to those that fell overboard. I ask him if I might be permitted to take photographs. Henry Tong shoots him an anxious, apologetic glance, but with an it's-all-right wave of the hand Chan consents. I snap away and, as if on safari, keep my camera out.

At the pre-Christmas peak, Po Sing employs around a hundred laborers, who, unless they are married, live in an adjoining dormitory. Today, during the post-Christmas lull, only forty or so are at work. I ask Chan where the other workers go during the slow season. Some, he says, have returned home early for Chinese New Year. Others have sought work elsewhere. If he has to lay workers off, Chan says, he gives them one month's pay as severance. One month's pay is eight hundred renminbi, around $115, the legal minimum wage in Dongguan. It's getting harder and harder to find and keep good workers, Chan tells me, now that the demand for labor in the Pearl River Delta has at last begun to exceed the supply. Many workers move from factory to factory until they find the job and boss they like best. As a result, Chan's labor costs have been creeping up, "but it's still cheap compared to the U.S.," he says, as if to reassure me.

Today, most of the remaining workers are upstairs, seated on long benches at tables strewn with yellow and red plastic tubes. They glance up when we enter, their expressions quizzical but emotionless, and then return silently, methodically, to their work. Clearly, they've received visits from strange white guys in sport coats before. With little bladed tools, the workers scrape away excess plastic by hand, one tube at a time. The tubes are components for a construction set, Chan explains. He plucks a yellow one, tugs at each end, and with a little crackle it lengthens, accordion style. He contracts it and expands it a few times for show. Snapped together, the tubes look like red and yellow sausage links. "You can build many different things with these," Chan says, "a home, a car, a spiderweb."

Finishing and packaging take place up here. Molding takes place downstairs, in a long corridor of a room where a couple of dozen machines are evenly spaced, a few feet apart, along either side of a central aisle. The machines look like antiquities, vestiges of a variety of industrialism that predates the space age and the information age. Some are wrought-iron black, others a metallic bronze green. They cast spindly shadows on the concrete floor. Only four are in operation today. At one, workers are "overmolding" handles for Dr. Brown–brand baby-bottle brushes. Overmolding, a technique for combining different colors of plastic into a single part, is a more time-consuming, labor-intensive process than regular molding, but it makes painting unnecessary. Inspecting the white plastic cores of brush handles cooling in a vat of water, I experience a shiver of recognition. "I own one of these brushes!" I tell Chan. "It's suctioned to the counter beside my sink back home!"

"So," he says, happily, eyes twinkling through his glasses, "you are one of my customers." He fishes out one of the little bobbing cylinders with his finger and, shaking the water off, presents it to me as a gift.

I'm not sure what to make of my own excitement. The thing is totally worthless, the cheap component of a cheap product that I already own, and yet for a moment it's as though some breach in my universe has been repaired, as if the arc between two oppositely charged poles has been jumped by an invisible surge. The air stills, the room grows quiet, even tensely ceremonial. Then I notice that the workers I've interrupted are watching me, a stranger mesmerized by a piece of plastic. It's clear that they understand nothing that Chan and I have said, and judging from their expressions, I'd guess that no one has explained to them who I am. Presumably, they mistake me for a customer, a toy importer from abroad. I nod hello. I try to say *doh je*. Either Cantonese is to their ears almost as foreign as it is to mine, or more likely, my pronunciation, despite Henry Tong's tutorials, still sucks: they respond by looking away.

At the fourth machine, an extrusion blow-molder, a worker is turning out yellow plastic ducks, one of which hangs from a wire in a plastic sack above his head. I climb onto a stool and peek into the hopper, an inverted pyramid that funnels yellow nurdles of LDPE (low-density polyethylene) resin down into the heated, spiraling barrel of the machine. From the barrel's other end emerges a yellow sock of goo, known in the trade as a parison. Seated there on a backless stool, the worker reminds me of a farmer milking his cow. Young, in his late teens or early twenties, he is dressed in a yellow T-shirt and shiny blue track pants with Adidas stripes running up the legs. Rhythmically, stoically, as if some strange American and his factory boss weren't ogling him, he milks his machine, yanking a lever that claps the alloy mold onto the parison, toeing a foot pedal that produces a blast of air, then yanking the mold back open. Out the yellow ducks drip, one by one. Every so often the worker pauses, tears off a ribbon of ducks, rips the remnant parison away, and tosses the ducks into the big cardboard box beside his stool. Three hours into the morning shift, the box is almost full. Considering its yellow contents, I find myself wishing that I could equip each toy with some sort of homing device and track their wayward journeys through the global economy, from this poorly lit factory floor to the slippery, hygienic bathtubs of America.

Tony Chan has a surprise for me. From storage, in anticipation of my visit, he's retrieved a zinc-alloy die-cast mold. He places it on the floor, open, so that the two halves mirror each other, one cross-section of a duck floating on its reflection. Chan's late father machine-tooled this particular mold. Machine-tooling molds is the most skilled and expensive part of the production process, Chan tells me. From my shoulder bag I brandish the yellow duck that Curtis Ebbesmeyer loaned me back in Seattle. Then I squat down and plug the thing into one of the zinc-alloy concavities in which, sixteen years before, with a single pneumatic blast, it had undergone a metamorphosis, transforming from yellow polyethylene goo into a hollow icon of childhood; from polymerized chains of carbon molecules into a marketable plaything. For a moment I half-expect some sort of cosmic magic to occur—rays of yellow light to come shooting from the mold, a portal to open in the space-time continuum. Instead, I just stand there muttering, idiotically, "Wow . . . Wow," while behind me Po Sing's perplexed managers look on and beside me an antiquated extrusion blow-molding machine operated by a youthful proletarian drips out new ducks one by one: *psssht, clamp*; *psssht, clamp*; *psssht, clamp*.

A zinc-alloy die-cast mold like this one, Chan explains, can produce 100,000 toys, three or four per minute, before it wears out. Steel molds, more expensive than alloy, can produce 500,000 pieces. Back in the autumn of 1991 when Po Sing received the order for that ill-fated shipment of Floatees, Chan's family made a profit margin of around 10 percent. Now the margin is "much, much less." Although he hasn't made Floatees for several years, he estimates that a set of four plastic animals blow-molded from LDPE would wholesale today for around eighty cents. More than forty cents of that would go toward raw materials. The cost of materials has doubled in the past five years, Chan says, due largely to the rising price of plastic resin, a derivative of oil. The packaging is now, in January 2008, sometimes more expensive to make than the product. A plastic blister package can eat up 20 percent of production costs; a plastic clamshell, 50 percent. Rent in Dongguan is also on the rise. Chan's family moved their plastics factory from Hong Kong to the mainland in the 1980s, but several years ago, when his lease expired, he had to relocate the factory again, to its current location. "First they invite us," he says. "Then they make us move."

Like his counterparts in Japan and Taiwan, Chan's grandfather got into the toy business for economic rather than sentimental reasons. Toys, plastic toys especially, can be produced easily by unskilled workers, and the start-up costs are minimal. But for the same reasons, toy-making and similar kinds of "light industry" will always be vulnerable to competition from less-developed, less-expensive countries, which is why the toy industry has been at the vanguard of globalization. Henry Tong is convinced that he and other Chinese toy exporters are now losing business to factories in Cambodia and Vietnam. To compete, he and his compatriots will need to start making products with greater "added value," products like electronic toys and video games.

Listening to him, I picture blow-molding machines and plastic ducks carried by a wave of outsourcing across the surface of the planet, from Massachusetts, where Ron Sidman's parents started turning out playthings a half century ago, to Japan, Taiwan, Hong Kong, then flooding the banks of the Pearl River Delta, before rippling northward to Shanghai and Beijing and southward toward Cambodia, Indonesia, and Vietnam, leaving behind it both greater prosperity and greater economic disparity, higher crime rates but also, perhaps, if recent developments in China are any indication, stronger labor and environmental laws. In June of 2007, sensing that the proletariat was growing restless, Beijing enacted the most sweeping labor reforms in the history of China's thirty-year experiment with capitalism, expanding legal protections for temporary workers and giving all workers the right to collective bargaining. In response, foreign companies threatened to take their business elsewhere, to more pliable labor markets.

Chan adheres strictly to China's laws, he assures me, as well as to the divergent product-safety codes of America and Europe. But with American importers refusing to pay more for Chinese toys, even as production costs rise, he is not surprised that some of his misguided competitors have given in to the temptation to cheat. "It takes a lot to meet all the safety standards," he says. "People complain about safety, but I don't think we kill anyone"—he hesitates a moment, studying his shoes, measuring the wisdom of his words—"not like the Americans with their bombs."

Given the openness with which he has welcomed me into his factory, I'm inclined to take Chan at his word. He seems like a sympathetic soul, trying to earn an honest living from his family's small business so that he can give his kids an education at least as good as the one he received in New York three decades ago. There are no obvious signs of malfeasance on display in his factory. No child laborers. No suspicious fumes. No lead paint. No paint at all, in fact, since Chan outsources any surface painting he needs done. But the truth is, I really have no way of knowing what goes on in Po Sing when strangers with notebooks aren't poking around.

Even if I could interview Chan's workers in private, I wouldn't know what to believe. In 2006, auditors from Wal-Mart visited a factory in Shenzhen that manufactures Bratz fashion dolls, the wildly popular, multiethnic, big-eyed, swag-loving contenders for Barbie's throne. In anticipation of the inspection, management at the plant gave workers a cheat sheet, later obtained by the National Labor Committee, that listed exemplary answers to fifty-four hypothetical questions of the sort the auditors might pose, questions and answers like

Q: Does management pay attention to problems that are raised?
A: Yes. For example, if it's too hot, the factory provides cold tea for the workers.
Q: Have you received or seen anyone receive unfair treatment? (Like fines, getting yelled at or hit?) How did it happen? Why was it unfair?
A: No.
Q: Is there anything else you would like to say?
A: No.[24]

Such documents alone do not incriminate China's toymakers, the great majority of whom may well be law-abiding businessmen. But they do reveal one of the least appreciated resources that the Pearl River Delta has for the past thirty years offered Western companies looking to outsource production: secrecy—secrecy amplified by oceanic distances, protected by multinational corporations and Chinese factory owners alike, fortified by a language barrier higher than most Westerners can surmount.

The little I know about Po Sing is still far more than I know about the shop to which Tony Chan subcontracts paint jobs, or the plant that mixes his resin dyes. Not even corporate auditors or labor-group investigators or Chinese regulators always succeed in unraveling the Delta's supply chain, a tangle of subcontractual relationships that can vanish into the warrenlike underground economy.

My tourist's visa permitted two border crossings, and two days after visiting Dongguan, I would return to China unchaperoned, taking a high-speed ferry through Hong Kong's seemingly endless port and then on, up the Pearl River.[25] On the outskirts of Guangzhou, accompanied

by a translator, I would pay an impromptu visit to a sofa factory in Long
Jiang with which one of my translator's former clients had done business. Like many so-called factories in China, this one was essentially
preindustrial, devoted to a form of manufacturing that verged on cottage industry. There were no machines to speak of. Workers sat on bare
floors hammering wooden sofa frames together or cutting out uphol-
stery with shears. Slats of wood had been nailed across a broken window. There were cracks in the walls, cracks in the concrete beams
overhead. Residents in a neighboring tenement had thrown trash into
the narrow courtyard that ran behind the shops. From a crack in the facade of one workshop, plants grew. The splintered ruins of a wooden
bench had been abandoned in a corner of the courtyard, and near it,
someone's laundry had been hung out to dry. I saw a mouse or else a
small rat disappear into a drain. The factory's owners, themselves mi-
grants from the north who'd come to the Pearl River Delta in pursuit of
prosperity, did their business in spare, dusty offices, the windows of
which seemed never to have been cleaned. The sunlight coming through
them made ocher squares on the floor. In one office, old, unused cubicles
were lined up against a wall, nested one into another like grocery carts,
their presence suggesting that their former occupants had been swept
away by a wave of administrative layoffs. Amy, my translator, introduced
me to the owner's son, James Liu, a chain-smoking twenty-six-year-old,
who was boyishly excited to meet me. In his office, beside his metal
desk, over which a blue ethernet cable ran from his laptop to an outlet in
the wall, was a brand-new leather sofa that, nice as it was, seemed out of
place. A customer had refused to pay for it, Liu explained. He had never
been outside China, not even to Hong Kong, but he nonetheless liked
America, he said, because American girls are beautiful.

Liu reminded me of something I'd read in Peter Hessler's book.
Those Chinese who had never visited America, Hessler writes, tend to
take one of two extreme views of the place: either America is "evil incar-
nate," or else it is the land of "wealth, opportunity, and freedom." But
there, in the littered courtyard of that sofa factory, trying to peer into a
world that was still difficult to comprehend even after I'd glimpsed it
with my own eyes, it occurred to me that American ideas about China
have of late grown similarly extreme, similarly confused.

Back in the eighties, the People's Republic played a supporting role in our popular culture. Soviet Russia, Nazi Germany, Vietnam, Tatooine—those were the foreign places my childish mind traveled to most often in movies and dreams. But then, sometime after Tiananmen Square and before 9/11, perhaps around 1998, when our growing trade deficit surpassed $50 billion, something changed. China began to cast a shadow and a spell over the American imagination. It opened its door, and our curiosity rushed in. Images of the new China proliferated in the media: futuristic skyscrapers and sporting arenas rose up among the sweatshops; SUVs and luxury sedans nosed like steel sharks through schools of bicycling proles whose clothes had turned suddenly colorful.

We no longer know quite what to make of China: Is it our ally or our enemy? Our rival or our doppelgänger? A repressive Communist oligarchy or the new century's new land of wealth and opportunity? In a way, thanks to China, we no longer quite know what to make of ourselves. Finances alone can't explain America's current crisis of confidence. Our anxieties about China are also to blame: China has tested our cherished belief that democracy and the free market are mutually necessary, symbiotic; the one is impossible without the other. Its rise has given other developing nations an alternative. It shines like a red lantern unto the world, a boomtown on a hill. With mounting curiosity, as well as anxiety, we gaze west across the Pacific, searching the horizon for signs of our influence and forebodings of our decline. Economically, the European Union may yet pose a greater threat to our supremacy than China, and OPEC a greater threat to our sovereignty, but in the markets of the imagination the renminbi is now mightier than the euro, and Thomas the Tank Engine more menacing to our children than a tanker of Arabian crude.

During the 2007 toy scare, when the Pearl River Delta suddenly stopped keeping its secrets, a disenchantment occurred. We were reminded, to our dismay, that there really is nothing brand-new under the sun. It wasn't merely the lead in the paint that scared us but the magnitude of our ignorance. In our children's vulnerability, we glimpsed an exaggerated version of our own. Banished from the garden of solicitous and magically affordable delights, we began to shop suspiciously, xenophobically, checking the labels. Even familiar American brands, those old friends, seemed to have turned against us. There were enemies in the pantry, sleeper cells in the toy box. The Consumer Products Safety Commission was the new Office of Homeland Security.

It didn't matter that American design flaws were to blame for most of the toy recalls, or that American toy companies, not Chinese factory owners, were the ones who'd profited most from China's regulatory neglect, or that we consumers were its beneficiaries, or that the Chinese workers spraying the lead paint were the ones exposed to the greatest risk, or that recalled products represented a minuscule percentage of the flood of Chinese goods. Nor did it matter that the only life claimed by the tainted toys, so far as we know, was that of Zhang Shuhong, the director of a factory implicated in one of Mattel's recalls, who one day in August, at the height of the scare, after instructing his managers to sell off all the machinery, ascended to his factory's third floor, locked himself in a workshop, and hanged himself. No, what mattered was that the material world was no longer under our control. It disturbed us how much we'd come to depend on the industry of shadowy strangers.

But during the time I spent in China, I couldn't shake the notion that I had traveled not into the future, which supposedly now belongs to China, but into some *Twilight Zone* version of America's economic past. Much as the Erie Canal once connected the Port of New York to the mill towns of the interior, so the Pearl River now connects the Port of Hong Kong to the factories of Shenzhen and Dongguan. Even if the scale of China's coastward migration is unprecedented, the phenomenon is not; similar economic forces sent immigrants and former slaves to America's Rust Belt. Much as competition from the Port of Newark eventually idled Manhattan's docks, so newer, cheaper ports up the Pearl River Delta have begun to draw business away from Hong Kong. It seemed to me more likely that a citizen of Hong Kong could catch a sneak preview of his future by visiting Manhattan, or a factory worker from Dongguan by visiting Akron (former capital of the rubber toy industry), than the other way around.

On the drive back to Hong Kong, Henry Tong tells me why he thinks his own economic future remains bright despite America's declining consumer confidence: the confidence of China's consumers is on the rise, and so is their disposable income. (Hong Kong's toy sales in China increased 369 percent between 2005 and 2007.) To capture a share of this emerging market, Tong and his brothers recently started a retail chain on the mainland called Baby Creations. Not long ago, Chinese parents spent as little as possible on their kids, but those who've gained access to China's burgeoning middle class "want to show off," Tong says, and so they've started outfitting their children with the same sorts of accessories American parents buy—new clothes, designer strollers, fancy baby bottles that will no doubt require special baby-bottle brushes. Unfortunately for Tong, there is one American buying habit most Chinese consumers have yet to learn. Worried about deteriorating plastic, American parents will replace a baby bottle's nipple every few months as instructed, Tong says. "But in China they'll buy one nipple and use it for a year." If only he can persuade them to throw away and replace, throw away and replace, the way Americans do. For now, more concerned about appearances than safety, they can be enticed by the fantasies that the bawds of consumerism peddle, but they remain indifferent to the fears. Soon, perhaps.

PLASTIC MAN

Back in the 1970s, when I was a child, rubber ducks were wilder than they are now. There was nothing iconic or nostalgic about them. The rubber ducks of the Nixon era were frequently white, some were calico. Some had swanlike necks and rosy circles on their cheeks. Some came with rococo feathers molded into their wings and tails. No one used them to sell baby clothes or soap. Normal adults did not give them to one another, or decorate their desks with them. So far as I can remember, no one I knew even owned a rubber duck. I did own one, however, on account of the pet name my mother had given me. Back then, before her breakdown and subsequent vanishing act, she liked to call my brother Benjamin Bunny. I, inevitably, was Donovan Duck.[26]

My own rubber duck was a somewhat hideous specimen, with white plumage, a green topcoat, a hydrocephalic head, and the gentlemanly posture of a penguin. It resembled a Hummel figurine that had sprouted a beak. In the bathtub, when squeezed, instead of squeaking, it shot water from the soles of its feet, a vaguely scatological feature that greatly amused my brother and me, though our preferred bath toys were Hot Wheels, which we'd drive in laps around the tub's rim and launch Evel Knievel–style off our wet knees. Most exotic varieties of rubber duck have since gone extinct—they are the dodos and carrier pigeons of the nursery—and what new ones have evolved share a single, yellow ancestor whose pop-cultural apotheosis was by my toddlerhood already under way.

It had begun in 1970, when an orange puppet named Ernie appeared on PBS and said, "Here I am in my tubby again. And my tubby's all filled with water and nice, fluffy suds. And I've got my soap and washcloth to wash myself. And I've got my nifty scrub brush to help me scrub my back. And I've got a big fluffy towel to dry myself when I'm done. But there's one other thing that makes tubby time the very best time of the whole day. And do you know what that is? It's a very special friend of mine. My very favorite little pal"—at which point Ernie reaches into the suds and, brandishing his yellow duck, bursts into song.

You can watch a video clip of the number online. A pink towel hangs from a wooden post at the left edge of the frame. The post looks like something out of an old western. There is no other scenery to speak of. Behind the bathtub—which is huge, presumably claw-footed, and decorated with three pink daisies—hangs a sky-blue backdrop. Bubbles of the sort you blow with a wand come floating up from the bottom of the screen, and the gurgle of water accompanies the music. Although I watched my share of *Sesame Street* as a child, I far preferred Super Grover's mock-heroic pratfalls to Ernie's snickering bonhomie, and I have no memory of the rubber duckie number. My wife, on the other hand, still knows the duckie song by heart.

Replaying the clip, I can't help imagining myself as a toddler sprawled on the shag rug, glassy-eyed and solitary before our old black-and-white Zenith, watching this solitary, football-headed, cross-eyed puppet—at once the child's alter ego and his imaginary friend, half preschooler, half possibly gay bachelor—as he serenades a squeak toy that is his own fabulous alter ego and imaginary friend. "Rubber Duckie, joy of joys," Ernie sings. "When I squeeze you, you make noise, / Rubber Duckie, you're my very best friend, it's true." It's all so synthetic, so lonely, so imaginary, so clean. And apparently children loved it. In the 1969 pilot episode of *Sesame Street*, in which a version of the rubber duckie song appeared, children in the test audience responded so enthusiastically to Ernie and Bert and so tepidly to segments featuring the live actors that the show's creators redesigned it, giving the puppets a starring role.

However novel the medium, however inventive Jim Henson's puppetry, Ernie's bathtub serenade draws on a history of representation that can be traced back to the eighteenth century, when British portraitists stopped painting children as diminutive adults and turned them into puppy-eyed personifications of Innocence. In the Romantic era, no longer was innocence merely the antithesis of guilt and childhood the antithesis of adulthood; innocent children were the antithesis of modernity, little noble savages. Childhood became a place as well as an age—a lost, imaginary, pastoral realm.[27]

It is striking how much the modern history of childhood resembles that of animals. "In the first stages of industrialism," John Berger writes, "animals were used as machines. As also were children." In the latter stages of industrialism, poor children who escaped the factory often took to the street, where they formed what social historians call "child societies," gangs of urchins who—like feral cats—invented a social order all their own. Partly in fear of child societies, middle-class parents of the Gilded Age began treating their children increasingly like pets. Nurseries and playrooms became more common, and toy chests began to overflow.

Then, in the early 1900s, came the crib, which unlike the cradle isolated infants in rooms of their own. According to Gary Cross, author of *Kids' Stuff: Toys and the Changing World of American Childhood*, in the padded cells of the modern nursery, "playthings served as 'antidotes for loneliness, . . . substitutes for the free play of the 'child society.'" Lonely little middle-class girls were encouraged to keep the company of dolls, and for lonely little middle-class boys (think of Christopher Robin), there were stuffed animals, a turn-of-the-century invention. By 1906 zoos were selling them as souvenirs, and later that same year, thanks in large part to conservationist in chief Teddy Roosevelt, the teddy bear fad hit.

Roosevelt doesn't deserve all the credit. Teddy bears and other stuffed animals, Peter Stearns explains in *Anxious Parents: A History of Modern Childrearing in America*, "were widely appealing at a time when parents were trying to facilitate new sleeping arrangements for babies and also to guard against unduly fervent emotional attachments to mothers. The decline in paid help for young children also opened the door to the use of toys as surrogate entertainment."

One commentator at the time speculated that toy animals helped children overcome primal fears of scary predators, turning lions and tigers and bears into snuggly sidekicks. But perhaps the most astute psychological profile of the child consumer appeared in *Playthings*, the brand-new trade journal of the American toy industry, which by 1912 was predicting that "the nervous temperament of the average American child and the rapidity with which it tires of things" would guarantee a never-ending bull market in toys.

Three other seemingly unrelated events coincided with the commercialization of childhood and the infantilization of animals: In 1871, a printer from Albany, New York, named John Wesley Hyatt added nitric acid to pulped cotton, thereby inventing celluloid. In 1873, the first Pekin ducks were imported to the United States from China. And in the 1880s, bathtubs began appearing in middle-class homes along with indoor plumbing. Celluloid eventually evolved into the plastics industry. The Pekin duck eventually became the preferred species of American duck breeders, making yellow ducklings a familiar symbol of birth and spring—familiar and far less alien than the Chinese themselves. And the average American bathroom, which had once consisted of a washtub and an outhouse, was consecrated as a temple of cleanliness.

Much as the modern nursery sheltered children from the social contamination of the street, so the modern bathroom protected their naked, slippery bodies from germs. In the first decades of the twentieth century, public health campaigns and soap advertisements—usually illustrated with pudgy little Pre-Raphaelite tots—exhorted parents to bathe their children often. Little boys, the thinking went, were naturally indisposed to bathing. Bath toys not only made hygiene boyishly fun; they helped overcome the naughty urges that bathing tended to arouse: "The baby will not spend much time handling his genitals if he has other interesting things to do," one government-issue child-care manual advised in 1942. "See to it that he has a toy to play with and he will not need to use his body as a plaything." Enter the rubber duck.

Ducklings are the aquatic equivalent of kittens and bunnies. In fact, it's hard to think of a smaller, cuddlier animal that can swim.

Most of the frogs and turtles of children's literature are middle-aged men, whereas even in nature ducklings are model offspring: obedient, dependent, vulnerable to predation, clumsy, soft, a little dumb. Just think of them waddling in a train behind a mother duck, a familiar image memorialized by Robert McCloskey's bestselling children's book *Make Way for Ducklings*. McCloskey's baby mallards, penciled in black and white, look like real baby mallards—a little stylized, but real. Like the ducks depicted in other venerable children's books, they bear little resemblance to Ernie's Day-Glo squeak toy. Beatrix Potter's Jemima is a white Pekin duck in a bonnet and shawl. Donald Duck, the most famous Anseriformes at midcentury, was also a white Pekin, and the most common toy duck was still the ancient bird-on-a-leash, a wooden pull toy with wheels instead of feet. Before the rubber duck could eclipse it, plastic had to replace wood as the preferred material for toys, which, following the technical innovations spurred by World War II, it did.

McCloskey published his book in 1941. That same year, at the beginning of the war, two British chemists, V. E. Yarsley and E. G. Couzens, prophesied with surprising accuracy and quaintly utopian innocence what middle-class childhood in the 1970s would be like. "Let us try to imagine a dweller in the 'Plastic Age,'" they wrote in the British magazine *Science Digest*.

> This creature of our imagination, this "Plastic Man," will come into a world of colour and bright shining surfaces, where childish hands find nothing to break, no sharp edges or corners to cut or graze, no crevices to harbour dirt or germs, because, being a child, his parents will see to it that he is surrounded on every side by this tough, safe, clean material which human thought has created. The walls of his nursery, all the articles of his bath and certain other necessities of his small life, all his toys, his cot, the moulded perambulator in which he takes the air, the teething ring he bites, the unbreakable bottle he feeds from . . . all will be plastic, brightly self-coloured and patterned with every design likely to please his childish mind.

Here, then, is one of the meanings of the duck. It represents this vision of childhood—the hygienic childhood, the safe childhood, the brightly colored childhood in which everything, even bathtub articles, has been designed to please the childish mind, much as the golden fruit in that most famous origin myth of paradise "was pleasant to the eyes" of childish Eve.

Yarsley and Couzens go on to imagine the rest of Plastic Man's life, and it is remarkable how little his adulthood differs from his childhood. When he grows up, Plastic Man will live in a house furnished with "beautiful, transparent, glass-like forms," he will play with plastic toys (tennis rackets and fishing tackle), he will "like a magician" be able to "make what he wants." And yet there is one imperfection, one run in this nylon dream. Plastic might make the pleasures of childhood last forever, but it could not make Plastic Man immortal. When he dies, he will sink "into his grave hygienically enclosed in a plastic coffin." The image must have been unsettling, even in 1941, that hygienically enclosed death too reminiscent of the hygienically enclosed life that preceded it. To banish the image of that plastic coffin from their readers' thoughts, the utopian chemists inject a little more Technicolor resin into their closing sentences. When "the dust and smoke" of war had cleared, plastic would deliver us "from moth and rust" into a world "full of colour . . . a new, brighter, cleaner, more beautiful world."

This chemical fantasy must have appealed to Londoners enduring the Blitz, and since plastic accommodates our dreams more readily than other substances, the fantasy in most respects came true. In *American Plastic: A Cultural History*, Jeffrey L. Meikle tells the story well. Just days before Japan surrendered, at a gathering of marketing executives, J. W. McCoy, a vice president at DuPont, delivered a sermon on mass consumption. The sacrifices consumers had made during the war economy had created "a great backlog of unfilled wants," McCoy said, and although this demand would send the peacetime economy into "an upward spiral of productivity," eventually, those wants would be satisfied, and "a satisfied people is a stagnant people."

To maintain growth, marketers would have to engineer perpetual dissatisfaction. Around the same time, a less enthusiastic prophet, the British design critic John Gloag, predicted that, after the asceticism of modernism and the austerities of war, consumers would indulge in "an orgy of ornament." Sure enough, by 1948 the *Architectural Review* was reporting that, as both Gloag and McCoy predicted, Americans had gone "on a baroque bender," gorging themselves on products that came "in a lurid rainbow of colors and a steadily changing array of styles." In the 1950s, when the patent on polyethylene expired, plastics became cheaper and thus more abundant than ever—so cheap that the only way polyethylene molders could turn a profit was by convincing consumers to start throwing plastics away.

"The future of plastics is in the garbage can," *Modern Plastics*, the trade journal of the industry, declared. The era of disposability was born. Polyethylene began filling grocery store shelves, and because children's desires are even more fleeting than those of adults, it also began filling the toy chests of the postwar baby boom. As Jeffrey Meikle observes, if the burgeoning market for cheap toys had not absorbed it, the excess supply of plastic resin "would have served witness to the folly of overproduction."

For new parents who had themselves grown up during the Depression and the war, the fantasy of childhood as consumer paradise exerted a powerful appeal. Browsing through issues of *Parents'* magazine from 1950, I came upon an ad campaign for Heinz baby food ("Scientific Cooking Gives Finer Flavor, Color and Texture to Heinz Strained Carrots"). In one Heinz ad targeted at new mothers, cartoon butterflies, fairies, and dolls encircle the photograph of a baby girl. "Wee elfin creatures go riding on butterfly wings," the copy reads, "dolls speak in a language all their own and something altogether new and wonderful happens everywhere a baby looks . . . your child lives in a magic world where everything's enchanted." Another ad featuring a baby boy replaces the talking dolls with teddy bears that "come mystically to life."

Then came television, enchantment in a box. Annual toy sales in America shot from $84 million in 1940 to $1.25 billion in 1960. Peg-and-socket pop beads sold to girls as costume jewelry consumed forty thousand pounds of polyethylene resin per month in 1956. In 1958, hula hoops and Frisbees consumed fifteen million pounds of the stuff. Polystyrene replaced balsa wood as the most popular material for model cars and planes. Plasticized polyvinyl chloride, the material from which the brand-new Barbie doll was made, provided a cheaper, more durable alternative to latex rubber, rendering traditional molded rubber animals and dolls obsolete except in name.

Now take another peek into Ernie's claw-footed tub. Here are Millais's wondrous bubbles, and here is the child accompanied by his small, cuddly pet that speaks a language all its own. Here is the antiquated stage set tinged with nostalgia (the wooden post, the claw-footed tub). Ernie is not dressed up in a costume (he's naked here), but in a way his orange, fuzzy pelt plays the same role as the Blue Boy's satin suit and the bubble boy's velvet green one; all three figures wear the colors of make-believe. Only now, in a kind of reversal of Pinocchio's metamorphosis, and a kind of culmination, the dream child has become a puppet, a toy brought "mystically to life," who himself plays puppeteer to his squeaky duck, the inanimate fakeness of which makes the puppet seem more real.

Meanwhile, the real flesh-and-blood child, the postmodern child, the Plastic Child has disappeared from view, banished to the other side of the glass screen. There, sprawled on the shag rug, his chin propped on his hands, or slumped into a beanbag chair, he reminds me of John Berger's zoo animals. "The space which they inhabit is artificial," Berger writes. "Hence their tendency to bundle towards the edge of it. (Beyond its edges there may be real space.) In some cages the light is equally artificial. In all cases the environment is illusory. Nothing surrounds them except their own lethargy or hyperactivity."

Not long after PBS first broadcast it, Ernie's rubber duckie song went to number sixteen on the Billboard charts. Radio stations were playing it, adults were buying it. And unlike the other *Sesame Street* characters, Ernie's rubber duck was not trademarked. Producers had picked up the prop at a local dime store, which meant that even as it became a recurring character and a pop music phenomenon, it remained in the public domain, free for the taking, no licensing fees required. In a way, its ascendancy confirms Daniel Boorstin's observation in *The Image: A Guide to Pseudo-Events in America* that in America we have arranged the world so that we do not have to experience it, replacing heroes with celebrities, actions with images—and, one might add, animals with television characters.

But does that mean that if Ernie had gone bathing with a white duck or a green one, our iconic ducks would be white or green? I'm not sure. The threads of chance and meaning are hard to disentangle. On the album cover of the LP single of the song, Ernie, for some reason, is holding a different duck, a calico duck, white with burnt-orange spots. Perhaps there is more to the message in this particular bottle than the medium. Perhaps celebrity alone does not explain the yellowness of the duck.

"Ideals of innocent beauty and the adorable have changed little in a hundred years or more," the historian Gary Cross writes. "Many today share with the Victorian middle class an attraction to the blond, blue-eyed, clear-skinned, and well-fed child and are appalled by, uninterested in, and even hostile to the dark, dirty, and emaciated child. Even when humanitarian groups try to shame us into giving money to support poor peoples far away, they usually show us an image of a smiling olive-skinned (not black) girl, a close copy of our ideal of innocence." So maybe it's just as Curtis Ebbesmeyer suspected. Maybe there is bigotry at play. Is it too much of a stretch to see in the yellowness of the rubber duck a visual reminder of that well-fed, blue-eyed, clear-skinned, yellow-haired Victorian ideal? After all, real ducklings have black, beady eyes, not blue ones like the ducks in Eric Carle's book.

Tuesdays during the seventh month of her pregnancy, my wife and I attended a prepared-childbirth class on the maternity ward of our hospital: linoleum, vinyl chairs, fluorescent lights, a tinted plate-glass window with a view of buildings, strangers, many with impressively inflated abdomens, some also with swollen breasts and feet. On one wall of the classroom hung a poster of an egg, mid-hatch. Contemplating it during the long, tedious hours of instruction, I began to wonder why this particular poster had been hung before our eyes. Was it meant to comfort us? Did we prefer the clean, white orb of an egg to the bloody mammalian mess of one body gushing forth from the wounded nethers of another? On the opposite wall of the classroom hung an enlarged sepia photograph of naked, racially diverse babies; aligned firing-squad-style along a fence over which they appeared to be attempting an escape, they displayed their wrinkly bums for our delight. Children are fundamentally the same, such images suggest, indistinguishable as ducklings despite the color of their skin. They inhabit a world before sex, before race, before history, before self, before humanity. Children, then, are beasts of burden, too—ducklings and bunnies of burden—asked to carry the needful daydreams of adults. The apotheosis of the rubber duck wouldn't be truly complete until the children who had watched that 1970 episode of *Sesame Street* grew old enough to look back forgetfully with longing and loss.

THE FIFTH CHASE

There are profitable ships and unprofitable ships. The profitable ship will carry a large load through all the hazards of the weather.

—Joseph Conrad, *The Mirror of the Sea*

[H]owever man may brag of his science and skill . . . for ever and for ever, to the crack of doom, the sea will insult and murder him, and pulverize the stateliest, stiffest frigate he can make.

—Herman Melville, *Moby-Dick*

AN UNPROFITABLE SHIP

In the fall of 1998, Paul Frankel, a furniture importer in Englewood, New Jersey, was waiting for his ship to come in, and on November 2 he began to wonder if it would. Really, it wasn't the ship Frankel was waiting for but one of the containers it was carrying, a jumbo-size forty-footer full of tables that his company, Collezione Europa, had purchased from a factory in the Philippines. A freight forwarder—the cargo industry's equivalent of a travel agent—had arranged to have Frankel's 237 tables shipped from Manila to Seattle aboard the APL *China*. In the fall of 1998, the APL *China* was one of the largest container ships plying the seas. It was also, so far as maritime engineers were concerned, one of the safest.

Like most successful importers in the age of containerization, Paul Frankel was not given to worrying whether his cargo would reach his warehouse intact, for a simple reason: it almost always did. Sure, every now and then a piece of furniture got scuffed or dinged in transit and had to be repaired by a furniture repairman (a furniture repairman very much like Charlie Moore, in his previous life), but it arrived nonetheless. What made Frankel start wondering now was a fax he received from the headquarters of APL, American President Lines, which, despite what its name suggests, belongs to a shipping company in Singapore. "Please be advised that the APL *China* v. 030 has been delayed due to severe weather encountered enroute to Seattle," the fax began. "The ship has suffered some weather damage, but we do not yet know the full extent."

To appreciate the full extent of the damage the APL *China* had suffered, one must first appreciate the full extent of the APL *China*. The *China* was a C-11-class post-Panamax ship, meaning that—at 906 feet long and 131 feet wide—it was too big for the locks of the Panama Canal. Standing on a dock beside it, you would have felt as though you were standing at the foot of an unnaturally smooth cliff, a palisade of steel. The carrying capacity of a container ship is measured in TEUs, or twenty-foot-equivalent units, because a standard shipping container is twenty feet long. One twenty-footer equals one TEU, a forty-footer, two. The *China* had a carrying capacity of 4,832 TEUs. (That of the Evergreen *Ever Laurel*, by contrast, was a mere 1,180 TEUs.)

Imagine a train pulling 4,832 boxcars: it would stretch for nineteen miles, from the southern tip of Manhattan up into Westchester. Or imagine this: both the Main Concourse of Grand Central Station and the Chartres Cathedral would have fit inside the *China*'s hull. Or imagine this: you could have fit Noah's ark inside its hull and had room for another ark of equal size. And the dimensions of the ark set down in Genesis were meant to suggest a vessel of fabulous grandeur, a vessel bigger than any ship built before the flood or since, a vessel—in short—of biblical proportions.

Now imagine if Noah had gone to sea in the APL *China* instead of his ark: A single twenty-foot-container could accommodate hundreds of thousands of insects, tens of thousands of frogs—plague in a box. A twenty-footer could comfortably house six horses, or, uncomfortably, a single elephant. The flood would have posed no danger to whales, but if Noah had wanted to, and if he'd had the skills and materials to carpenter a really big fish tank, he could have squeezed a gray whale into a jumbo-size forty-foot container like the one Paul Frankel was waiting for. If Noah had sailed in the *China* instead of his ark, he might not have been able to save every species on earth, but he could have come close. On the other hand, if Noah's luck were as bad as that of Parvez Guard, captain of the *China*, thousands of the animals he was trying to save would have been lost at sea. (The whale, at least, might have survived.)

Paul Frankel wasn't alone in his wondering. All across North America that autumn, people were waiting for the *China* to come in, and all across North America people began to wonder if it would. Not long after APL's fax arrived at the offices of Collezione Europa, an even more worrisome message reached the offices of Consolidated Transportation Services Inc., a freight forwarder in Carson, California. The message concerned two containers of clothing that the company had arranged to be shipped to New Jersey from Palau, an island five hundred miles east of the Philippines. "Both containers have gone overboard," the message reported. A few weeks later, another freight forwarder, H. W. Robinson & Co., based near JFK Airport in New York, also received a worrisome fax. "As you may know, the APL *China* encountered heavy weather in the Pacific Ocean on or around October 26, on its voyage from the Far East," this fax began. "The information we have received thus far indicates the storm was unusually severe with wave heights exceeding fifty feet."

People at the headquarters of Eddie Bauer began wondering what had happened to the 1,583 cartons of cotton dresses they'd ordered from Sri Lanka. People at Schwinn Cycling & Fitness wondered about the bicycles they'd ordered from Taiwan. People at Pier 1 Imports were wondering about their 236 cartons of acrylic bottle stoppers. And at Azuma Foods, people wondered what fate had befallen 940 cartons of seafood that had been packed onto the *China* inside special, refrigerated containers known as reefers. Most worried of all, perhaps, were executives at Toshiba, who were waiting for nine containers of cordless phones, $4.2 million worth. Paul Frankel's tables, by comparison, had cost him a mere $7,000.

The *China* had staggered into port in Seattle on November 1. It was a bluebird day, the sun bright and sparkly on the water of Puget Sound. A container ship crossing a harbor on such a day, passing among fishing boats and yachts with the imperturbable majesty of a zeppelin among gulls, is a beautiful sight. The sight of such a ship on such a day suggests that human ingenuity has succeeded in taming the sea, turning Melville's "watery wilderness" into so many watery highways, along which freighters travel as routinely as eighteen-wheelers travel the roads.

A Seattle longshoreman named Rich Austin was dispatched to the *China* salvage operation that day. As he drove into port, the sight wasn't picturesque at all; it was, he remembers, "ominous." From bow to stern, stacks of containers had toppled like dominoes, some to starboard, some to port. Some containers were crumpled up like wads of aluminum foil, others were pancaked flat. One of the *China's* sixty-two cargo bays gaped like a missing tooth; an entire row of containers, stacked six high and sixteen across, had been swept away.

Later that afternoon, hoisted up in a man basket by a crane, Rich

Austin got a better view of the devastation. "It looked almost like a landfill in some areas," he remembers. Containers had split like dropped melons, spewing cargo: remote-control boats, golf clubs, frozen lobster tails, thousands of plastic air fresheners. Many photographs of the ravaged *China* were taken that week. Photos of bay 1 show boxes stuffed with clothing labels yet to be stitched on. With a magnifying glass, you can read what the labels say: DANNY & NICOLE. Scrolled bolts of cloth protrude from the ruins of a container in bay 15. Striped, child-size shirts of the sort favored by *Sesame Street*'s Ernie drape the scaffolding in bay 17. In photos of bay 36, you can see packages of frozen shrimp; in those of bay 58, Kenwood bookshelf stereo systems. Perhaps most impressive of all are the photos of bay 59, which show a smorgasbord of consumer goods—stainless steel pots, bouquets of plastic flowers, white sneakers, gray trousers from the Gap, all intermingled and strewn about. One pair of white sneakers hangs from a rail by its tied laces, like shoes thrown over a telephone wire. "Whatever Americans were consuming at that time, there it was," another longshoreman, Dan McKisson, told me. "There was Christmas, laying on the deck."

Salvaging this wreckage was slow, dangerous work, like playing pick-up sticks with thirty-ton sticks. While a pair of cranes lifted stevedores up in man baskets, another crane swung four steel hooks out to them. The stevedores would catch the hooks, latch them one by one into a container's corner castings, the man basket would swing clear, and up and away the snared container would go.

At least that's how it was supposed to work. Sometimes to reach the corner castings, a stevedore would have to climb out of the man basket onto a tipped container. Sometimes when a crane lifted one container, those leaning against it would topple. Sometimes when the hooks rose, the container would sunder, spilling cargo onto the decks and docks. There was melting seafood everywhere, and after a couple of days the smell was so bad that Austin started swiping spilled air fresheners and rubbing the fragrance onto his mustache. "Other guys put earplugs in their nostrils," he recalls. Dan McKisson remembers that as one container rose into the air, its contents suddenly shifted. There was a crash. Then the steel wall of the container gave way at one end, opening like a hatch. Down came "a rain shower" of cardboard boxes, some of which tore open when they hit the decks. Inside? Bicycles. The people at Schwinn could stop wondering.

Austin and McKisson had seen cargo losses before. Almost every winter at least one container ship turns up in Seattle with containers damaged or missing. But they'd never seen devastation like this. Neither had their foreman, Don Minnekan, and Minnekan was nearing retirement. Neither had any other longshoreman, ever. What the longshoremen bore witness to that November morning was, in monetary terms, the worst shipping disaster in maritime history. Of the 1,300 containers the *China* had been carrying when it departed Taiwan for Seattle eleven days before, 407 had been lost at sea. Of those remaining onboard, another few hundred had been damaged or destroyed.

"I've been out at sea on tugboats and fishing boats. When it's snotty out, it's no fun," Austin told me. "But I can't imagine what it would be like to be in something that would do that much damage to a ship like that."

By the time I made my pilgrimage to the toy factory in Dongguan, I'd gone to sea several times—on the *Malaspina*, the *Opus*, the *Alguita*. And yet I could no more imagine what it would be like to ride a container ship through snotty weather than Rich Austin could. Nor could I imagine what it would take to make containers tumble overboard. The accident that had set the toys adrift remained to me mysterious. By process of elimination, after contacting the Port of Tacoma and consulting old shipping schedules in the *Journal of Commerce*, I managed to identify the ship—the Evergreen *Ever Laurel*—from which the toys I was chasing had fallen. But there was no mention of the accident anywhere in the public record, not in *Lloyd's List*, not in the *Journal of Commerce*, not in the Port of Tacoma's archives. The Coast Guard's records did contain an inspection report, which I acquired under the Freedom of Information Act, but all it said was that the *Ever Laurel*, on January 16, 1992, while in Tacoma, had been subjected to a Coast Guard inspection, which it had passed.

Except under extraordinary circumstances—if a law has been broken in American waters, for instance, or a hazardous substance has spilled— the U.S. Coast Guard does not investigate shipping accidents. Most of the time, it's left to lawyers and insurance adjusters to reconstruct the sequence of events and assign blame. If small sums of money are involved, the ship's owners and underwriters will dispense with the investigation, accept liability, and settle the claims. The case of the *China* was different. The damages were too costly for APL to absorb; 361 claimants represented by twenty-five lawyers would eventually file for damages exceeding $100 million, more than the ship itself was worth, and because APL decided to contest their claims, the details of the *China* disaster are part of the public record. If you visit the federal courthouse in the Southern District of Manhattan and give the file clerk the correct docket number, she will emerge from the archives wheeling a cartload of legal files in which you will find bills of lading, invoices, faxes, memos, claims for damages, motions and counter-motions, thousands of documents—the sort of archival mother lode for which, in the case of the *Ever Laurel*, I'd searched in vain.

I'd heard about the *China* from a number of sources. A maritime lawyer named Geoff Gill had told me about it. Charlie Moore had told me about it. Mariners I'd met had told me about it. The *China* disaster had become as legendary as the rubber ducks lost at sea. Perversely perhaps, listening to the legend, I wanted more than ever, if only briefly, to join the merchant marine, but joining the merchant marine turns out to be far harder than it used to be. You can't just show up at a port and offer your services as a cabin boy—at least not as a cabin boy on a container ship; on a tramp steamer, maybe. To join the crew of a container ship you have to go to school and secure a license, and even then, in America, whose merchant fleet has dwindled, atrophied by globalization, licensed mariners have trouble finding work. I asked a few different shipping lines—including Evergreen—if they'd let me hitch a ride. Those that returned my calls said no. I offered to pay. Still, the answer was no. Taking on passengers is more trouble than it's worth, I was informed, especially with all the security restrictions put in place after 9/11. Then someone told me about a German shipping line called Reederei NSB, NSB for short, one of the last shipping lines that still does a side business in tourism. So long as I was willing to attain a doctor's note attesting to my good health, sign a legal waiver six pages long, and fork over a couple grand, I could learn "first hand what it means to 'sail the seven seas' aboard an ocean-going giant," the NSB website promised. I would return from my voyage "able to tell no end of sailor's yarns."

That sounded pretty good to me. I've always been a sucker for sailor's yarns. And ever since I was a kid growing up in San Francisco, I've wondered what it would be like to ship out on one of those oceangoing giants, which, on clear days, I could see from my childhood bedroom, out on San Francisco Bay, going to and from the docks in Alameda, transacting their mysterious business with the faraway. In mid-January, when I'd be returning from China, there were no NSB ships departing from Hong Kong for Seattle, but there was one ship—the Hanjin *Ottawa*— departing from Pusan, South Korea, for Seattle, following roughly the same route the *Ever Laurel* had taken sixteen years before, and roughly the same route the *China* had taken in 1998. Along this route, the toys had broken free, changing from containerized cargo into legendary characters. Along this route, some oceanic force had beaten a post-Panamax ship to ribbons. Now, from Pusan, I'd travel this route.

I also had other, vaguer, more philosophical reasons for shipping out, reasons that the actuarial phrase "act of God" helps explain. I didn't expect an ocean crossing to restore my faith in God, exactly—at least not in a biblical God; I lost that irretrievably long ago. But I did hope that it might refresh my capacity for awe. Rich Austin thought the sight of the devastated *China* was ominous. I agreed. The toy spill was ominous, too. What these omens seemed to portend was this: that the high seas may yet be the wildest wilderness in the world after all; that despite the cargo industry's best, most technologically advanced efforts to turn shipping lanes into watery highways, there remain on this diminishing, warming, terraqueous globe of ours tropics of mystery, seasons of astonishment, zones of the sublime, which not even vast expenditures of capital and ingenuity may ever fully tame. After exploring the Pearl River Delta, I set out to test this augury of mine, this wish.

PORT OF CALL. PUSAN. 35°04'N, 129°06'E.

It seems that my ship, the Hanjin *Ottawa*—a 5,618-TEU post-Panamax box boat built right here, in the South Korean shipyards—has been delayed by dirty weather in Shanghai. I've called Mr. Shin every day for the past three days, and every day he tells me the same thing: call back tomorrow. Mr. Shin is a freight forwarder, whose name and number my travel agent, a woman with a Slavic accent and a mailing address in up-state New York, sent me along with my ticket and my six-page legal waiver. I was to contact Mr. Shin upon my arrival in Pusan.

Flying from Hong Kong to Pusan was like flying from summer into winter. While waiting, I've been living at the Hotel Phoenix, down in the city's Cinema District, where pedestrians in down coats, breath steaming, gather around pushcarts to eat fried pancakes full of bean paste from little folded circles of paper. My first day here, I bought one for five hundred won, a pittance, and watched the woman turn it three times in the hot oil with her tongs, before dropping it into its little paper pocket. The first bite burned my mouth, but I didn't mind because I was cold and short on cash and it was so sweet and greasy and cheap that when I finished it, I bought another. That's all I ate my first day in South Korea, waiting for my ship to come in: bean paste pancakes.

In the top floor of the building across the street from the Hotel Phoenix is a grooming parlor for poodles. From my room, you can catch glimpses of the poodles through the parlor's windows, poodles almost epileptic with fear, claws skittering on the stainless steel tables, while disembodied hands tease their forelocks into poofs. Beyond the grooming parlor is the Jagalchi fish market, the largest fish market I have ever seen. Yesterday, I walked through it, biding time.

The muddy, bloody streets were lined with fishmongers, all women—the wives and daughters of fishermen, I presumed. Their heads covered in neat little kerchiefs, they crouched beside basins and plastic laundry baskets in which much slithering and crawling and dying was taking place. Other women in kerchiefs pushed wooden carts heaped high with assorted vegetables and crustaceans, shouting like carnival barkers as they went, and down along the water, inside what I at first mistook for a convention center or an opera house—a three-story edifice gleaming with architectural ambition, its silhouette suggestive of billowing spin-nakers and silvery waves—were still more fishmongers, selling octo-puses puddled atop beds of ice and flayed fish drying on racks like zombie laundry.

This seemed to me an auspicious beginning to my transpacific voy-age. In Manhattan, the once vibrant downtown waterfront has been

turned into Stamford, Connecticut. They've upholstered one pier in Astroturf, built a playground on another, while of others all that remains
is a grid of moldering piles, past which joggers on the meticulously landscaped promenade go bouncing along, listening to their iPods and enjoying the view of condominiums in Hoboken.

The opening chapter of *Moby-Dick* describes what New York's waterfront used to be like. "There now is your insular city of the Manhattoes, belted round by wharves as Indian isles by coral reefs—commerce
surrounds it with her surf," Ishmael writes. Promenaders on the Battery
could peep "over the bulwarks of ships from China." Now if an insular
Manhatto wants a good look at a ship from China, he has to take the
A train to Queens. Gaze out to the horizon from Rockaway Beach,
where on summer days the urban masses lounge and frolic among the
orange umbrellas of the New York City Parks Department and pigeons
fraternize with the gulls, and you will see them there, the great box
boats, converging out of the haze on the Port of Newark, their decks
loaded with the provender of the world. They process so silently and
slowly, they can almost seem stationary, like oil rigs viewed from the
Gulf Coast.

It is at first pleasurable to find oneself marooned by happenstance in
a foreign city one knows absolutely nothing about. All around you people are going about their daily lives, rushing to catch commuter trains
that will spirit them away to neighborhoods you cannot imagine, while
you bob among them like flotsam in a current, aimless, borne along. But
after three days the novelty begins to wear off and loneliness sets in.
Amazement at the varieties of human strangeness gives way to sentimental thoughts of home.

BAD WEATHER. NIGHT. 35°04'N, 129°06'E.

The sun has sunk behind the mountains and on the blackening foothills
the million lit windows look like stars. The Hanjin *Ottawa* finally arrived this morning at dawn, and an hour or so later Mr. Shin woke me
with a phone call. I was to bring myself and my luggage to the customs
house by 10 A.M. so that he could usher me through immigration. A day
one of his ships comes in is for Mr. Shin a busy one. He had more im-

portant things to worry about, more valuable things, than some American idiot willing to throw away good money to cross the North Pacific aboard a container ship at the height of the winter storm season. But my mysterious travel agent had enlisted Mr. Shin's services, and he'd accepted her payment, and now he was stuck with me.

BEAUTIFUL WONDER KOREA / IMMIGRATION SMART SERVICE, a sign outside the immigration office read. Inside, two uniformed bureaucrats sat behind a long desk under a wall clock. While Mr. Shin glanced frequently at the clock with an air of frantic impatience, the bureaucrat inspecting my passport glanced frequently at me with an air of bemused curiosity. He and Mr. Shin conversed in Korean, and at one point, something Mr. Shin said made them both laugh. Judging from the looks they gave me, I was the joke. At last, with a flurry of rubber stamping, the bureaucrat returned my documents to Mr. Shin, who spun on his heel and out the door to his paneled van, tugging on his leather driving gloves and beckoning me to follow.

Speeding along the docks, through a maze of interchanges and feeder roads, past acres of containers, above which poked the red booms of gantry cranes, we spoke only once, when out of nowhere, Mr. Shin said, "May I ask a question? Why you take the ship? Long time! Ten days!"

It was a good question, to which I had a number of complicated answers, but even if I could have made myself understood, there wasn't enough time to explain to Mr. Shin about the legend of the rubber ducks lost at sea, or about turning a map into a world, or about my philosophical interest in the wilderness of water, so instead I told him that I was a writer, as if that explained anything. He shrugged and shook his head, as he did again upon delivering me to the *Ottawa*'s gangway, where a Filipino oiler named Marco Aaron descended the long, clattering flight of metal stairs to relieve me of my suitcase. "Safe voyage!" Mr. Shin exclaimed, still shaking his head. Then he sprang into his van and sped away.

Aaron handed me and my suitcase off to Joe the Messman, a cherubically youthful Filipino steward dressed in white, who led me down a narrow linoleum corridor. Entering a ship is always disorienting, I've found. All the corridors look the same, all the doors look the same. Even after a few days aboard, you still find yourself forgetting which way is froward, which way aft. Rising eight stories from the deck to the bridge, the house of the *Ottawa*—the habitable part—is a bit like a floating business-class hotel, a business-class hotel with a space-age control tower on top (the bridge), a factory (the engine room) in the basement, and lots of big red axes bracketed to the walls. Like a business-class hotel, the *Ottawa* has a gym, an elevator, even a swimming pool— empty now because it has to be pumped full of seawater, and the *Ottawa* has been traveling through frigid latitudes.

There are two lounges, one for the officers, one for the crew. Though similarly appointed—with wet bars, home theater systems, couches— the officers' lounge is more ornately decorated. It has a painting of a mountain, a long box of surprisingly lifelike plastic plants, and a big glass case displaying a Japanese doll, a white-faced geisha in a kimono. In the crew's lounge, by contrast, the only decoration is a poster-size pinup calendar, compliments of United Shipchandlers Ltd., on which a golden-haired bathing beauty can be seen lifting up her halter-top, of- fering the sex-starved Filipino oilers and deckhands a charitable peek at her shiny, suntanned, and preternaturally orbicular hooters.

In light of Mr. Shin's bafflement, I was surprised to learn—from Joe the Messman—that I wasn't the *Ottawa*'s only passenger. A retired cou- ple named Bob and Claire are taking the round-trip from Seattle to East Asia and back. As soon as Joe had delivered me via elevator to my cabin—I have my own desk, television, DVD player, bathroom, and enough glassware to host a cocktail party—Bob and Claire turned up to welcome me aboard. They'd decided to take this voyage on a whim, Claire explained. Before marrying Bob, a divorcé with grown children, Claire had spent a few years living on a boat docked near Portland—not Portland, Maine, mind you, Portland, Oregon, which is where she and Bob live, though not in Portland, exactly, in a small town on its out- skirts, a really lovely town, which Claire likes quite a lot. Anyway, as she was saying, she lived on this boat, this antique boat, that she'd spent all her free time restoring, a beautiful wooden boat, but it wasn't really sea- worthy, and she wasn't really any kind of sailor, though living on that boat made her want to take an ocean voyage someday. As for Bob, well, until he retired two years ago, Bob was an engineer, and feats of engi- neering like container ships always interested him, and furthermore his grandfather had been a captain in the merchant marine during the last

MOBY-DUCK

years of the age of sail. Claire had heard about this travel agency that arranged voyages on container ships, and so, well, next thing you knew, they were embarking from Seattle aboard the *Ottawa*, and now, two weeks later, after stopping in Yokohama and Shanghai, here they were in Pusan. So far they hadn't been disappointed. The eastbound trip had been just lovely, a little rocky at first, but mostly lovely, Claire said, and the cook, who was getting off here in Pusan, sadly—sadly for us, not him, of course; he'd been at sea seven months, pour soul—the cook was absolutely wonderful, just wonderful, it was like eating in a fancy restaurant. She favors the first person plural ("We had to wait at anchor off Shanghai for three days because of a storm; it was just lovely"), whereas he prefers the all-knowing third person ("The storm had already passed, but they were backed up in the port").

There's nothing to dispel one's fantasies of adventure like a retired couple on holiday. It's as though, having attained base camp on Everest, one were to find a pair of senior citizens lounging in lawn chairs beneath the awning of a Winnebago.

One of my cabin's two portholes looks out at the end of a maroon container, the other at the end of a white one. They are both the uppermost containers in their respective stacks, and they are close enough to my cabin that I could poke them with a broom handle if I had both the desire and a broom handle. Between the containers and the house is a sort of crevasse several feet wide and fifty feet deep. I undogged one of the portholes and, peeping down into the crevasse, felt a bit like a prisoner in a medieval dungeon.

Unlike my cabin, the officers' lounge affords an excellent view of the forecastle deck, where two gantry cranes were at work, loading and unloading containers; a third was at work astern. I spent the afternoon watching them. Most of the time you see a gantry crane, it's at rest, its boom pointing toward the sky, so that it resembles a steel brontosaurus. After a container ship has finished docking and the longshoremen have rolled the cranes into position on their enormous tires, the booms come groaning down.

At the shoreward end of the boom, where eighteen-wheelers had queued up, the operator sent down a kind of rectangular hoist called a "spreader." It grasped a container by the corners, plucked it daintily from the trailer bed of an eighteen-wheeler, hoisted it 160 feet into the air, and slung it out to the ship. A skilled crane operator can move forty containers an hour. It is surprisingly graceful, this logistical trapeze, this ballet for heavy machinery, one crane operator zipping out as another zips back toward shore, one container dropping on steel cables like a spider on its thread, as another comes rising up out of the hold's shadowy depths. A single container can weigh as much as forty tons, and whenever a crane lowers one into place, there is a great metallic *kaboom* that makes the bulkheads shudder.

The cranes do not merely unload the containers bound for Pusan and load the new ones bound for North America. They also rearrange containers according to a computer-generated stowage plan that takes into account a host of variables—the shape of the ship's hull, the total weight of the cargo, the size and location of the ballast tanks. The calculations that the stowage software performs are complex, but their purpose is not. Their purpose, like that of everything in the shipping industry, is to maximize efficiency and minimize risk.[28]

The most important variable in the stowage software's calculations is something naval architects call metacentric height, which is determined by the ship's buoyancy and center of gravity. Think of an old-fashioned, windup metronome, an upside-down pendulum with a weight that slides up and down the stem. Slide the weight to the bottom of the pendulum and the pendulum will rock back and forth in short, quick, allegro arcs: *tictoctictoctictoc*. Slide the weight up to the pendulum's tip and it will sweep from side to side in wide, slow andante arcs: *tic . . . toc . . . tic . . . toc*. A ship rolls the way an upside-down pendulum swings, with the keel at the base of the pendulum and the crow's nest at its tip.

A vessel's center of gravity is analogous to the metronome's sliding weight. You can adjust it by adding or discharging cargo or ballast water. Make the center of gravity too low and the ship will be "stiff"; a stiff ship is stable, but like a metronome set to allegro, it will jerk violently back and forth in short, quick rolls. Make the center of gravity too high and the ship will be "tender," rolling steeply, righting itself slowly. With every roll the crow's nest will swing way out over the water to starboard, then way out over the water to port, describing long, gut-wrenching arcs.

Once the crane operators finish their work, the lashing crew will have to finish theirs, and then the lashings and hatch covers will have to be inspected, and all this must be done according to strict stowage guidelines issued by the classification society, Germanischer Lloyd, that has certified the *Ottawa* as seaworthy. Classification societies, of which there are only several in the world, exist to minimize risks, but also to minimize liability. Underwriters will not insure an unclassified ship. If the officers of the *Ottawa* do not adhere strictly to Germanischer Lloyd's stowage guidelines and something goes wrong at sea, the *Ottawa*'s owners rather than its underwriters could be held liable. Preventing liability is the second most important part of the captain's job, after ensuring the safety of his crew. This night will be for the *Ottawa*'s captain a long one. It could be dawn before we get under way, he said.

Busy with preparations for our departure, he came to dinner late, just as the other officers were leaving, and, alone at his table, a linen napkin tucked into his collar, ate his pork stew in preoccupied silence, until Claire decided to bother him. His name is Uwe Jakubowski. A German of Russian descent, he is a big-shouldered, soft-voiced, white-haired man with a gap between his front teeth and a beaky wedge of a nose. His cheeks are ruddy with burst capillaries. Donning wire-rim glasses to study a logbook or an instrument, he resembles a professor reviewing his lecture notes.

Like all the officers, and most of the crew, Captain Jakubowski speaks English with an accent, but speaks it well. A modern container ship is a polyglot place. With the exception of the ship's German mechanic, Klaus Scharmach, the *Ottawa* has an entirely Filipino crew. With the exception of the Filipino third mate, Ricardo Salva, and the Finnish second mate, Fredrik Nystrom, the officers are all German. You'll hear Filipino deckhands speaking Tagalog or officers speaking German, but the lingua franca is English.

Claire asked the captain when he thought we'd reach Seattle, now that our departure from Pusan had been delayed. "They have us coming in on the 26th," he said, untucking his napkin and pushing back his chair, "but I don't think we'll make it, because of the weather. It's winter after all. Who knows? We may get lucky. But it looks like we will have bad weather." Then, with a courteous, distracted gap-toothed smile, he excused himself to catch a few hours' sleep.

This afternoon, up on the bridge, I checked the latest forecast, a litany of warnings. There are storms and gales expected in the western Aleutians, the waters east of Hokkaido, and the Sea of Okhotsk. Although it forebodes sleepless nights for the captain, this is good news for me. So long as I survive to tell the tale, I want to see just what the North Pacific can dish out. Odds are I will survive it, all too easily, but the odds are also good that we'll see a little action, a little *Sturm* and possibly some *Drang*, before we make Seattle.

"For what is the array of the strongest ropes, the tallest spars, and the stoutest canvas against the mighty breath of the infinite," Joseph Conrad asks in *The Mirror of the Sea*, "but thistle stalks, cobwebs, and gossamer?" Of course, a modern freighter like the *Ottawa*, or the *China*, or the *Ever Laurel* resembles a sailboat about as much as a 747 does a hang glider. Spars and canvas have given way to seventy-thousand-horsepower engines that burn two hundred metric tons of fuel per day. Wooden hulls have given way to steel, the astrolabe and sextant to gyrocompasses and satellites. And yet, today's cargo vessels also take riskier routes.

When the trucking magnate Malcolm McLean perfected the humble intermodal shipping container in the early fifties, he revolutionized the stolid shipping industry. Containerization introduced efficiencies and economies of scale that made shipping fees plummet. The only way to make more money was to increase volume by making bigger vessels deliver more cargo faster. Hulls had to be enlarged—by 2006 they would exceed 1,300 feet in length, 340 feet longer than the *QE2*. Port times and transit times had to be shortened. Now, instead of keeping well away from winter storms, freighters travel between them, like cyclists riding the drafts of semis. And storms, like sleepy truckers, can sometimes take sudden, unforeseen turns.

This is one reason merchant seafaring is still, by some accounts, the world's second most dangerous occupation, after commercial fishing. According to Imperial College London, two hundred supertankers and container ships have sunk in the past two decades because of weather. Wolfgang Rosenthal, a scientist at the European Space Agency, which studies sea conditions via satellite, estimates that two "large ships" sink every week on average. Most of these, he says, "simply get put down to 'bad weather.'"

"The shipping industry is decades behind the airline industry" in its management of risk, says Geoffrey Gill, the maritime attorney I'd spoken with. Why?

"Because there are no passengers, and because most merchant mariners these days are Filipino. A lot of people don't seem to care if twenty-five Filipino sailors drown."

And drown they do. How many, exactly? Nobody knows for sure, but the number of accidental seafaring fatalities appears to exceed one thousand lives per year, and the number-one cause of death is believed to be drowning. Maritime losses—of cargo, vessels, digits, limbs, life—are enough to fill a few pages of the *Lloyd's List* weekly Casualty Report. There are accounts of collisions, of fires, of piracy. Most of the casualties can be attributed to mechanical failures, human nefariousness, or human error, but around 10 percent of shipping casualties are indeed ascribed to bad weather and left otherwise unexplained. And dozens of the most catastrophic weather-related mishaps to have befallen container ships—such as the 2003 Maersk *Carolina* disaster, the 2006 *P&O Nedlloyd Genoa* disaster, the 2006 *CMA CGM Otello* disaster—bear a mysterious resemblance to the *China* disaster.

There's no doubt that the *China* had encountered severe weather, as APL claimed, but severe weather at sea is routine during winter, when winds between Taiwan and Seattle often attain hurricane force, and waves routinely exceed thirty feet or, less routinely, forty to fifty feet. Sometimes, under the worst conditions, waves as high as seventy, eighty, ninety, even a hundred feet can loom up out of nowhere, spelling catastrophe. These great and sudden waves have seized the popular imagination, sinking a cruiser called the *Poseidon* in not one but two cheesy films. Their names make them sound as fabulous as the kraken—freak waves, rogue waves, extreme waves. My personal favorite is the German term *Monsterwellen*.

Was it a *Monsterwelle* that had ravaged the Evergreen *Ever Laurel*? Probably not. According to Curtis Ebbesmeyer, the ship had encountered hurricane-force winds and forty-foot waves—rough conditions, to be sure, but altogether typical of the North Pacific in winter, and twelve containers overboard isn't, comparatively speaking, all that much; not enough to upset the underwriters. What about the *China*? To people like the longshoreman Rich Austin, it sure as hell looked like something monstrous had happened out there in the Graveyard of the Pacific. Still, even when a captain, with help from weather-routing services that recommend course changes via fax, can't avoid severe weather, the ship is supposed to survive it. If the cargo has been properly lashed and the hatch covers tightly battened, if the engines have not failed and the helmsman has time to take evasive action, not even eighty-foot waves are supposed to send stacks of containers tumbling over the rails—certainly not 407 of them in a single night.

Even before the *China* entered Puget Sound, the speculating and finger-pointing had begun. To limit its liability, APL had to prove that the accident had been a so-called act of God. They appeared to have a strong case. Weather reports pointed to a prime superhuman suspect: Super Typhoon Babs. In late October 1998, when the *China* was at port in Taiwan, Babs was laying waste to villages in the Philippines. Early news reports assumed that Babs had laid waste to the *China* too. But as weather records reveal, the *China* and Babs had not crossed paths. Reports to the contrary were mistaken. The ship had departed Taiwan on October 21, six days before the remnants of Babs hit.

When lawyers questioned the officers in Seattle, what they heard strained credulity. The scuttlebutt at the longshoremen's union hall was that the ship had lost power and gotten caught in the trough between waves. But the officers claimed that the engines had failed after they watched the container stacks fall. Before it ever lost power, the ship had rolled wildly, inexplicably, they said, despite attempts to take evasive action, and at the worst of it, they'd seen, at bridge level, "green water"—nautical-speak for a wave tall enough to wash over the main deck. In calm seas, the main deck would be about four stories above waterline, and the bridge eight stories above that. Most of the lawyers listening to this tale took it to be a tall one, literally. After all, sailors always exaggerate, especially when trying to exonerate themselves, and such giant waves have long been the stuff of sailors' lore.

Wrote Melville, "In maritime life, far more than in that of terra firma, wild rumors abound, wherever there is any adequate reality for them to cling to."

NAUFRAGIA. 38°14'N, 134°41'E.

Steaming east through the snowy Sea of Japan, birthplace of storms. Last night sometime after four, trying to remain awake for our departure, I nodded off with my shoes on reading *Typhoon*, Conrad's novella about the ill-fated steamship *Nan-Shan*, commanded by the unlucky yet miraculously competent Captain MacWhirr, of whom, when the barometer suddenly plummets, Conrad writes, "Omens were as nothing to him, and he was unable to discover the message of a prophecy till the fulfillment had brought it home to his very door."

When I woke up this morning, I sensed from the motion of the ship and the throb of the engine that we were already under way. I rushed outside and up three flights of stairs to the bridge deck, where I was hit by a cold blast of headwind. It wasn't all that windy; mostly it was the ship's speed—twenty-four knots—that knocked me back. Ahead, the Pacific was lost in fog. Behind us, Pusan was vanishing into it. I could hardly feel the roll of the ship, which didn't seem to be steaming through the waves so much as steamrolling over them.

On the bridge, sleep deprived but nattily attired in a navy sweater with black-and-gold epaulets, the captain was doing his paperwork. Jakubowski has been a merchant mariner for forty-four years, since the age of eighteen, when, against his parents' wishes, he shipped out on a Baltic Sea break-bulk freighter. I asked whether he'd ever lost containers overboard. "Never," he said, but he'd come close. Once, near the Azores, the ship he was commanding was struck by a wave sixty-six feet tall; and on this very trip, westbound from Seattle, the *Ottawa* had rolled 20 degrees to starboard, 26 to port. "Here, you can still see." He pointed above the helm, to the clinometer, which had yet to be reset.

Like Europe-bound flights that arc north over Greenland, a container ship from Asia usually describes an arc—the Great Circle route— toward the Aleutians, passing through the Graveyard of the Pacific. Because of the stormy forecast, the weather-routing service recommends we take a northerly detour, straying from the Great Circle route and into the Bering Sea, seeking shelter in the lee of the Aleutians. "Will we see any ice?" I asked, hopefully.

"We want to get close to it," the captain said, "but not that close."

Mariners of the age of sail avoided the Great Circle route in winter. It's easy to understand why: Back then the only way to forecast the weather was to watch the barometer, study the portents, and heed the wisdom of the almanacs. Every year sailing ships nevertheless sank by the dozens, in bad years by the hundreds, which is why both underwriters and classification societies had to be invented. British shipowners of the age of sail lost 10 to 20 percent of their revenues to shipwrecks, and accounts written by the survivors were so numerous and so popular they constituted a veritable literary genre—naufragia, it was called, after the French, *naufrage*, for shipwreck. The authors of these accounts, perhaps appealing to the sensibilities of their landlubber readers, tended to interpret their harrowing experiences as religious allegories, and their publishers tended to favor subtitles the length of paragraphs, as in, *God's Wonders in the Great Deep, or A narrative of the shipwreck of the briguntine* Alida and Catharine, *Joseph Bailey, master, on the 27th of December 1749. Bound from New-York for Antigua. Wherein the wonderful mercy of the divine Providence is displayed in the preservation of the said master, with all his men, from the time of the said vessel's over-setting to the time of their being taken up by a vessel bound from Boston to Surranam, on the 3d of January following; all which time, being seven nights, they were in the most imminent danger & distress.*

CABIN FEVER. 43°74'N, 149°16'E.

It's snowing. At sea you can see a snowstorm from a long ways off, a white disturbance in the air. Now it's swirling all around us, accumulating on the bridge wings, dusting the corrugated tops of the stacked containers with white triangles from which you can tell the direction of the wind. Way out beyond the stacked containers, the water looks like a tarnished mirror, shiny in some patches, leaden in others. Up on the bridge, there are little icicles on the windshield wipers.

Two days out and I'm already experiencing symptoms of cabin fever.

I take half-mile laps around the cargo decks, read Conrad in bed, dine with Bob and Claire at the passengers' table, visit the bridge, chat with the officer on watch, admire the 360-degree view. So far there has been little to see, other than changes of weather and light—few other vessels, no wildlife besides gulls. There is almost never more than one officer on the bridge, and most of the officers do not greet my questions as amiably as the captain.

This morning, second mate Fredrik Nystrom was at the helm, paging through the Nautical Institute's newsletter, reading the accident reports sent in by seafarers. I asked Nystrom if he had any idea what was inside all those colorful boxes out there. "We have a manifest somewhere," he said, "but in the end it's not that interesting."

Then he resumed his reading until, a few moments later, the VHF crackled and over the airwaves came a man's voice singing—in a language neither Nystrom nor I recognized, Japanese, presumably, or Korean—a joyous, possibly intoxicated chantey. "Not quite what you're supposed to do," Nystrom said, smirking. Then we stopped talking and listened and gazed out into the swirling snow.

Tonight when I entered the bridge, Chief Officer Hermann Josef Bollig was standing watch, as he always does between 8 P.M. and midnight. A bearded German giant, Bollig scares me a little. He seemed to scowl when he saw who it was. At night, to maximize visibility, the bridge is kept dark. As the ship goes autopiloting along, Bollig sits there, his face lit by the lunar glow of computer screens, surveying the seas for fishing boats. If a small one should stray into our path, the *Ottawa* would plow right over it, leaving behind little but a trail of flotsam and sticks. Fully loaded, the *Ottawa* weighs more than 140 million pounds. At twenty-four knots, the forward momentum of that much weight through water is almost planetary, and difficult to stop, even with the engine in reverse.

If I've read the charts right, in three days, near the international date line, we'll come close to the site of the *China* disaster and even closer to those coordinates—44.7°N, 178.1°E—that three years ago, late one night while my pregnant wife slept in the other room, I marked in my *Atlas of the World*.

———

Six years after the *China* dropped its cargo into the sea, Space Agency spent three weeks studying winter waves Waves of more than one hundred feet are altogether real, confirmed. There's evidence, in fact, that as the planet warms bers may be rising, and ESA scientists believe that their data may help solve unsolved maritime mysteries.

One of the most famous such mysteries is the sinking of the MS *München*, in 1978. The *München*, at the time a year-old state-of-the-art barge carrier, vanished while crossing the North Atlantic. Its disappearance has never been explained, but many nautical detectives suspect that it and its crew of twenty-seven fell victim to a *Monsterwelle*. A true *Monsterwelle* is defined not by size but by its mathematical improbability, an example of chaos theory in action. The seafarer's handbook *Heavy Weather Sailing* notes that "whilst one wave in twenty-three is over twice the height of the average wave, and one in 1,175 is over three times the average height, only one in over 300,000 exceeds four times the average height." If the average waves in a particular sea are only three feet tall, and a twelve-foot breaker were to suddenly rear up in their midst, it would be a little monster. What really makes the fearsome *Monsterwellen* fearsome isn't size; it's the element of surprise. They can come out of nowhere, even in calm seas, overwhelming a ship before the helmsman has time to escape it.

Scientists are still trying to explain monster waves. One leading theory is that they arise when a strong current—like the Gulf Stream, or the Kuroshio, or the Algu-has, down at the Horn of Africa—collides with a countervailing storm swell, or when a deep-ocean swell collides with a shallow shelf. Another theory, called "constructive interference" or "random superposition," holds that in chaotic seas two wave trains with identical wave periods and crest heights can ever so rarely combine into a kind of super train, producing monsters. Which might explain why such waves often seem to come in sets of three, a phenomenon known to sailors as "the three sisters." Three sisters forty or fifty feet tall can be more dangerous than a single one-hundred-foot giant. The first monstrous sister rocks the boat mightily before plunging it into a trough so deep, one eyewitness account compares it to a "hole in the ocean." Then the second or third wave sinks it.

Nonetheless, sailors are often mistaken when they identify mon-

..ously large waves as freaks or rogues. The tsunamis that struck land all around the Indian Ocean in 2004 were well over a hundred feet, but they were the result of an enormous undersea earthquake, not chaotic hydrodynamics. Typhoons and hurricanes regularly whip up waves over seventy feet high, but evidence suggests that you're more likely to encounter a *Monsterwelle* in *lower* sea states than in tall ones.

The peril of the sea that modern merchant mariners are most likely to face, navigational wisdom holds, isn't a *Monsterwelle* but something known as "synchronous rolling," so called because the natural roll period of the ship falls in sync with that of the waves. I first learned about synchronous rolling from Beau Gouig, an oiler I met on a research vessel. Gouig explained it to me this way: "Basically what happens is when the wave crests get far enough apart, the ship starts rolling, and with each roll it takes longer to recover, so you're going farther to starboard, and then farther to port."

Small container spills happen all the time, Gouig told me. "Nobody really cares—except the environmentalists," he said. The usual cause, according to him: synchronous rolling, a phenomenon he'd experienced firsthand. "I've been through two typhoons," he said. "The last one, we almost lost the ship." Gouig and I were in the research vessel's mess, at a Formica table bolted to the floor beneath a porthole, in swiveling plastic chairs, also bolted to the floor. I fetched a refill of coffee and took out my notebook.

"It was in January," Gouig began. "January and February are the bad months for typhoons. We were sailing from Japan to L.A. The last port we visited was Yokohama. We were watching the storm on the weather maps, plotting two charts, one for the storm and one for the ship. A typhoon can go in two directions. Either it goes up toward China"—he drew invisible maps on the table with his fingers and pantomimed the storm, his hand cutting a line past the west coast of a coffee mug up into the China Sea—"or it brushes past Japan and spins out across the Pacific." His fist swept toward the table's edge.

"All the predictions were saying this typhoon was headed this way, to China. But our first mate knew something was wrong. At Yokohama they stuck us at the dock and made us wait three days. This mate, he said, 'All hands on deck! Double-lash all the containers, tighten every twistlock, batten down all the hatches.' We had all the ABs out there"— all the able-bodied seamen.

"We left at around four in the morning. By eight or nine o'clock we were getting hammered. After ten hours we tried to turn back. Eventually you don't care where it was you were supposed to be going. You're just trying to get out of the way. We kept sailing south. A little farther and we would have made it to Hawaii. This was an eight-hundred-foot container ship, and that storm blew us four days off course. We keep making the ships bigger, but we're fooling ourselves if we think they're safe from the sea. Big ships still sink." He took a sip of coffee, then continued.

"I was up on the bridge. We were in autopilot, and we were getting hammered. We did three or four 50-degree rolls, buried the bow three times. Waves swept all the lifeboat ladders off deck. I said to the captain, 'These wave crests are getting far apart.' That's basically why you're up there, to watch for synchronous rolling. Suddenly everything on the bridge just goes *voom*."

The ship's captain did exactly what all captains are trained to do: he hove to, immediately turning into the oncoming waves and slowing down, letting the ship ride up and over. "All the books say that you've got only two minutes to break that cycle. You've got to make a hard turn and get the bow up on a wave or down into a trough. Another thirty seconds and I think we would have rolled right over."

The crew were all amazed, Gouig said, that no containers had been lost. "Spills happen all the time," he said, again. Then he told me the story of what he believed to be the most famous case of synchronous rolling, the case of the APL *China*.

GRAVEYARD OF THE PACIFIC. WEST OF THE INTERNATIONAL DATE LINE.

Two days ago, in the Tsugaru Strait, the snow let up, and through the fog could be glimpsed the mountains of Hokkaido to the north, the mountains of mainland Japan to the south, their black ridges striped with snow. Claire and Bob came bustling into the bridge to peep at them through binoculars, she exclaiming breathlessly, "Oh, aren't they just breathtaking!" The next land we'd see would be American.

The weather service recommended the Great Circle route after all. We're now way out in the Graveyard of the Pacific, and you can tell, you can feel the giant swell moving under the hull. Not sure exactly how much we're rolling, but it feels like a lot—enough to make a port glass in my cabinet slip loose from the rack and go tinkling rhythmically around: *tink,* roll, *tink*, roll, *tink*. Last night, it woke me up—it and the wind howling at my portholes.

This evening, I ran into the captain as he was returning from an inspection of the outer decks, dressed in a red jumpsuit, carrying a flashlight and walkie-talkie. "A little snow," he told me, smiling his courteous, gap-toothed smile, "is nothing to worry about." Feeling cabin-feverish, I decided to see what it felt like to walk among the containers in a snowstorm. Gave myself something of a scare.

Back inside the *Ottawa*'s eight-story house, the habitable part of the ship, I learned from an oiler named Joel Nipales that solitary nocturnal circumambulations of the main deck are strictly forbidden. If an officer or deckhand has to go out at night, he alerts the bridge, puts on foul-weather gear, and brings a walkie-talkie. It wouldn't have taken much to knock me overboard: a stumble, a snowy gust. No one would have discovered my absence until morning. Cast away in dark, cold water four miles deep, deeper even than the deep water I swam in off the coast of Hawaii, I would have watched the *Ottawa*'s running lights appear and disappear beyond the crests, as if blinking on and off, dwindling and dimming into the swirling snow.

The weather report on the morning of the *China*'s departure from Taiwan wasn't auspicious, but it wasn't ominous either, calling for winds registering 6 or 7 on the Beaufort scale. Four days out, six hundred miles east of Tokyo, the *China*'s weather-routing service recommended an easterly change of course. Two low-pressure systems following in the ship's wake had unexpectedly merged, developing into what climatologists used to call a "meteorological bomb" but which they now, far less entertainingly, call a "rapidly intensifying low." Also unexpectedly, this weather bomb had moved in an easterly, rather than northeasterly, direction, toward the *China* instead of away from it.

Two days later, on October 26, by noon local time, the storm had drawn to within 120 nautical miles. The *China* once again changed course, veering farther south. Nine hours later it veered south yet again, onto an almost due-easterly bearing. Weirdly, instead of continuing on its track, the storm veered too, as if in pursuit, and the *China* suddenly found itself in what meteorologists call the most dangerous quarter of the storm. If the master had ignored the weather service and stayed on his original course, he might have escaped. Instead, in perfect darkness, at 3 A.M. on October 27, disaster struck.

The winds were now Beaufort force 11: "exceptionally high (30–45 ft) waves, foam patches cover sea, visibility more reduced." Analysis of the historical weather data would eventually reveal that the conditions were even worse than Beaufort had described. The tallest of the normal, unfreakish waves likely attained heights of seventy or eighty feet. Rogue waves are by definition unlikely, but if one had arisen under these conditions, it would have been, wave-height distribution statistics suggest, 105.5 feet tall. It was an hour and a half before the fury of the storm began to subside. The sun rose on a ruin.

Exactly what happened during that nightmarish hour and a half is difficult to say. Among the cartload of legal files archived at the federal courthouse in the Southern District of Manhattan you will not find transcripts of the depositions taken from the *China*'s crew, which remain sealed. I asked APL for access to them. They denied my request. I asked for access to the mariners themselves. Again, no.

One of the lawyers deposing the mariners in Seattle in November of 1998 was Bill France, of Healy & Baillee, a once venerable, now defunct New York firm hired to represent APL and its underwriters. France, who grew up miles from the sea, on a mink farm in Chippewa Falls, Wisconsin, is not your typical maritime lawyer. When I called him to ask about the APL *China*, the deep-voiced person who answered agreed to meet with me but, before hanging up, said there were two things I should know: First, in honor of attorney-client confidentiality, we could discuss only what was already in the public record; second, the fifty-nine-year-old person I'd be meeting was no longer a man named Bill but a tall, transgendered, "somewhat mannish woman" called Willa.

Not only is Willa France transgendered; she is a poet, a student of Jewish theology, and a certified naval architect.[29] France knows about the mysticism of Buber and the prosody of Frost. She knows about the physics of waves and the applied science of engineering. I asked whether she wanted me to identify her as Willa in print, and she said yes and, in a characteristically eloquent e-mail, compared her transgendering to the change the *China* had undergone on its fateful ocean crossing. Not that it had been for her a personal disaster; quite the opposite. But it had been overwhelming, "not unlike the consequence of an elemental and unpredictable force." Although no actual detectives were assigned to the case of the APL *China*, of all the lawyers involved, France, with her expertise in naval architecture, came closest to playing the part. She is this mystery's Sherlock.

Like the lawyers representing the cargo owners and their underwriters, France was at first skeptical about the story recounted by the *China*'s crew. The details didn't make scientific sense. The ship's anemometer, which measures wind speed, had been on the fritz, so the sea conditions recorded in the logbook were estimates made in darkness, in wildly "confused seas," by sleep-deprived, tempest-tossed mortals, and the ship had been yawing hard, off its intended course. In confused seas, the waves move every which way, and the prevailing direction is difficult to determine. It would be understandable if the officers had been wildly confused too.

"It was only after listening to four or five guys that I began to take them seriously," France told me. Their stories all matched, and France discovered forensic evidence confirming some of the details—green water inside a running light up on the bridge, wave damage to the outermost containers in stacks still standing, a dent on the bow's protective steel bulwark. But what finally made France a believer was the testimony of the *China*'s master, Parvez Guard.

A seasoned mariner from India who'd been captaining container ships for fifteen years, Guard was an exceptionally expert witness, France said. In a deposition lasting three of the six days that the *China* spent in Seattle, Guard reconstructed the voyage day by day, then, as the time of the disaster neared, hour by hour, then minute by minute, corroborating his testimony with entries in the logbook. While that testimony isn't in the public record, one very telling quote from it is. Just before the containers began to fall, the ship had suddenly become "uncontrollable," Guard testified, "as if there were a devil in it."

NOT DOWN IN ANY MAP. 47°6'N, 178°1'E.

Today, a little while ago, we came as close to the site of the toy spill as we'll come, about 180 nautical miles north of the 45th parallel. The conditions were fairly calm. The sky was white, the water gray. There was a light roll to the ship. Wind out of the southeast, but we didn't know how strong because the *Ottawa*'s anemometer, like that of the *China* a decade ago, is on the fritz. Fredrik Nystrom, second mate, was on watch. "You are lucky," I told him upon entering the bridge after lunch, a bit giddy, and he looked at me with a skeptical smile. "You will be present for an exciting moment!" I said. I'd already recounted to him the legend of the rubber ducks lost at sea.

"Oh, yes?" he said. Humoring me, he calculated as precisely as possible when we would cross the longitudinal line I was waiting for, 178.1° E. For almost three years now I've been contemplating this anonymous place on the map, this nowhere, this freak coincidence of coordinates where there are no landmarks to be seen. "An event *took place*," we say, and in taking place here, in taking this particular place, in tumbling overboard here, or near here, the toys transformed this middle of nowhere into the middle of somewhere, at least in my imagination. On a computer screen, Nystrom and I watched the degrees tick closer, 177.8, 177.9, 178.0. And then we were there.

I rushed out onto a bridge wing, then down the metal staircases, six flights, to the main deck, all the while thinking of that day—or night— sixteen years ago, all the while wondering: Which twelve containers fell? Did two columns fall? Or containers from different columns? Were they to starboard or to port? Fore? Aft?

As we steamed on at twenty-three knots, I felt as though the toys were still out there, as though the point I'd marked on my map were already slipping away, as if I were leaving the toys behind. I made my way to the *Ottawa*'s stern, where—below the aftmost rows of containers, stacked overhead, water raining down between them—there was a great cavern. Standing at the taffrail, I studied the horizon to the southeast, the horizon toward which the *Ottawa*'s wake foamed and frothed and stretched.

The containers would still have been afloat. Perhaps, standing at the taffrail of the *Ever Laurel* on that day—if it had been day still and not night—you could have seen the colorful boxes, appearing and disappearing from behind the waves. Had the one containing the toys burst open yet? Had the brown cardboard boxes inside yet escaped? Perhaps. And perhaps even one or two of the cardboard boxes had already opened. Doubtful, but possible, depending on the physics of the spill and the qualities of the glue. Perhaps the little packages had already come bobbing to the surface and begun to drift apart.

I had my own yellow duck in the pocket of my windbreaker, the duck Ebbesmeyer had given me, not the one I'd retrieved from Gore Point, and seriously considered chucking him overboard just to see with my own eyes that image I had been imagining, the image of a lone duck on the open sea. No one was around to stop me. I put my hand in my pocket. I felt the little plastic ball of its head. How easy it would be.

The movement of water along the hull of a boat is deceiving. With no fixed point, it's hard to separate the speed of the ship from the movement of the waves, which is why sailors in the past would throw a log overboard and count the knots of rope that unraveled behind it in order to calculate speed. How would the duck move across the chaos of the surface? I wondered. The *Ottawa*, big as it was, seemed hardly to move at all, but surely a little plastic duck would change the scale, making the ocean seem all the grander. Surely it would topple about, overturning and righting and overturning again, though I supposed it was possible that it might ride upright over the waves, the way we would like to picture it, sliding backward up to the crest, and then tipping as it slid down the other side, the waves moving beneath it. I withdrew my duck from my pocket, held it the way you'd hold a baseball, and leaned over the rail.

But I hesitated. Jettisoning it seemed wrong, and not simply because I wanted to keep it, nor because I had promised to return it to Curtis Ebbesmeyer when I was done with it and still meant to keep my promise, nor because by jettisoning it I would be violating international law and making an infinitesimal contribution to the pollution of the sea. It seemed wrong because Moby Duck is and has always been a dream. The story I and others were enchanted by was enchanting because it was illusory, and no matter how much forensic evidence I assembled, it would remain illusory. A reenactment would get me no closer to that event. That event had taken place, but that place could not be visited, because not even a container ship can subdue the seas of time. I knew that if I threw my duck overboard it would fall far short of the dream. I knew that if I threw it overboard, watching that yellow evanescent dot drift off, I'd be filled not with exhilaration but with disappointment and regret.

Pocketing my duck, I opened my notebook. What to write? I should have prepared something! "Fare thee well," I began, then scribbled it out, remembering the line from *Moby-Dick*. "Good-bye, Moby Duck!" I began again. "Thou cans't never return! God keep thee!" Writing this, I wasn't even sure what I meant by this. To what would Moby Duck wish to return?

I tore the page free and folded it into quarters. On one side I drew a little duck, inscribing THE FIRST YEARS across its wing. On the other side, I wrote, QUACK! Then over the port rail of the deck I flung this little folded farewell, this prayer, the wind snatching it away faster than I'd anticipated. I lost sight of it the instant it left my fingers, and my heart sank. I had wanted to watch it drift away. But then, running to the taffrail, I saw it, leaping and spinning and diving over the boiling churn, blown open, fluttering, borne aloft by gusts of turbulence as if it were alive, dancing in the air. Then it dove suddenly down onto the white wake and was gone.

I stood there, inside the thunderous cavern of the *Ottawa*'s stern, gazing at the spot while water dripped down onto my head from the containers above, and my hands went numb on the cold rail, and my ears ached in the wind, and I felt sad—not unsatisfied. Not wishing I could do something more. Not wishing I could tip 28,800 toys into the sea to watch them float away, or uncork a bottle of champagne. Not disappointed, exactly, but sad, as though I had lost something.

I'm back in my little cabin now. It's a slushy, half-frozen rain that I can see falling at an angle past my two portholes, the one looking out onto the maroon container, the other onto the white one. Although we are five days from Seattle, I think this trip may already be over. I just realized that we are, quite possibly at this very moment, crossing the international date line, which means that although yesterday was Tuesday, today it is Tuesday again, and for the first time since I flew to Hong

Kong two weeks ago, it is the same day here as it is in New York. A minute ago, I was more than a day ahead of Beth and Bruno. Now I am six hours behind them. Only six more time zones to go.

I've begun to wonder if the greatest peril modern-day merchant mariners face isn't the life-threatening *Monsterwellen* but the mind-numbing boredom. The other night on the bridge Chief Officer Hermann Josef Bollig pointed out that there were "blue skies," meaning clear ones. Stars were shining, the first we'd seen in days. Hanging above the horizon like an incandescent peach was the moon, bigger and yellower than it's been. "That's quite a moon," I said. And Bollig said, "There's a rumor they put a man on it." He reminisced about watching the moonwalk on television as a boy. I asked if he'd wanted to be an astronaut. "Oh, no, I couldn't," he said. "I'm only intelligent enough to drive this ship."

Bollig, it turns out, is a perfectly friendly German giant, who has the charming habit of wearing his glasses on the back of his head when he doesn't need them. Twice now he has delivered what is obviously a favorite line. Last night the line went like this: "The children they want the champagne and caviar." Tonight it was "caviar and chocolate." He has children, and so he rides the sea. That is all there is to it. Beside him, swaying with the rolls, listening to the ninety-five-revolutions-per-minute throb of the engine far below, I asked him, "Do you still find it beautiful? The sea?"

"No," he said, still staring straight ahead. The one good thing about his job is that after four months away, he gets to spend three months at home in Saxony, lying in his hammock in his garden, his garden of apples and plums.

All the *Ottawa*'s mariners seem to have terrestrial dreams. Captain Jakubowski dreams of New Zealand, where he hopes to vacation soon with his wife and two daughters. This morning during his coffee break in the ship's office, he told me that in his youth he had read Conrad, and Melville, all the classics of naufragia, and they had fueled in him a yearning for the age of sail and for portages many weeks long and for the freedom shipmasters formerly enjoyed. Someday, he thought back when he was young, he'd buy a small freighter of his own, so that he could be both commander and boss, but in the sixties, when he first shipped out, such a dream had already been foredoomed by the container. "And so," he said, "I have always been an employee."

"Who aint a slave?" Ishmael famously asks, but on a container ship one encounters degrees of slavery, as well as of freedom, six of them, at least. Ships have always been microcosms of the world onshore. The *Ottawa* is a little chip of Europe afloat on the North Pacific. Here, too, there are the executives and middle managers and menial laborers of a darker race, and even a few American tourists seeking pleasure, or thrills, or truth, or God, or wildness, or a hollow plastic duck.

The other night, after taking my forbidden nocturnal walk, I spent an hour in the crew's lounge drinking cheap red with the oilers Marco Aaron and Joel Nipales. On his last trip, Aaron's ship, one even larger than the *Ottawa*, failed to escape a hurricane. "You know the distance from one wave to another wave?" he said. "It's four hundred meters. Our ship's four hundred meters also. It was rolling, pitching, everything." Belowdecks in a storm like that, "every time you walk you have to carry your empty glass or empty bucket. So that you not throw up anywhere. As long as the engine's running, nothing can happen to you. But once the engine stops, you have to pray. You have to call all the saints."

Seasick as he was, the homesickness may be worse. Every day at sea, Aaron misses his wife and baby daughter. She'll be turning one next week, and he's planning a long phone call as soon as we reach Seattle. He's not sure he'll last as long as Nipales, who, at forty-two, is eleven years older. But at sea, an oiler can make a decent living. "For one month we get $1,300 U.S.," Aaron said. "But in the Philippines, if you're going to work there, the maximum for the beginners, you'll only earn $200." Unlike the officers, the oilers and deckhands are not employees of NSB but temp workers subcontracted by an agency in Cyprus. The officers do four months on, three months off, whereas the oilers and deckhands ship out for seven months at a time; at the end of those seven months many sign on for seven more, and in some ports of call there's not even time for shore leave. "It's very hard. Seven months is too much," Aaron said. "Almost 70 percent of your life you spend on a ship." Nipales once spent twenty-six consecutive months at sea. Even more than what it would take to send the *Ever Laurel's* twelve containers overboard, let alone the *China's* 407, that's something I have difficulty imagining.

No one besides the mariners who were there will likely ever know exactly what happened aboard the Evergreen *Ever Laurel* that stormy day or night in January of 1992. Likewise no one will ever know what Parvez Guard and his crew went through on that stormy October night in 1998 aboard the *China*. Demonic possession, however, would never stand up in court. Given the stakes involved, APL subsidized an expensive forensic investigation led by Willa France.

France hired three meteorological consultants to hindcast the sea conditions with computers. Next, she hired oceanographers to computeranimate what would happen to a C-11 container ship under the conditions the *China* encountered, and the Maritime Research Institute Netherlands to conduct model tests in a wave tank three times as long as an Olympic swimming pool.

If Parvez Guard was to be believed, the *China*, already hove to, couldn't have fallen prey to synchronous rolling. Furthermore, the waves were too close together to sync with the ship's roll period. But in 1973, experiments conducted with a scale model revealed that the hull shape of container ships made them vulnerable to a kind of rolling rarer and quite possibly more dangerous than synchronous rolling: "parametric rolling," so called because it occurs not when a ship's roll period is in sync with the waves but when the waves come exactly twice as fast as the ship rolls. Those model tests proved that parametric rolling could be excited by stern-quartering seas—that is, when the waves hit the ship askance on either corner of the stern. France set out to prove that it could also occur in bow-quartering seas.

Months after my trip aboard the *Ottawa*, she invited me to the East Harlem brownstone where she still lives with her wife of thirty-five years. While we ate sandwiches and pickles off silver plastic trays, France played me footage of the 1973 model tests. I watched as the toy boat first yawed a few degrees off its bearing, then suddenly began rolling hard—so hard the damn thing keeled right over. Next came the computer animations of a C-11 container ship hove to in bow-quartering seas. Away the *China*'s avatar went, pitching merrily along, up and down, over a red grid of giant waves until, suddenly, for no perceptible reason, something changed. The digital ship rolled. At first a little.

Then more. Then more, until the bridge wings were ticktocking like a metronome.

Then came the wave-tank experiment. Here again was a model boat, a replica of the *China*, which also went pitching merrily along in confused seas, actual ones this time, created by hydraulic paddles. For a minute it seemed as though the experiment would fail. "Nothing is happening," France said, narrating for me, remembering for herself. "I'm beginning to bite my nails, because we've invested so much money in this and we're not seeing what the computer program has predicted." She stood next to the screen now to deliver her closing argument, pointing out details, like a weatherman delivering the forecast. "Now watch," she said. As in the simulations, something subtly changed. The toy boat rolled a little to starboard as the bow pitched down, a little to port as the stern rocked back. Steadily, the amplitudes steepened. Finally there it was: full-on demonic possession, green water over the rails, waves swiping at the stacks.

Would any of these waves have qualified as genuine *Monsterwellen*? Probably not. The wave the officers described was monstrously large (taller than France's brownstone) but, computerized hindcasts showed, not statistically improbable. The ship was rolling so heavily, dipping its bridge wings so close to the water as it pitched into the troughs, that even a fifty-foot wave could have splashed them. Officers reported rolls as steep as 40 degrees—steeper and more violent than the steel lashings had been designed to endure. At 40 degrees, the stacks of containers would have been almost as horizontal as vertical, and a mariner standing on a bridge wing would have been staring into the abyss. The experiment proved that, under such conditions, if Parvez Guard had done everything he'd been trained to do, if he'd hove to and decelerated and the engines had not yet failed, the accident still would have occurred. And ironically, if Guard had not done what he'd been trained to do—if he'd maintained his speed, for instance—disaster might never have struck.

No judge ever decided whether France's explanation solved the mystery. APL settled out of court for an undisclosed amount. But France's discoveries did help limit APL's liability. "We got a very favorable settlement," France told me. Not all of France's colleagues understood her fascination with the *China* disaster any better than they understood her poetry, or the mysticism of Buber, or, later, her metamorphosis, but in 2003, after the legal proceedings had all been settled, France published her findings in the journal of the Society of Naval Architects and Marine Engineers. A few months later, under the headline "Parametric Rolling Will Rock Insurers," *Lloyd's List* warned the marine-insurance industry of this "alarming new danger," which appears to be an unintended consequence of the oversize, U-shaped, post-Panamax hull. France's findings not only helped explain the mystery of the *China;* they would later help explain what had happened to the Maersk *Carolina*, the *P&O Nedlloyd Genoa*, the *CMA CGM Otello*, and an unknown number of other maritime mysteries. In July of 2008 the American Bureau of Shipping and the Polish Registry of Shipping announced that they were embarking on a "multi-year, joint research and development program" to find technology that would help prevent parametric rolling in "extreme wave conditions," hoping to exorcise once and for all the devil that possessed the *China*.

LANDFALL. SUNSET. 48°28'N, 124°60'W.

Yesterday we saw the first blue daytime sky in almost a week—just a portal of it, like a window atop a dome. Now we're close to the fictitious Strait of Juan de Fuca, so close that gulls have begun visiting us and cell phones have reactivated. Late last night in the crew's lounge, there was a party. Bob and Claire, who appear to be teetotalers, did not attend, but Chief Officer Bollig and I did. Ronaldo Cuevas, the bosun—chief of the deckhands, Bacchus of the lounge, a big, round-faced dude wearing a muscle tee and a Fu Manchu—seemed intent on getting me drunk. He refilled my tumbler of wine even when I told him not to. At least I managed to get Joe the Messman to play "The Boxer" on his guitar.

I don't remember much of last night's party, but there is one night from this long, uneventful voyage that I think I will never forget, the one when I took my forbidden nocturnal walk. As I made my way along a catwalk glazed with sea spray and snow, I had to take cover behind a bulkhead every few yards to warm my ears and hands. Above me the containers creaked and moaned and clanged, straining at their lashings with every roll. I had intended to walk all the way to the stern, where cataracts of water rain down from the container lids overhead and the roar of the propeller makes the whole place thunder like a cavern in an old myth. But then I thought better of it. The winds were at Beaufort force 7, near gale, and just as Sir Beaufort promised, the sea had heaped up and foam was streaking off the crests of breakers thirteen to twenty feet tall. You could see only the waves that came heaving and hissing along the hull, blue and foamy and luminous in the house light, but you could sense the rest of the Pacific, and if you looked hard, you could vaguely distinguish the greater darkness of the ocean from the lesser darkness of the starless sky. Standing at the starboard rail—facing south toward the spot five hundred miles away where, on a night far stormier than this one, containers full of clothes and toys and phones and fish and tables had gone crashing overboard—I gazed a long while into both varieties of darkness, the watery and the ethereal one, as if into the tenebrous seas and mists of time. Then, ears aching, hands numb, I turned around, leaned into the wind, and staggered to the crew's lounge for a little human company, a little Filipino television, and a tumbler of cheap red.

THE LAST CHASE, PART ONE

To see! To see!—this is the craving of the sailor, as of the rest of blind humanity. To have his path made clear for him is the aspiration of every human being in our beclouded and tempestuous existence.

—Joseph Conrad, *The Mirror of the Sea*

THROUGH THE BERING STRAIT

Picture a red beaver, a blue turtle, a green frog, a yellow duck—one set of toys hatched from the same plastic shell. It is February 1992, a month after the spill. The cardboard backs of the packages inscribed with colorful copy have long since dissolved. Carried east by a current known as the North Pacific Drift, the four castaway toys have traveled over two hundred nautical miles along the northern edge of the 45th parallel, the northern edge of the subarctic front, the boundary between the Subpolar and the Subtropical gyres.

The four animals have remained close together, and relatively close to the other 28,796 castaway toys, traveling the current in a diffuse flock, smiling refugees on a possibly interminable road, prisoners in the labyrinth of drift, a labyrinth that is the collaborative work of invisible architects—the spinning of the planet and the heat of the sun, the saltness of the sea and the influence of the winds. When the seas are calm, the zooplanktonic spawn of flying fish try to raft on the toys, as do the spawn of gooseneck barnacles. On the hunt for nutrients, albatrosses, riding the same winds that generate the waves, swoop down. For now the four toys are probably too big for an albatross to swallow.

Squalls come and go. Snow falls in swirling flurries, melting on contact with the ocean's surface, which out here in the Graveyard of the Pacific is cold but not freezing, 5 degrees Celsius, 41 degrees Fahrenheit. Overboard in these waters, a strong swimmer might last a half hour. Though stormy, the North Pacific is in February 1992 calmer than it usually is in winter. The moon is growing full, and on those nights when the skies clear, it paints on the wave crests a triangle of white lines that diminish toward a vanishing point on the horizon. Viewed from the deck of a nearby container ship, the lines of moonshine on the black water seem to climb into the black sky like the rungs of Jacob's ladder.

It is the incongruity of the toys that above all things enchants me.

They are incongruously small compared with the deep; incongruously colorful in that gray-green seascape, the red beaver and the yellow duck especially since they have only just begun to fade; incongruously cheerful under circumstances that would drive mad a castaway both sentient and mortal; incongruously human, and childish, and domestic, and pastoral.

Months pass. Winter turns to spring. The days lengthen, the nights shorten, the weather moderates. Storms still roll through, but less often now and with less violence. Passing cargo ships have grown rarer, too; this is the slow season for trade with the Far East. Up and down goes the price of oil, up and down go the profits of toymakers and shipping lines, up and down go the four toys on the Ferris wheel of waves, aspin in space and time.

In June, 690 miles southwest of Sitka, for the first time since the spill, dramatic complications occur. As it collides with the continental shelf and then with the freshwater gushing out of the rainforests of the coastal mountains, and then with the coast, the North Pacific Drift loses its coherence, crazies, sends out fractal meanders and eddies and tendrils that tease the four voyagers apart. We don't know for certain what happens next, but statistical models suggest that at least one of the four voyagers I'm imagining—the frog, let's pretend—will turn south, carried by an eddy or a meander into the California Current, which will likely deliver it, after many months, into the North Pacific Subtropical Gyre.

You may now forget about the frog. We already know its story—how, as it disintegrates, it will contribute a few tablespoons of plastic to the Garbage Patch, or to Hawaii's Plastic Beach, or to the dinner of an albatross, or to a sample collected in the codpiece of Charlie Moore's manta trawl. *Bon voyage, petite grenouille!* May the fatal threads of the ocean currents spare you from a plundering albatross and spare an albatross chick from you!

The other three toys turn north. In November 1992, the Alaska Coastal Current carries one of them—the turtle, let's pretend—onto Kruzof Island, where it will remain, caught in the jackstraw, until Tyler Orbison comes sloshing ashore the following summer. You may now forget about the turtle.

Hundreds of other toys make landfall on the Alaska Peninsula that November, as Eben Punderson will later report in the Weekend section of the *Daily Sitka Sentinel.* But thousands, including the duck and the beaver I have asked you to imagine, keep bobbing along on a northerly bearing. At Kayak Island, north of Juneau, the currents make a sharp, westerly left. By January 10, the first anniversary of the spill, the beaver and duck have reached the entrance to Prince William Sound, passing Bligh Reef, named for the notorious sea captain, made famous by the notorious tanker that, running aground, spilled its oil there just three years before the Evergreen *Ever Laurel* spilled its toys.

A few weeks later, traveling the same route, roughly, as Exxon's crude—the same route, roughly, that Pallister and I would take aboard the *Opus*—the toys reach Gore Point's windward shore. A southeaster blows up and a storm surge lobs the beaver over the driftwood berm. Hurricane-force winds lift it aloft, then send it tumbling across the forest floor, then pluck it up again, drop it again, until finally it comes to rest in the lee of a spruce. For fourteen years, there it sits, atop the pine needles, among fishing floats and water bottles, until on July 16, 2007, along comes an erstwhile schoolteacher turned amateur driftologist. He bends back a leaf of devil's club with the toe of his rubber boot and, careful not to strain his surgically repaired lumbar region, crouches down, hollering with ridiculous excitement. You may now forget about the beaver. We know how its story will end—as a lab animal, inside a mass spectrometer, contributing a few tablespoons of data to the oceans of information.

You may not forget about the duck, however. At least not if you're me. If you're me, prey to nocturnal flights of fancy, you will spend many years thinking about that stupid, sun-bleached, ghostly, hollow, iconic, and altogether hypothetical plaything with a body the size of a bar of soap and a head the size of a Ping-Pong ball.

Out on the Alaskan Current, it has escaped the witch's finger of Gore Point and is now headed, along with thousands of other toys, west by southwest toward the Aleutians. Most of those toys, those puerile satellites, will remain, the simulations of OSCURS predict, in the Subpolar Gyre's orbit, completing a lap once every few years, their numbers diminishing with every lap. Some will end up on the rocky shores of the mostly uninhabited Aleutians, others in Siberia. Others, after completing a lap or two, will return to Alaska and join some farrago of wrack in some out-of-the-way cave of ocean. And some—not many, OSCURS tells us, but some—will slip between the Aleutian Islands and into the Bering Sea, where under favorable circumstances an even smaller number—including, let's pretend, the once yellow, now pale duck I'm imagining—will pass through the Bering Strait and into the Arctic. There OSCURS had lost them. And so had I.

Truth be told, by the time the Hanjin *Ottawa* passed through the Strait of Juan de Fuca, I'd had enough of seafaring. At the entrance to Puget Sound a launch motored out, a harbor pilot came aboard. "A Departure, the last professional sight of land, is always good, or at least good enough," Conrad writes in *The Mirror of the Sea*. "For, even if the weather be thick, it does not matter much to a ship having all the open sea before her bows." Landfall, in Conrad's time as in ours, is another matter. It cannot be entrusted to the autopilot, nor even to the helmsman or master, which is why the harbor pilot, who knew Puget Sound by heart, appeared on the *Ottawa*'s bridge carrying a duffel bag of charts. Occasionally consulting the radar screen, mostly studying the scenery ahead, he called out numerical bearings and the helmsmen called the bearings back in confirmation. The rest of us, even Captain Jakubowski, stood by, gazing through the glass in passive silence, faces luminously masked by the glowing screens, waiting for greater Seattle to appear on the dark horizon, and when it finally did appear on the horizon, like some resurrected twenty-first-century Atlantis rising up out of the water into the night sky, which it polluted deliciously with light, an electrical dawn, my spirits rose too.

It was well after midnight by the time we reached pier 5, Hanjin's container terminal in Seattle. Bob and Claire and much of the crew had gone to sleep, content to disembark the following morning. Not me. As a tugboat nosed the colossal ship toward the dock, I stood at the starboard rail, bags already packed. Ashore, a car and a truck materialized, and longshoremen climbed out. In hard hats and life vests, they took up their positions beside the yellow bollards, waiting for deckhands to throw them the lines. As soon as the ship had tied up, customs agents came aboard, and as soon as they cleared me, I rushed down the gangway onto the deserted sodium-lit pier.[30] Among containers stacked six high in rows, I wandered, searching for the exit.

I'd had enough of seafaring. I was ready to fly—not sail, carbon footprint be damned—to the insular city of the Manhattoes and stay there a good long while. Ready to pend myself up in lath and plaster, tie myself to a counter, nail myself to a bench or clinch myself to a desk, so long as said pending, tying, nailing, or clinching came with decent health benefits and a retirement plan. Ready to circumambulate the city of a dreamy Sabbath afternoon in the company of my long-suffering wife and my paternally neglected son.

To my paternally neglected son, when I got home, I'd read the opening chapters of *The Wind and the Willows*, and then, some Sabbath afternoon, I'd take him rowboating not on the life-threatening waters of Resurrection Bay but on the pond in Central Park, just as, in the very first chapter of *The Wind and the Willows*, River Rat takes landlocked Mole rowboating on the downstream backwaters of the river Thames. I, the oarsman, would play the part of River Rat. Young Bruno would be landlocked Mole. As he leaned back in his seat and felt the boat sway lightly under him, I would recite to him, dreamily, in character, the famous line: "Believe me, my young friend, there is *nothing*—absolutely nothing—half so much worth doing as simply messing about in boats."

In short, I was ready to come home. But I couldn't. Not yet. What I'd gone searching for had yet to be found. I had one last riddle to puzzle out, one last journey to take—or rather to finish. It had begun one September morning, before I went sailing with Charlie Moore or visited the Po Sing plastics works or crossed the Pacific on the Hanjin *Ottawa* at the height of the winter storm season. It would end, months later, after a long hiatus and numerous detours, deep in the interior of the Canadian Arctic on the icy shores of the Northwest Passage.

THE BLIND OCEANOGRAPHER

At the main dock behind Smith Laboratory on Water Street in downtown Woods Hole, the research vessel *Knorr* was preparing to depart. Forklifts zipped around, beeping. Stevedores and deckhands walked the aluminum gangway, busy as leafcutter ants, loading and stowing cargo—provisions, instruments, a big cardboard box containing a new Nordic-Track treadmill for the *Knorr*'s onboard gym. Through the main dock's chain link security gates, a silver minivan arrived, and a woman descended from the passenger seat. Forty-six years old, she was on this mid-September morning girlishly attired—as if for an afternoon of yachting on Vineyard Sound—in a rose-and-white waterproof jacket, white shorts, white sneakers. She had a pair of sunglasses pushed up into her coppery, shoulder-length hair and clutched a black leather pouch in her left hand. This was Amy Bower, a senior scientist in the Department of Physical Oceanography at the Woods Hole Oceanographic Institution and the chief scientist on the first leg of the voyage that was about to begin—voyage 192 it was officially and forgettably called. A photographer asked Bower to pose for a few publicity shots, dockside, against the picturesque backdrop of the *Knorr*.

Since its maiden voyage, in 1968, the *Knorr* had carried scientists to every corner of the ocean, from the Arctic to the Antarctic, the Gulf of Alaska to the Gulf of Maine, the Bay of Bengal to the Bay of Fundy. Aboard the *Knorr* scientists had collected the first images of the wrecked *Titanic*. They'd revealed the secrets of plate tectonics at the Mid-Atlantic Ridge. In February 1977, above the Galápagos Rift, they'd sent a camera 8,250 feet down and discovered bizarre organisms improbably well adapted to the infernally hot, infernally sulfurous environs of volcanic sea vents—giant albino clams, giant albino mussels, giant albino crabs, giant albino tube worms growing in thickets, like Martian bamboo, red obscene tulips of flesh blossoming from their tall white stalks. Some scientists speculate that it was not at the stormy surface of the ocean, or in Darwin's lightning-struck pond, but out of such volcanic vents—black smokers, they're called—that life first arose.

In the previous two years alone, the *Knorr*'s itinerary had included stops in San Diego; the Galápagos Islands; Valparaíso, Chile; Buenos Aires; Reykjavík; the Caribbean (Bermuda, Martinique, St. John's); and Nuuk, capital of Greenland, which is where the first leg of voyage 192 would officially end. Tourist maps from these ports decorated the hallway outside the ship's mess—"Feel the warmth and softness of the Icelandic wool," enticed an ad for Helga's Wool Market on the map of Reykjavík. Now nearing retirement, the *Knorr* was nevertheless a beautiful ship, high-prowed like a navy cutter. Its hull, 279 feet long, was the deep blue of the ocean on a world map, its upper decks, white as a wedding cake. A light-blue stripe beribboned its single white stack, and the white faceted sphere of a satellite antenna rose amidships.

On the *Knorr*'s fantail, beside the starboard rail, looming above Bower's shoulder as she posed for photographs, appeared another, smaller sphere, painted the yellow of a rubber duck. The size of a wrecking ball, made of syntactic foam interlarded with hollow glass orbs, it sat atop a big steel trellis. Across its northern latitudes, in black block letters, the following message ran.

IF FOUND ADRIFT CONTACT
MOORING OPERATIONS GROUP
WOODS HOLE OCEANOGRAPHIC
INSTITUTION
WOODS HOLE, MA 02543

The photo shoot over, Bower extracted from her black leather pouch a collapsible cane. With an expert flourish of her wrist, like a magic trick, she made the cane spring forth and went tap, tap, tapping up the gangway.

The week before, I'd driven to Cape Cod in a rented Chevy and checked myself in to the Sands of Time motel. After a fitful night beneath a floral print bedspread in the motel's basement suite, my mind sluggish with caffeine withdrawal but abrim with boyish curiosity, I'd made my way on foot to what had once been a church of the classic New England sort—white siding, white bell tower adorning a peaked roof. From outside it resembled a Puritan chapel whose cross and steeple had been dismasted by an atheistic wind. The interior of the building conjured forth altogether different associations.

Up front, where perhaps a preacher in a black cassock had once detailed the dangers awaiting sinners in the hands of an angry god, there now protruded the flukes of a whale. The beast seemed to have been snared in Sheetrock while attempting to escape. From a nearby computer mounted on a pedestal emanated the otherworldly ululations of humpbacks, and in a glass case beneath a window could be seen replicas of the tube worms that thirty years ago scientists on the *Knorr* had discovered growing along the sea vents of the Galápagos Rift.

The former church now serves as the gift shop and exhibit center of the Woods Hole Oceanographic Institution, whose initials, WHOI, its employees charmingly pronounce *HOO-ee*. Woods Hole, I learned during the week I spent there, is a marvelous place, a veritable distillery of marvels. Passing through it on your way to catch the ferry to Martha's Vineyard, you'd never suspect that this sleepy seaside village—also home to the Marine Biological Laboratory, a Coast Guard station, and branch offices of both NOAA and the U.S. Geological Survey—had once been the Houston of deep-sea exploration and the Los Alamos of submarine warfare.

With remote-controlled vehicles resembling torpedoes, Woods Hole oceanographers have looked for cracks and leaks in the forty-five-mile-long network of aqueducts that deliver drinking water to New York City. Some can read the history of the planet in tubes of sediment, thousands of which are kept, carefully archived, in a climate-controlled warehouse, a sort of library of dirt, whose contents date back decades and may well contain auguries of decades to come. With a mass spectrometer, they can analyze the isotopes in a baleen frond and tell you that the whale to whom it formerly belonged wintered in the tropics and summered in the Arctic. Others are experts on sand, which is more interesting than you'd think.[31]

I hadn't come to Woods Hole seeking wonders, however. I'd come seeking a guide, some wayfinding oceanographer willing to help me follow the trail of the toys into and out of the Arctic's icy maze. I didn't find one, but I did find Amy Bower, who was willing—so long as I packed quickly and could afford a last-minute plane ticket home from Greenland—to take me to the Arctic's brink. There, at 60.6°N, 52.4°W, just south of the Davis Strait, in the northeastern waters of the Labrador Sea, gateway to the Northwest Passage, birthplace of icebergs, we would deploy a "densely instrumented mooring," a sort of underwater weather vane almost two miles long in waters two miles deep. The purpose of the mooring was to gather intelligence on Irminger Rings, a variety of "mesoscale eddy" spawned by the Irminger Current, a remnant of the Gulf Stream.

The Labrador Sea, I'd learned by then, is the source of the Labrador

Current, which transports cold Arctic water south along the coasts of Newfoundland and Maine. If those beachcombers in Kennebunkport and Curtis Ebbesmeyer were right, if one of the castaway ducks had made it to Gooch's Beach by the summer of 2003, if the duck I'd spent years chasing was in fact a data point rather than a will-o'-the-wisp, it would have ridden the Labrador Current to get there. For me voyage 192 was a free ride, a voyage of opportunity. When I accepted Bower's last-minute invitation, I had no particular interest in Irminger Rings and mesoscale eddies. I had no interest in them because I'd never heard of them. In the dank, dim, florally themed basement suite of the Sands of Time motel, I consulted the oceanographic textbooks I'd stuffed into my suitcase. "Mesoscale eddies are the oceanic analogues of weather systems in the atmosphere," one textbook said. They are also, it said, "an exciting and relatively new discovery."

At nine thirty sharp, the provisions all stowed, the equipment all lashed, the stevedores slipped the lines, and with a blast of its horn, the *Knorr* drifted from the dock. At the starboard rail Bower and I and the four other members of her scientific team waved at the crowd that had come to send us off. In the aimless manner of a councilman on a parade float, I waved at figures I could see but did not know. Bower waved, aimlessly, at figures she knew and loved but could not see.

A seasoned seafarer, on previous research cruises she'd braved winter storms on the North Atlantic and Somali pirates in the Gulf of Aden—pirates, armed with grenade launchers, whom the research vessel's crew had managed to repel with a high-powered hose. From her colleagues her fearlessness had earned her the nickname "Hurricane Amy." She'd published dozens of scientific papers in academic journals, papers with abstruse titles like "Structure of the Mediterranean Undercurrent and Mediterranean Water Spreading Around the Southwestern Iberian Peninsula." She also happened to have, while performing these feats of seamanship and scholarship, gone almost totally blind.

She first learned she was losing her sight in her early twenties, when, driving to the University of Rhode Island from her mother's house in Maine, aware that her night vision was poor, she'd turned down the dash lights. With the dash lights off, it was easier to see the road ahead, but with the dash lights off, she couldn't read the speedometer. Going seventy on a two-lane highway, she'd come up too fast on a truck and skidded into it. She'd escaped the wreck unscathed, but not her next visit to the ophthalmologist. He informed her that she'd developed a blind spot. It was eventually determined that she was suffering from not one but two congenital diseases, macular degeneration and retinitis pigmentosa.

The macula, near the center of the retina, "is what you use for fine vision," Bower told me. When it degenerates, "everything you look at," everything you try to focus on, "disappears." Retinitis pigmentosa, meanwhile, attacks your peripheral vision. Bower's loss of sight resembled the meticulous restoration of a painting, only in reverse—a meticulous defacement. Slowly, little by little, from the center and from the edges, her vision was being rubbed away as if by a rag dipped in turpentine. When she looks at you with what remains of her vision, she doesn't appear to be looking at you. She appears to be looking at something to one side of you.

She could sense this morning that the sky was overcast. That over there, to aft, were the gray waters of Vineyard Sound, and over there, to fore, the village of Woods Hole, a blur of Cape Cod clapboard and academic brick. But when she looked straight ahead, at the crowd gathered on the dock, whatever she tried to focus on disappeared. In the crowd were Bower's husband, David Fisichella, and, in his arms, dressed in pink leggings and a fleece sweatshirt, their adopted Guatemalan daughter, five-year-old, black-haired Sara. It was to Sara that Bower blindly waved. That morning Sara had been "acting out" in protest of her mother's desertion, or so Bower believed. "She wanted to wear a bunny costume to bed last night, and when she woke up one of the socks was missing, and she had a total meltdown," Bower said.

Sending forth from its stack a yellowy stocking of diesel fumes, the *Knorr* circled about and steamed north, past the wooded coast of Martha's Vineyard. From the *Knorr*'s mast an American flag snapped in the headwind. Atop the bridge the white bar of the LORAN lazily spun. At the starboard rail, Bower reported the day's forecast: "Four- to five-foot waves out of the northeast. Sounds good to me."

Meanwhile, 828 miles above the lowering clouds, hurtling through space at 17.4 times the speed of sound, a NASA satellite named Jason-1 was beaming microwaves silently and invisibly down, and eight hundred miles north of Woods Hole a mesoscale eddy was invisibly gathering in the depths of the Labrador Sea.

"Mesoscale eddies are like watery storms, kind of like tornadoes, only much slower," Bower tells me that first morning in the *Knorr*'s main lab, as we're steaming past Nantucket, making twelve knots. The main lab is a long room lit by portholes and fluorescent lights and furnished with galvanized metal workbenches surfaced in plywood and bolted to the floor. We're seated at one of these workbenches, in front of our computers, beneath a pair of portholes through which can be seen—though not by Bower—the leaping shapes of the waves. As the ship rolls to port, the portholes seem to fill, just a little, with water. As it rolls to starboard, they seem to drain. Plugged into power strips bracketed to the ceiling, the cords of our computers sway, and from belowdecks come the rumble and throb of the *Knorr*'s four engines.

Not only are mesoscale eddies like watery storms, Bower explains; from the viewpoint of a physicist, they are watery storms—not storms of wind and waves, rain and lightning, all the usual atmospheric jazz, but storms of spinning water. To a physicist, water and air are both turbulent fluids. "Mesoscale" means that, relative to other climatological phenomena, these eddies we're hunting are pretty big—dozens of miles wide—but relative to others, not that big, not megascale big, not thousands of miles wide; not as big as the great oceanic gyres Curtis Ebbesmeyer had taught me about.

At the other extreme from the megascale gyres are "microscale eddies," eddies the size of dance floors or dimes, and if you would like to see one, drag your hand through a bathtub and watch. There they go, swirling away, as ephemeral as they are small. Throw a rubber duck in and you can watch it swirl away too. If you were a god dragging your divine hand through the ocean basins, that's what mesoscale eddies would be like. Underwater storms are slower than atmospheric ones. The watery winds of an Irminger Ring attain a "swirl speed," Bower said, of around one mile per hour—the speed, in other words, not of a hurricane or a gale but of a breeze.

Their slowness belies their strength. In their watery coils they can transport up to 1.95 trillion cubic meters of water, along with flora and fauna and flotsam, seaweed and krill, driftwood and rubber ducks. They can also, if warmer than the water through which they swirl, as Irminger Rings are, transport heat, how much no one precisely knows. A few years ago, a Seaglider—a kind of motorized remote-controlled underwater drone—strayed into a gathering Irminger Ring while exploring the Irminger Current, which winds westward around Greenland's continental shelf. Caught in an underwater storm, it took the motorized glider fourteen days to break free and resume its preprogrammed route.

Underwater storms are also smaller than atmospheric ones— Irminger Rings measure, on average, thirty miles across. They're also denser, of course, which explains their sluggishness, and their sluggishness in turn explains this: much as mammals with slow metabolisms tend to have long life spans, so the longevity of underwater storms tends to exceed that of their atmospheric counterparts. Hurricanes decay and vanish just days after meteorologists name them. A mesoscale eddy of water by contrast can last—or "live," as Bower likes to say—for many months.

Perhaps the most meaningful difference between atmospheric storms and underwater ones from a terrestrial point of view is this: you can't see mesoscale eddies, or feel them. You could be sailing on calm seas, at Beaufort force 0 ("sea like a mirror"), and the underwater storm of the century could be swirling slowly beneath you. Nobody gives them names, or watches them on the Weather Channel. Nevertheless they are as much a cause and effect of the climate as rogue waves and hurricanes. What made the "relatively new discovery" of mesoscale eddies so exciting? Until recently, no one, not sailors or scientists, knew that the world below the waves was such a stormy place.

By sunset we are somewhere east of Boston, no land in sight. The skies have begun to clear. A crescent moon rises to starboard. Through a porthole in the main lab I watch the black shape of a cargo ship—a post-Panamax container ship, by the look of it—cross the sunset in silhouette. After a dinner of buttery fish and rice in the mess, the *Knorr*'s off-duty oilers and deckhands gather in a lounge to watch a movie about an assassination plot. Long after the portholes have all gone black, through the lounge's closed door, the muffled sound of gunfire can be heard. On the bridge one officer and one able-bodied seaman stand watch in darkness, their faces, lit by screens, like those of sleepy revenants. In the main lab, Bower stays up late working at her computer, which blows documents up to a scale she can decipher. Special software speed-reads text aloud in a robotic voice as nonsensical to my ears as the chattering of a telegraph. Taking a break from her work, she clicks through the library of music she's brought along. "I always loved this song," she announces, and a moment later, somewhere out on the Gulf of Maine, inside the *Knorr*'s main lab, UB40's "Red Red Wine" begins to play.

In my windowless cabin belowdecks, I stay up late reading about the history of oceanography. When I turn out my bunkside light, the darkness is absolute, as black a darkness as I've experienced, and I wonder if this is what it's like to be totally blind. I have been, since the age of four or five, acutely myopic. I spent my childhood wearing—and losing, and replacing, and lashing to my head with unflattering elastic accoutrements—spectacles with Coke-bottle lenses and breakproof plastic frames made, unconvincingly, to resemble tortoiseshell. From such eyewear I was liberated, as a teenager, by the contact lens, to the inventors of which I will remain eternally grateful, whatever my misgivings about disposable plastics. Without glasses or contact lenses I too would be prohibited from driving. Without glasses or contacts, I should probably be prohibited from walking. I tried it once on the sidewalks of New York, circumambulating by night while visually impaired. Traffic lights turned into colorful chandeliers, pedestrians into shadows that caught me by surprise. To read a book without contact lenses or glasses, I have to bury my face between the pages so that I appear to be snorting meanings through my nostrils.

Aboard the *Knorr*, in the darkness of my submarine cabin, I lie awake awhile, eyes open, contacts out, glasses off, wondering what Beth and Bruno are doing, listening to the pulsing RPMs of the engines below and to the amniotic hiss and whoosh of the waves above. Then I close my eyes and fall asleep thinking how strange it is that I am below the surface of the high Atlantic. Inches from my pillow, for all I know, fish swim.

THE MYSTERY OF OCEAN CURRENTS

In 323 B.C., the last year of his life, charged with philosophical heresies by the pagan inquisitors of Athens, Aristotle—generally regarded by scientific historians as the forefather of oceanography—fled to his family's country house in Chalcis on the Greek island of Euboea. Chalcis happens to be one of the best spots in the Mediterranean to observe the tides, a phenomenon of which the Greeks of Aristotle's time were largely unaware, for good reason. Tides in the shallow, enclosed waters of the Mediterranean are weak, so weak that on some shores near Athens, the difference between flood and ebb measures less than a centimeter. If you were an ancient Athenian, you too would be unaware of the tides. Unless, that is, you'd visited a place like Chalcis.

There the Euripus Strait, separating the island of Euboea from the Greek mainland, narrows into a channel only 130 feet wide. There even the weak tides of the Mediterranean make the funneled water diurnally rush, frothing and churning, first in one direction, then the other. This twice-daily riot was something undreamt of in Aristotle's philosophy. In fact, Aristotle believed that since water seeks its level and in the ocean finds it, below its windy surface the ocean must be still, as if an ocean basin were a kind of giant cistern. A few months after retiring to Chalcis, the forefather of oceanography died. The cause of death?

According to a legend that persisted into the seventeenth century: drowning. Suicidal drowning. Suicidal drowning brought on by despair. Despair brought on by confusion. Confusion induced by the sea. Confronted with the tumultuous, inexplicable waters of the Euripus Strait, Aristotle supposedly hurled himself into them. "The great Master of Philosophy drowned himself," the seventeenth-century British scientist Richard Bolland wrote, "because he could not apprehend the Cause of Tydes." So begins the history of physical oceanography. Except that it's mostly fictional. Aristotle did indeed flee to Chalcis, but the actual cause of his death was an undiagnosed gastrointestinal ailment, brought on, probably, by a funky oyster. In Aristotle's suicidal confusion, it seems, Bolland and his seventeenth-century oceanographic colleagues saw their own.

Just twelve years after Bolland repeated the apocryphal story of Aristotle's suicide, Isaac Newton finally determined the lunar and solar
"Cause of Tydes." The currents, however, would prove to be an even more
maddeningly intractable riddle. "The secrets of the currents in the seas,"
Melville observes in *Moby-Dick*, which unlike most novels includes cita-
tions to actual scientific papers, "have never yet been divulged, even to
the most erudite research." By the time Melville died, in 1891, most of the
major surface currents had been charted, or at least sketched, but they
had yet to be explained. "We are now becoming acquainted with only the
roughest features of the oceanic circulations," the oceanographer Harald
Sverdrup observed in a 1929 article called "The Mystery of Ocean Currents." Three years later the British mathematician Horace Lamb famously remarked, "I am an old man now, and when I die and go to
Heaven there are two matters on which I hope for enlightenment. One
is quantum electrodynamics, and the other is the turbulent motion of
fluids. And about the former I am really rather optimistic."

By the 1950s, oceanographers, many of them at Woods Hole, seemed
finally on the verge of solving the mystery of ocean currents and had
even begun to illuminate what the "scientifics" of the *Challenger* expedition had called "the dismal abyss." What the space race would later do
for astronomy, the submarine warfare of World War II did for ocean science. Deep-sea explorers were the astronauts of the postwar years, the
abyss the final frontier. No longer a howling waste, or a great blankness,
the ocean had become the latest locus of an old, elusive dream: the
dream of a new world. "In this Kingdom most of the plants are animals,
the fish are friends, colors are unearthly in their shift and delicacy," the
deep-sea explorer William Beebe had written in the 1930s. "Here miracles become marvels, and marvels recurring wonders." Now, in the
1950s, with the advent of underwater cinematography, we could all be
armchair deep-sea explorers of this magical kingdom. Jacques Cousteau
was a celebrity and Rachel Carson's *The Sea Around Us* a bestseller. A
former director of WHOI appeared on the cover of *Time*. In his first
State of the Union address, Kennedy declared, in a passage that would
now seem anachronistic, "We have neglected oceanography, saline water
conversion, and the basic research that lies at the root of progress." With
instruments developed for the U.S. Navy, aboard research vessels the
Pentagon had subsidized, marine geographers mapped the seafloor in

exquisite detail, supporting the then controversial theory of plate tectonics. Currents flowed through the earth's mantle, their soundings revealed, as well as through water and air. Everything was adrift, even
continents. The only difference was the rate of flow.

In the last chapter of *The Sea Around Us,* published in 1950, Rachel
Carson informed her readers that the benighted cartographers of the
Middle Ages had thought of the ocean as the "dread Sea of Darkness."
Over the centuries, little by little, explorers and then scientists had
pulled the veil of darkness back. "Here and there, in a few out-of-the-
way places, the darkness of antiquity still lingers over the surface of the
waters," Carson wrote. "But it is rapidly being dispelled, and most of the
length and breadth of the ocean is known; it is only in thinking of its
third dimension that we can still apply the concept of the Sea of Darkness. It took centuries to chart the surface of the sea; our progress in de-
lineating the unseen world beneath it seems by comparison phenomenally
rapid."

If Carson had stopped there, if she'd made of oceanography yet another chapter in the triumphant march of progress, implying that soon
there would be no watery mysteries left, the ocean would have made of
her a fool, as it had of Aristotle. But Carson didn't end her story on a triumphant note. Prophetically, she added this: "Even with all our modern
instruments for probing and sampling the deep ocean, no one now can
say that we shall ever resolve the last, the ultimate mysteries of the sea."
Which brings us to mesoscale eddies.

In 1960, aboard a one-hundred-foot ketch called the *Aries,* a British
oceanographer named John Swallow sailed northeast out of Bermuda
into the Sargasso Sea in search of a vast, deep, and altogether hypothetical northerly current that he was confident he would find. He was confident because a scientist at Woods Hole, Henry Stommel, the Isaac
Newton of physical oceanographers, had deduced its existence. Most
oceanographers sailed out, collected data, and then interpreted it. Stommel made his most important discoveries on land, with pencil and paper.
In 1947, scribbling on a place mat at a roadside diner, he'd mathemati-
cally explained the physics of the Gulf Stream. And in the midfifties,
once again with pencil and paper, for sound mathematical reasons, he'd

hypothesized that beneath the northerly Gulf Stream there must be an-other fast-moving western boundary current—the Gulf Stream's shad-owy twin, its abyssal, invisible doppelgänger—flowing in the opposite direction, south, toward the equator.

How to prove this hypothesis? On the far side of the Atlantic, John Swallow had found a way. To follow a deep current, Swallow realized, what you needed was a neutrally buoyant float, ballasted to sink toward but not quite to the ocean floor. If you equipped such a float with a battery-powered transducer that could send forth pings, oceanographers aboard a research vessel could track it—not with terrestrial eyes, but with aquified ears. In March 1957, west of Bermuda, aboard the British research vessel *Discovery II*, Swallow had launched three of his experi-mental floats—Swallow floats, they're now called—into rough seas. One collided, ruinously, with the *Discovery II*. One wandered aimlessly about. The third Swallow failed to track. Launching eight more floats, Swallow's experiment met with success: one was lost, but seven were carried south, below the Gulf Stream, by the unseen current whose ex-istence Henry Stommel had divined.

Now, three years later, in 1960, aboard the *Aries*, Swallow expected to vindicate Stommel once again. Again he cast his floats overboard. Then, from the deck of the ketch, he lowered his hydrophone into the water and listened for pings. This time his floats sauntered extrava-gantly, first one way, then another. They spun. They described arcs. They meandered. They doubled back. Not one of them drifted steadily north, as Stommel had predicted. Even more than those of the castaway toys, their drift routes seemed hand-drawn by a cartographer with palsy, a drunken cartographer with palsy. Examining his chaotic data, Swallow began to discern a vortical method to this oceanic madness. His floats hadn't charted rivers in the sea. Instead they'd revealed watery breezes, watery winds, watery storms, and this revelation fundamentally changed the way scientists think of the sea.

There are still oceanographers preoccupied by the ocean's third di-mension, but a growing number—perhaps even most—are thinking about its fourth and darkest dimension: time. Just as geographers and cartographers mapped the earth, so the physical oceanographers of the nineteenth century set out to map the sea—to chart its currents, to fathom its depths. As Swallow's discovery made clear, the geographic analogy turned out to be a poor one. The ocean was not so much a place as a kind of weather. What we think of as the surface of the sea oceanographers now think of as the "ocean-atmosphere interface," a membrane more permeable than it looks. The "hydrosphere," one of Swallow's contemporaries proposed we rename the ocean. The climate of the planet stretches twenty-five miles, from the depths of the Mariana Trench to the ozone layer, oceanographers tell us.

No wonder so much of my *Atlas of the World* is blue and blank. Map the ocean! You might as well map the clouds. Chase a castaway duck! You might as well chase the wind. Caught in a mesoscale eddy like a seagull in a hurricane, who knew where a castaway duck might go? Not Curtis Ebbesmeyer. Certainly not Charlie Moore or Chris Pallister. Not OSCURS. Not I.

ARROGANCE IS A BAD THING

The skies had cleared. The sun was nearing its meridian. Three gulls were gliding in our draft, occasionally climbing up or dropping down, but mostly just hovering in place, resting in the air, their wings locked. We were due east of Gooch's Beach, south of Nova Scotia, in the temperate climate of the Gulf Stream, no land in sight. The seas were, as the deck boss, Will Ostrom, put it, "flat-ass calm."

It was the second day of our voyage, and out on the *Knorr*'s fantail Ostrom was conducting a crash course in mooring deployment. A fantail is just what its name suggests. Instead of tapering aerodynamically, the *Knorr*'s tail fanned wide, wide but also low, so that the deck was just a yard or two above the water, so low that even in calm seas the wave crests poked above the bulwarks. A white A-frame crane straddled the stern. An A-frame crane is also what its name suggests—a crane shaped like the letter *A*, only, in this case, without the crossbar. Gazing between its steel legs, you could see the *Knorr*'s wake fading away toward a vanishing point on the horizon. Bolted to the deck of the fantail were wire cages full of "hard hats." In the language of oceanography—which pleasurably mixes the arcane, Latinate, acronym-laden jargon of science with the salty, poetic argot of sailors—"hard hats" are glass floats encased in yellow plastic shells, and from the outside they look very much like what their name suggests, two hard hats sealed together. Also bolted to the deck was an enormous diesel-electric winch, onto the drum of which Ostrom had already spooled approximately two miles of steel cable sheathed in black plastic.

Working with a skilled, trained team of specialists from the Woods Hole Mooring Operations Engineering and Field Support Group, Ostrom can deploy ten moorings in a single day, one after another, even in rough seas. But Bower couldn't afford to pay the usual mooring ops team to spend nine days in transit and one day at work, deploying a single mooring. She could barely afford Ostrom, who was therefore obliged to press-gang the available able-bodied members of our unusually small scientific crew. There were only three of us: me, a thirty-five-year-old writer with a bad back; Kate Fraser, a matronly science teacher from the Perkins School for the Blind, in Watertown, Massachusetts, who was helping Bower develop oceanographic curricula for visually disabled teenagers; and Dave Sutherland, a tall, blond, twenty-eight-year-old doctoral candidate from North Carolina who'd volunteered to assist Bower, sans compensation. Sutherland had high cheekbones and close-set eyes that gave him a look of puzzlement, and he was kind and polite in a slightly sheepish, slightly innocent way. When I noted that he would soon be able to call himself Dr. Sutherland, he smiled, embarrassed, and said, "Yeah, I know. Pretty wild."

"I've been doing this for thirty-four years," Ostrom said to the three of us out on the fantail. "When I started, I was twenty-one years old, twenty years younger than most guys. I was this little guy." At fifty-five, he was still a little guy, littler even than Chris Pallister, with windblown white hair and a two-tone beard, white on his cheeks, black around his mouth. His wardrobe seemed mostly to consist of old blue jeans and old flannel shirts, the sleeves of which he wore rolled up. Deep crow's-feet had formed at the corners of his eyes, and black crescents of grease had collected—permanently, it seemed—under his fingernails. "The guys who taught me," he continued, "most of them were military. They'd been in World War II, Korea. Now I've got to train guys like you"—he shot a sardonic glance first at Sutherland and then at me—"so that (a) you won't kill me and (b) you won't lose the mooring."

Over dinner in the mess our first night at sea, at a Formica table, beneath a wall-mounted rack of condiments, Sutherland had revealed that he was engaged to a marine biologist. I'd offered my congratulations. Ostrom had offered his condolences, and advised the young scientist to start saving for alimony now—a lesson he'd learned the hard way, Ostrom said. He referred to his children as "brats." He warned the rest of us never to take solitary nocturnal walks on the fantail, then added, "*I will, because I don't care.*" He liked to act grumpy, even when he wasn't—mock grumpy. With Ishmael, he seemed to concur that going to sea was the best psychotherapy, the only substitute for pistol and ball. He went to sea every chance he got, and because his skills were both valuable and rare, he was offered many chances. Aboard research vessels, he'd traveled the world. He'd walked the frozen seas of the Arctic, gone marlin fishing in the Caribbean. More than once, he'd played golf at the old military base on Midway back before the base was closed and the island designated an albatross sanctuary. Golfing on Midway back then, you had to watch out for the albatross nests as well as the sand traps, Ostrom said. When condemned to spend time at home, he occupied himself by doing a side business, repairing the moorings of yachts.

The mooring of a yacht resembles the high-tech underwater weather vane that we were soon to deploy about as much as a yacht resembles the R/V *Knorr*. Bower's mooring wasn't only high-tech. Like John Swallow's neutrally buoyant floats, it was experimental. In the half century since Swallow set sail on the *Aries*, oceanographers had used one of two methods to study underwater storms. Some, following Swallow's lead, had sailed out, searched for a mesoscale eddy, and, if they were lucky enough to find one, tossed a float overboard. This is known as the Lagrangian method. Others had anchored data-collecting moorings in waters where eddies are known to propagate and waited for one to swirl past. This is known as the Eulerian method. Two years ago, over lunch in the WHOI cafeteria, Amy Bower had struck on a novel way to combine the two methods. "Suppose we were to launch a float from a mooring?" she'd wondered aloud.

As it so happened, one of Bower's colleagues at Woods Hole, a physical oceanographer named David Fratantoni, had recently invented a device that could do just that—a submerged autonomous launch platform, or SALP, Fratantoni called it. He'd tested his invention off the coast of Bermuda, but it was still a novelty. "That was the selling point of the proposal," Bower told me aboard the *Knorr*, "that we were trying to do something new. People don't seem to recognize the risk. I hope they continue not to recognize the risk."

At Woods Hole, I'd attended a lecture during which Bower had played a digital animation of the risky experiment she'd spent two years orchestrating. On the lecture hall screen, the mooring appeared, anchored to the floor of a transparent sea by a five-ton cylinder of solid iron, tugged toward but not quite to the surface by a yellow sphere, a subsurface buoy (avatar of the actual yellow sphere now lashed to the deck of the *Knorr*). Strung onto a half-inch-thick steel cable between anchor and sphere, like big, data-collecting beads, were assorted gadgets and instruments with magical names—eight Aanderaa current meters, nine Seabird SBE 37-SM MicroCATs (high-tech thermometers that also measure salinity), a pair of SALPs custom-built at Woods Hole by engineer Jim Valdes in a machine shop the size of an airplane hangar.

The actual SALPs, now lashed to the *Knorr's* fantail, were the size of elevators. Each one resembled the chamber of a colossal six-shooter, a six-shooter of the sort Jupiter might whip from his cosmic holster in a duel with Neptune. Into the twelve chambers of the two SALPs we were to load twelve profiling floats, nifty gizmos that show well just how far Lagrangian technology has come since John Swallow cobbled together his neutrally buoyant novelties out of scrap metal and navy-surplus electronics. Profiling floats resemble torpedoes but behave like hot-air balloons. Drifting upright, noses pointed to the sky, they can, by filling or emptying a hydraulic bladder, recalibrate their own buoyancy, sinking to the ocean floor or ascending to the surface, analyzing—or "profiling"—the water column as they go. When they surface they can beam data to shore via satellite and download into their circuitry any new directives their oceanographic masters choose to beam back.

Onto the lecture hall screen there appeared a lethargic tornado of blue water, a cartoon Irminger Ring. It caught the yellow sphere and, towing it along, made the mooring acutely lean, like a palm tree in a hurricane—or like, Bower suggested in pantomime, a red-and-white cane tilted by a blind oceanographer on the stage of a lecture hall. Pressure gauges on the SALPs registered this disturbance. Another sensor registered temperature anomalies indicative of Irminger Rings, which are one to two degrees warmer than the frigid water through which they swirl. At the behest of a "decision algorithm" programmed into its computerized brain, the uppermost SALP tripped a burn wire and, a moment later, spat out a profiling float painted the bright, innocent yellow of a rubber duck.

Down the yellow float swooped into the eddy's watery coils. There, for the next several months, it would stay, like a weather balloon launched into a hurricane's eye, traveling wherever the storm carried it, surfacing once every several days to beam its findings home. Bower's mooring would allow her to study Irminger Rings remotely for two years, even during the brutal Labrador winter, when sea spray will freeze on contact to a ship's bulwarks, shellacking them in ice. With the data her profiling floats collected, she'd be able to do for an Irminger Ring what I, so many months ago, had dreamed of doing for a castaway duck—tell its story, from birth to death.

Assuming, that is, that the experiment worked. To her office door in Woods Hole, alongside a Monet print (woman with parasol, in a field speckled with red poppies) and snapshots of her Guatemalan daughter, Bower had taped a slip of paper bearing a typewritten motto: "Theory is when you know everything but nothing works. Experiment is when everything works but you know nothing. Most of the time nothing works and no one knows why." Deck boss Will Ostrom, whose job was to make sure that everything worked and if it didn't that someone knew why, also had a motto: "Arrogance is a bad thing."

Our first morning at sea, in the main lab, using a diagram as a visual aid, Ostrom had explained how the mooring deployment should, in theory, go. As soon as the *Knorr* was "on station," we'd "do a little bathymetry." The mooring had been cut, precisely, for waters 3,200 meters deep. "Obviously we don't want to make this"—he'd jabbed a greasy fingertip at the yellow sphere on the diagram—"a hundred meters higher. It would be on the surface, right?" At the surface it could get fouled in the screw of a passing boat. "So we want it right here," he'd said, "ninety-five meters down." The anchor would go over last, the sphere first. Once the sphere was in the water someone with good penmanship would have to keep a meticulous log—"a diary of the balloon" Ostrom called it— because we would be measuring out the mooring in increments of time. "Okay, like a 480-meter shot of wire should take about twenty minutes," Ostrom said. "So what if instead it takes ten? That means it's too short." When we shackled an instrument onto the cable, someone would have to check the cotter pins. "The smallest detail, and this thing will fail, will come apart. I don't know what the bottom-line cost of this mooring is but—"

"A lot," Bower interjected. The bottom-line cost of the mooring (labor, equipment, ship time, R & D), I knew by then, was $1.2 million.

"Beaucoup, okay?" Ostrom said, pronouncing beaucoup *bookoo*. "I want to be right up front. If there is anything that you don't understand and you don't feel comfortable doing, just say so. But it is a problem if you are a person that says, 'I can do everything,' and does a really lousy job. Arrogance is a bad thing."

Now, out on the fantail, Ostrom played variations on this theme. "You can see scars on my eyes when I close them, right there," he said, exhibiting his eyelids. "I've been hit in the face with chains. Twice." That big yellow sphere sitting over there by the starboard rail weighed 2,600 pounds. If a line fouled on a rail as we sent the sphere overboard, it would rip the rail right out. "And that's bad," Ostrom said, "because now you've got to steer a 2,600-pound sphere that's loose and coming back at you." Once, on the University of Rhode Island's research vessel, an eye bolt in the deck had sheared off and gone flying into the bosun's leg, breaking it "like that," Ostrom said and snapped his fingers. Just the other day at Woods Hole, the counterweight on a core-sampling apparatus had come crashing onto the docks. "It was an act of God that it didn't kill someone," Ostrom said. "And it happened just like that." He snapped his fingers. Another time, on another voyage, on this very ship, right here, on the fantail, right after a pretty yellow sphere had gone overboard, the cable had caught a deckhand in the face, and flipped him. "Tore his mouth, all the skin on his lip," Ostrom said. "He was a real talker too, and couldn't talk anymore for the rest of the trip." Arrogance is a bad thing. Another time Ostrom was trying to free a fouled line with a six-foot crowbar when a "big, perky" seaman came running to his aid. Right then Ostrom had "popped" the line, which twanged like a giant bowstring into the big, perky seaman's arm— "fractured it in four places, just like that," Ostrom said and snapped his fingers. "It really screwed him up. I mean, we carted him off and shot him full of morphine, put him on another ship that night. The point

here is, no freelancing, okay?" The injury had screwed up the perky seaman so badly that he quit seafaring and became an English teacher.
This was Ostrom's cardinal rule: No freelancing. Do as you've been told
but nothing else. He had other rules: Don't get underneath a crane
load. Keep out of the bight. Wear a hard hat. If you're not working, stay
back.

For the next hour, like a director rehearsing a troupe of talentless actors, as we rocked about at the center of the circle described by the horizon, Ostrom put us through our paces, teaching us how to cleat and
"clear" a tag line—a rope that keeps hoisted objects from swinging
wildly about; how to tie a bowline and a clove hitch; how to coil a rope
properly, letting it drip from your fingers to the deck in a clockwise el-
lipse; how to operate the diesel-electric winch, and to follow commands
flashed in the semaphoric language of crane signals. "If you're at the
winch, I will point to you and then I'll point down. What does down
mean? Pay it out. Up is in. Down is out. And this means what?" He held
up a fist.

"Stop," Sutherland and I said, almost in unison.

Out here in the sunlit, seasickening arena of his life's work, Ostrom's
world-weary air seemed to fall away. You could tell that despite or probably because of the risks, he loved his work. Adored it. Although there
was no glory or fame in doing what he did (deck bosses never appear on
the cover of *Time* or even on the cover of *Oceanus*, WHOI's in-house
organ), Ostrom knew he was good at what he did, as good at it or better
than anyone alive. He knew moorings. Moorings were just about all he
knew well. A computer could autopilot a ship. A computerized glider or
float could explore the deep. But no computer knew what Ostrom knew.
He could cleat and clear a slip line single-handedly with the legerde-
main of a rodeo cowboy lassoing a post. In the small world of ocean science, he'd gained a certain notoriety. The bosun on a NOAA research
vessel once drew a caricature of Ostrom, depicting him as a little devil
in a red jumpsuit, pricking deckhands in the ass with a pitchfork. Ostrom loved this drawing. During the course of our voyage, David
Sutherland would name a photo of Ostrom that he uploaded onto the
Knorr's computer server "will_overseer."

NIGHT VISION

For the next three days, the seas are flat-ass calm. Ostrom and the engineer, Jim Valdes, unpack instruments from crates and begin hitching them together. Dave Sutherland sits at his laptop beneath the portholes in the main lab, working on his dissertation, a study of the East Greenland Coastal Current. Amy Bower and Kate Fraser post daily "audio postcards" on a website for the blind and take questions from Fraser's students by satellite phone:

How deep is the water?

The depth varies, but in the Labrador Sea, where we will deploy the mooring, the water is about two miles deep.

How cold is the air?

Fifty-five degrees Fahrenheit here in the moderating influence of the Gulf Stream, but soon, in the Labrador Sea, it will fall to just a few degrees above freezing.

How do you make the batteries in the profiling floats last two years?

Lithium.

Can you bring a seeing-eye dog onto a research vessel?

No, but Bower planned to adopt one shortly after returning to Woods Hole. From Fraser's students this news elicits raucous cheers.[32]

At sea, Bower has no need for a seeing-eye dog. She finds life on a ship easier than life at home, where her daughter strews toys in her path and her husband misplaces the remote control. On a ship, everything is where it's supposed to be. She navigates the corridors with her cane, detecting the thresholds of the watertight doors. She memorizes the number of stairs between the main deck and her cabin on the upper decks and counts them carefully off as she ascends and descends. We've all been assigned a labeled coffee mug—mine is labeled sci 6, since I'm the sixth and lowliest member of Bower's team—that we are responsible for washing; around hers, labeled sci 1, Bower has snapped a knotted rubber band so that on the wooden rack in the mess where the mugs are kept she can find it by touch.

When Bower first learned that she was losing her sight she felt foredoomed to darkness and to failure, to a life of dependence and disability checks, and in another century, she would have been. Even in another discipline she would have been. Marine biologists, for instance, have to see the specimens they dissect. Paradoxically, since physical oceanographers study invisible phenomena, since all physical oceanographers are in effect partially blind, Bower's blindness proved less disabling than she'd feared. When it comes to looking at the ocean, her vision is far superior to mine. The papers she's published are filled with colorful maps and charts that make watery winds and storms as visible as atmospheric ones. She has, with the help of her expensive instruments, acquired aqueous powers of perception, a sense of oxygen, a sense of isotopes, a sense of current, a sense of salt.

As we pass through the Strait of Belle Isle and emerge from the shelter of Newfoundland's lee, we're struck, and rocked, by northwesterly winds. The first mate pipes instructions from the *Knorr*'s bridge: we are to lash down or stow all belongings. Rough weather ahead. Rough weather and icebergs, forty-one of them, according to Canadian ice charts. Seated before an array of glowing screens in the main lab, on which various data appear, spectacles perched on the tip of his nose, the chief technician, Robbie Laird, says of the officers on the bridge, "They can see icebergs fine. The only thing we have to worry about are bergy bits." Silly as it sounds, "bergy bit" is the technical name for a little iceberg, a molehill rather than a mountain of ice. "A bergy bit probably wouldn't sink this ship," Laird says, "but it could do some damage and end this cruise."

On the bridge two ABs begin standing watch instead of one, scanning the seas with night-vision goggles. Night-vision goggles work best in total darkness. Second officer Mark Maloof passes through the main lab, extinguishing the portholes. The other night, in the mess, while I was trying my best to explain my weird quest to a table of mariners, Maloof said, "Kind of like *Moby-Dick*!" He, too, had fallen under Melville's spell. Now whenever we pass each other in the corridors or in the mess, he greets me with Ahab's famous question: "Hast seen the white whale?" To which I reply, "Hast seen the yellow duck?" He hasn't, but some of the other mariners on the *Knorr* have heard the tale. In fact, one afternoon on the bridge a balding piratical seaman named Kevin, who sports a hoop earring and a handlebar mustache, decided to share with me an incredible tale. He commenced to recount the legend of the rubber ducks lost at sea. Interrupting him, I brandished my yellow duck and explained why I was here. When Maloof has finished clamping shut the metal lids of the portholes in the main lab, he moves on to the mess. The lit windows of the *Knorr* wink out, one by one, and the ship moves stealthily through the dark.

Overnight, autumn turns to winter. The temperature drops from 55 degrees Fahrenheit to 41 degrees. In an icy rain we wallow through gray swells marbled with foam. Fearful that her experimental mooring might fail, Bower has made a change of plans. She's decided to keep one profiling float in reserve. Before or after we deploy the mooring, depending on the weather, we'll investigate Irminger Rings the old-fashioned way, the Swallow way—by hunting one down and hurling a float into it. Hunting for Irminger Rings, I'm pleased to hear, requires a good deal of fancy detective work. "If you just go out and poke around looking for an eddy, you're not going to find one," Bower says.

Beneath the shuttered portholes, she downloads onto her computer's magnifying nineteen-inch monitor images of red and orange blobs in a field of yellow. Nose almost touching the screen, pecking at her mouse, robotic software chattering away, she studies the images for clues. Generated from data beamed down twenty-four hours ago by Jason-1, the NASA satellite speeding overhead, the images on Bower's computer are altimetric maps of the Labrador Sea—maps that measure subtle variations in the ocean's topography, variations as small as three centimeters. The red blobs represent Irminger Rings, the circling currents of which raise a bump on the ocean's surface, a bump just fifteen centimeters taller than the surrounding water. Spread out across a ring's thirty-mile diameter, that fifteen-centimeter elevation is invisible to the naked eye, but not to the satellite-corrected eyes of Amy Bower.

Unfortunately, Bower's maps are already out of date, blurred by the passage of time. They show us where rings were yesterday, and suggest where we should look for them now, but not where we'll be sure to find them. The easiest way to confirm the presence of an underwater storm is to measure the speed of its watery winds with something called an acoustic Doppler current profiler, or ADCP. The *Knorr* has an ADCP, but it's on the fritz. We have no choice but to resort to more time-

consuming methods, taking the water's temperature not with a ther-
mometer but with yet another acronymic piece of electronic equipment,
an expendable bathythermograph, or XBT. An XBT looks, deceptively,
like a black poster tube.

At 59.5°N, 52.4°W, off Greenland's tip, Bower phones the bridge. The
helmsman powers down the engines. Chief Technician Laird pulls on
his parka, loads an XTB into an orange metal launcher shaped like a
spackle gun, and teeters out onto the fantail, trailing a cord behind him.
David Sutherland and I follow. At the stern, as the *Knorr* idles, Laird
leans over the rail and aims the XBT at the gray water. I lean over too.
Looking at the ocean from the deck of a ship, I've found, it's hard not to
think of drowning. Not just think of it, imagine it—the shock of the
cold you'd feel hitting the surface, the weight of your sodden clothes
pulling you down, the salty water rushing into your nostrils and lungs,
the darkness below, the diminishing light above. Laird pulls the trigger.
I expect some sort of charge to fire, the XBT to go whizzing harpoon-
like through the air. Instead it plops into the *Knorr*'s loud wake, a copper wire thinner than fishing line paying out behind it.

Along that wire travel data, into the orange launcher, through the
cord, back to a computer in the main lab, where we gather around a
monitor to watch. As the little expendable instrument sinks—fifty meters, one hundred, three hundred, a thousand—it draws two lines down
a graph, one for temperature, one for depth. Where it crosses the boundary between the cold, fresh surface waters to the warmer, saltier deep
water below, the temperature line abruptly jogs then resumes its linear
descent. By warmer I mean 3.5 degrees centigrade, just a degree or two
warmer than the surface waters above, into which thousands of icebergs
and floes have dissolved.

No eddy here. In the core of an eddy, the water temperature should
rise to a balmy 5 degrees centigrade. But the absence of an eddy is good
news: Bower didn't expect to find one here. She is collecting control
data, profiling the water's background temperature and depth. Meanwhile, the XBT keeps sinking—1,300 feet, 1,500, 1,800. Then the fila-
ment connecting it to the surface runs out and snaps. This explains the
X in XBT. Expendable bathythermographs can't be retrieved. Oblivi-
ously now, it continues to sink to the seafloor, connected to us only by

filaments of the imagination. An hour later we launch another XBT. This time I get to pull the trigger. Out drops the black tube. In comes the data. Still no eddy.

When oceanographers talk about climatological models, they themselves draw the visual analogy, comparing their computerized simulations to cameras, their satellites to eyes. The better the model, the higher "the resolution," they say. Although they are still the best crystal balls oceanographers have, the resolution of current computer models, Bower told me, is still "very low." She emphasized the word *very*. The only way to increase the resolution is to gather more data—data finer in scale, not megascale data but mesoscale or even microscale data, and if you were deciding where to begin collecting such data, the Labrador Sea would be a good place to start.

Along the Labrador Sea's western edge, the cold water of the Arctic flows south into the North Atlantic. Along its eastern edge the comparatively warm remnants of the Gulf Stream flow north toward the Arctic. At the heart of the Labrador Sea, chilled by Canadian winds, as temperatures drop and ice forms, making the water saltier, the converging remnants of the Gulf Stream grow dense enough to sink. Hundreds of meters down, they begin the long, slow journey back toward the equator. There are only two spots in the entire Northern Hemisphere where this sort of sinking occurs, here in the Labrador Sea, and just east of Greenland. "Circulation pumps," such spots have been called. The pumping isn't constant. It happens only in winter, and some years the pumps are more active than others. But for at least ten thousand years, since the end of the last ice age, the intermittent pumps have kept the North Atlantic surface currents flowing to subarctic latitudes, delivering a trillion kilowatts of heat or more (scientists have yet to calculate the precise figure) to the Northern Hemisphere, heat that munificent breezes and winds carry east around the globe.

That, at least, is how the low-res megascale climatological model of the Labrador Sea works, and oceanographers have understood the low-res megascale model ever since Henry Stommel puzzled it out on a notepad a half century ago. The mesoscale model of the Labrador Sea, however, is still blurry. Oceanographers like Bower are now trying to bring it into focus, trying to do what Emerson required of the poet—"magnify the small," "micrify the great."

DIARY OF THE BALLOON

Amy Bower has received some very exciting news. The latest altimetric map shows an Irminger Ring right near the mooring site. The news from the bridge is also exciting, but in a bad way: rapidly, sickeningly, sea conditions have continued to deteriorate. Low clouds of drizzling rain have shut down around us. The waters above and the waters below seem one. The ocean-atmosphere interface has vanished in the mist. It's doubtful that any fishing boats are still out. We haven't seen one in days. To be safe, every two minutes, the *Knorr*'s foghorn booms out its two-toned song. Big waves come rolling in from the east, hitting us broad-side. The portholes in the main lab seem to fill and empty with every roll: nothing but water, then nothing but fog. Seasick, Kate Fraser retreats to her cabin belowdecks. Bower remains at her computer. A whiteboard keeps leaning away from the wall, then smacking back into place. A yellow hard hat—an actual hard hat, not a float—zigs and zags across a workbench. Tools clank around inside Willy Ostrom's tool chest. Walking from starboard to port feels like walking on a seesaw. You're leaning into a hill one moment, then running down it the next.

By 2100 hours—nine o'clock—we will reach our destination, just west of Greenland's continental shelf, and therefore have a decision to make: deploy the mooring tonight, in rough seas and darkness, or wait for daylight. "The weather is the driving force," Ostrom says. "If the weather's going to blow up before tomorrow morning then it makes sense to do it now."

For the first time since we left Woods Hole six days ago, the *Knorr*'s captain, Kent Sheasley, appears in the main lab, along with first mate Mark Maloof. Sheasley, age thirty-six, is young for a captain. A man of few words and few displays of emotion, he dresses casually, in tattered khakis and an old rugby shirt. Maloof spreads out the latest weather maps on a workbench and we gather around. There's a low-pressure system in the area. The winds right now are blowing 31.5 knots out of the north—out of, that is, the Arctic. Through the portholes we can hear them scream. By tomorrow morning, according to the forecasts, the winds will be forty knots—gale force. "At least we know there'll be no terrorists trying to come aboard," Ostrom cracks.

Since the *Knorr* has to make Nuuk, the capital of Greenland, in just three days, there's no time for us to heave to and wait out the storm. This is Sheasley's recommendation: we deploy tonight, as soon as we reach the mooring site.

"You okay with that?" Bower, determined but worried, and looking at him askance, maximizing her vision, asks Ostrom.

"Look, it's crappy out there," Ostrom says. "But this is what you're going to see for the next three days. The sooner we do this the better. This sea state's just going to keep filling over the next five hours. I mean it's not going down."

Later, up on the bridge, looking out into the rainy mist, Captain Sheasley says to his third mate, "Once we're on station, we've got to check the motion. These guys are going to be out there setting things off the stern. We don't want these guys out there pitching, getting stern slapped."

On the fantail, Ostrom, Valdes, and Sutherland, all bundled up in parkas, start loading the yellow floats into the SALPs. Ostrom has on a red, fur-lined hunter's cap. "Fascinating isn't it," he says of the SALPs in a tone of utter boredom, his cap's earflaps flapping in the wind. "Ninety percent of science is appearance. Doesn't work, but hey, it looks cool. No data, but that sure was cool."

In the main lab, Bower posts an audio postcard on the website for the visually disabled: "Two hours from the mooring site," her podcast begins. "We're in a race with the weather."

At 2200 hours, up on the bridge, Captain Sheasley instructed the helmsman to idle the engines and turn on the exterior lights, which gave the aft deck a theatrical brightness. Beneath the A-frame crane there gaped an opening in the bulwarks. Until an hour ago, a safety chain had stretched across it. Now the chain had been undone, and there was nothing between the deck and the sea; walk through the A-frame portal of the crane, step into the yawing darkness, and you'd fall into water two miles deep. Every so often a wave came crashing up over the stern, glazing the deck with a slippery veneer that went rippling out the scuppers. This was the stage, the turbulent precipice, on which the night's action would play out.

It was raining, not heavily, but painfully—a cold, needling rain. The deckhands all wore rubber boots with steel toes, and hard hats and orange, insulated waterproof jumpsuits decorated with hi-viz reflective tape. I wore a hard hat too, and waterproof pants that I'd purchased from an army surplus store on Cape Cod. I had on my Sitka sneakers, and my yellow PVC slicker over another raincoat made out of recycled plastic bottles, and under that, an insulated coat, and over the slicker a life vest, and on my hands, gloves, and under the hard hat, a woolen one. The gloves, it was now painfully clear, weren't waterproof.

Bower stood by, in the lee of the main lab, helping Kate Fraser enter data into the mooring log—"the diary of the balloon." Will Ostrom was in charge now. Even Captain Sheasley was under Ostrom's command. In the staging area between the winch and the stern, Ostrom transformed into an angry dynamo of action, darting about the heaving deck with the agility of a gymnast, barking out orders and flashing crane signals to the deckhands manning the ropes and rigging and heavy machinery with which, expensive instrument by expensive instrument, meter by meter, the mooring would be assembled and unspooled: "Tie that down!" Ostrom shouted at Dave Sutherland, who was manning a slip line. "Hold it fast! Clear it! I said clear it!" He pointed at me: "You! On the winch!" I climbed up to the control panel and watched Ostrom through a grid of steel, there to protect me from the backlash of a snapping cable. "Pay it out! Pay it out!" Ostrom boomed, pointing down. With cold fingers I pushed the little toggle forward, and with a diesel-electric moan the wire on the drum began to unspool. "Stop!" Ostrom commanded, flashing a fist, but I was slow to react. "For fuck's sake, I said stop!"

Over the next several hours, I would come to appreciate that cartoon depicting Ostrom as a devil in a jumpsuit, and why it was Sutherland had called that photo "will_overseer." There was, nevertheless, something beautiful about Ostrom's grace and expertise. He seemed like a conductor with his symphony in full swell, or a chef in his kitchen, or a quarterback in motion. He was an adept of improvisational physics, judging by sight the weight of an object that had been hoisted into the air, anticipating its swing.

We began with the great yellow sphere that sat on its trellis by the starboard rail. It measured sixty-four inches across. Its buoyancy would exert 2,832 pounds of upward pressure. Atop it, at its north pole, was a satellite transponder and a beacon. Sutherland climbed up on the trellis and switched the transponder and beacon on. The latter began to flash, once every ten seconds. Deckhands undid the sphere's lashing. From the winch, the cable now ran up through the A-frame's block and then out around the starboard rail, where the bosun cotter-pinned it to a grommet at the sphere's south pole. Sutherland, obviously better at this job than me, took over on the winch. My new job was to keep the line from fouling on the rail. As I leaned against the cold steel, and stretched my arms out over the water in a seemingly beseeching gesture, cable slack across my numb and pruning palms, the waves heaving past seemed close enough to touch. They had been gray all day long, but here, up close, in the glare of the deck lights, they were turquoise and crystalline, shot through with light, and tempting somehow   so close, so beautiful, so cold, so deep. Out beyond the edge of the light, a glaucous gull floated contentedly on a swell, a white dot of sentience in the icy dark.

Now a deck crane lifted the yellow sphere into the air, dangling it like a colossal Christmas tree ornament or the yo-yo of a god, then swung it over the side, and let it fall. The splash was tremendous, a great diadem of droplets bursting from beneath it. The cable came alive in my hands, and I pulled it taut. It felt as though I'd hooked an enormous fish, an enormous, yellow, spherical fish. The sphere drifted out and then aft, and I drifted with it, following it to the stern, keeping the cable clear of the rail, until Ostrom gave the yell, "Let it go!" and I set the cable free. Away the sphere went over the waves, its beacon flashing, and my heart leapt with a curious exhilaration. Its yellow form was bright against the blue waves, which is why, of course, oceanographers paint their floats and buoys the color of a rubber duck, the color that in the gray-green welter of the sea is easiest to spot. It was the only bright thing out there, the only symmetrical thing out there, the only human thing—or at least the only human thing visible to the naked eye.

Measured in meters and minutes, the night slowly unspooled. My hands grew numb, the sphere's flashing beacon fainter and fainter as it receded behind the intervening waves, a distant star that you had to wait to catch a glimpse of. And then it was gone, and with it went the last remnants of my exhilaration. It was impossible to stay warm and dry. The cold rain and the cold wind sneaked in at my collar and cuffs. Besides the rain, there was the spindrift. Everything was wet, and blurry. Water blurred my vision, fatigue blurred my mind. I wished I'd taken a nap that afternoon.

A foot or two from the stern's edge, the men in jumpsuits wrestled instruments onto the black cable—the current meters, the thermometers, the SALPs loaded with yellow floats—and then coaxed them overboard one by one. It seemed a wonder that none of them tumbled after. Every so often, Ostrom would yell commands in my direction— "Donovan, man the air tugger!" "Donovan, up on the winch, give Dave a break!" "When I point down, you pay it out, goddammit!" These were the easy jobs, but in my numb, sleep-deprived inexperience, they seemed plenty hard, and grew harder as the night wore on. At the winch, I stood braced with worry, muttering to myself, "Down means out, down means out," trying my best to keep my mind from wandering when all it wanted to do was wander off to sleep.

Late—how late exactly I don't know, at three in the morning, or maybe four—Ostrom pointed at a rope and commanded me to take it over there, to the corner, behind the deck crane, and coil it up, neatly, the way he'd shown us that sunlit afternoon our second day at sea. The rope was a tangled mess, and that corner of the deck was slippery with grease. I hydroplaned around in my boots. Time seemed to slow, and the world to diminish, until there was nothing but me and that damn rope, my adversary, locked in slippery battle. It took me forever to untangle the thing. I kept falling down, stumbling into the crane, or into the starboard rail, where the waves no longer seemed tempting, but menacing. I was shivering uncontrollably by then, feeling the first symptoms of hypothermia. At last there lay before me on the greasy deck something resembling a coil—a mess, a sloppy pile of only vaguely concentric circles. Fearful of Ostrom's wrath, like some marlinspike Sisyphus, I started over again. My second attempt improved only slightly on my first. Good enough, I decided. But what was I supposed to do with the rope's tail? There was a trick Ostrom had showed us. You wrap it around the coil two times, or was it three? Then what? *Fuck it.* I gathered the sloppy coil into my arms, dropped it into the wooden chest where the ropes were kept, and deserted to the mess for a mug of hot coffee.

Unlike the rest of us, Amy Bower remained exhilarated throughout the long night. In the lee of the main lab, in turtleneck and parka, happily sipping hot cocoa from her mug marked with a knotted rubber band, she asked Fraser questions—"The second SALP is over now?"—and Fraser, as best she could, narrated the action playing out before Bower's blind eyes. I was standing beside her, sipping my coffee, when Bower made a chipper remark: "Look at that," she said. "It's first light!"

She was right. Without my noticing, the sky had gone from black to charcoal gray. A moment later you could begin to distinguish the grayness of the water from the grayness of the sky. I was cold as ever, tired as ever, but when Ostrom commanded me to take another turn at the winch, I didn't mind as much now that day had begun to dawn. When, a little before 0600 hours, the mooring's anchor tumbled overboard, there was no sense of finale, only relief. Ostrom continued to shout commands: coil that rope, pick up that wrench. At last, he dismissed us, and the moment he did he underwent a kind of metamorphosis. A devil no more, he became his former amiable if curmudgeonly self.

Out of gratitude and pride, I stayed up to help Bower enter data from the mooring log into her computer, reading it aloud. The winds were now gale force, just as the forecast had predicted. Under steam again, the *Knorr* was rolling more steeply than ever. When the portholes in the main lab filled, it took a surprisingly long while before they began to empty. Even there, inside, I couldn't stop shivering, and my head had begun to throb. Bower was chipper as ever.

"I think I need to sleep," I told her, and she apologized for keeping me.

For the first time on that rocky voyage, I stumbled to the head and vomited. Then, fully clothed, I crashed onto my bunk and slept fourteen hours straight. When I woke up, the storm was mostly over, and it was night again.

Our last day at sea, we found our elusive Irminger Ring, or so Jason-1 and an expendable bathythermograph alleged. The discovery was anticlimactic. The seas had calmed. And yet below us, if the data were to be believed, there raged a watery storm. We human beings are such visual creatures that for a semi–scientifically literate layperson like me, believing in invisible if observable phenomena—mesoscale eddies; rising CO_2 levels measured in parts per billion; rising sea levels measured in millimeters; electrons, dark matter, quarks; the waves through which cell phones and satellites communicate—requires a leap of faith, or at least a leap of trust.

Out there on the fantail of the *Knorr*, on the last day of our voyage, in seas far less stormy than they'd been the day before, trying in vain to perceive some trace or sign of the watery storm below, I couldn't help but feel a bit envious of the naturalists of centuries past, those scientific voyeurs who, with microscopes and telescopes, made discoveries everywhere they looked, perceiving ecosystems in drops of water, cosmologies in the dying rays of intergalactic light. Several years ago, reading Darwin's *Journal of Researches,* I was struck by how anachronistic—how innocent, even—the god-toppling biologist's exuberant curiosity seemed. His journal entries were rhapsodies of descriptive prose. Forms of the words "interesting" and "surprising" toll among his sentences like a refrain of wonderment. "I was much surprised to find particles of stone above the thousandth of an inch square, mixed with finer matter," he writes of a handful of dust.

The relationship between seeing and knowing helps explain, I think, why it is I was compelled to chase the yellow ducks lost at sea. In following their trail I've strived to raise, if only by a megapixel or two, the resolution of my own mental model of the world. Of course, as most oceanographers will tell you, a rubber duck isn't a very sensitive instrument. You can't follow it by satellite. It can collect PCBs and other POPs, which is of some scientific use, but it can't measure salinity, or levels of dissolved pollutants like CO_2 and mercury, or take the water's temperature. "In fact, this data is not very good," a Woods Hole oceanographer named John Toole told me, a bit apologetically, when I described for him Ebbesmeyer's accidental flotsam studies. Bower's dozen yellow floats would tell us far more about the ocean than a million castaway toys ever could, and her floats, as I learned from John Toole, are just a small part of a global fleet. Since the year 2000 oceanographers have seeded the oceans of the world with more than three thousand of these underwater robots. You can follow their peregrinations online. If the Evergreen *Ever Laurel* had spilled a shipment of profiling floats, we'd know their fates.

Profiling floats can't descend below two thousand meters, however. And unlike polyethylene ducks, they can't ascend into Arctic latitudes, where sea ice makes it impossible for them to communicate via satellite. North of Canada and Siberia are what oceanographers call the Canadian Hole and the Russian Hole. These holes aren't holes in the ice, or holes in the seafloor. They are giant holes in the climatological record. The abyss and the poles remain the last redoubts of oceanic darkness. The data that profiling floats beam home won't banish darkness from the deep once and for all, or resolve the ultimate mysteries of the sea. But slowly, over the next several decades, they may, one hopes, shed a little more light on the ocean's fourth and darkest dimension.

With a big heave-ho, Ostrom and Valdes hurled Bower's last float into the *Knorr*'s wake, where it swirled about in the roiling foam, then righted itself and went bobbing away. In a week or two it would, as Bower liked to say, "phone home." Months later, I would visit Bower at her office on Cape Cod, and she would pull onto her big computer screen maps on which her floats— all but one successfully launched from her experimental mooring—had traced their wayward routes.

Our last night at sea, shivering on the *Knorr*'s bow, Sutherland, Ostrom, Maloof, and I watched the northern lights flicker green and psychedelic across the sky. "Aliens in September," Maloof said, and Ostrom said, "It's Elvis. I'm telling you that's where he went. He's with us."

The *Knorr* docked in Nuuk a little after dawn. Waking in my cabin belowdecks, I sensed a strange stillness. We weren't rolling, or pitching, or yawing or indulging in any of the six degrees of freedom. The *Knorr*'s engines had fallen silent. And when I ascended to the main lab and looked out the portholes, I'd seen an astonishment—dark brown mountains, frosted with snow: Greenland! Greenland, that white, icy island, huge as a continent, misnamed by that real estate developer Erik the Red. Greenland, which on world maps resembles a wordless thought bubble floating up from the coast of Labrador, as if Canada's mind had gone blank. Bower and I and most of the scientific team spent the day wandering around Nuuk, sampling the reindeer soup, admiring a pair of bergy bits stranded in a fjord. With a startling crack, loud as a rifle report, one of them split in two. The two halves rolled over. Slush sizzled into the water. The next day, on the first of three connecting flights, we flew over the Greenland ice cap, that white Sahara. The flight lasted an hour, and for all but a few minutes of it there was nothing to see from my window seat but ice—no sign of life, no trace of color except the occasional pool of blue melt.

THE LAST CHASE,
PART TWO

*It's a Hyperborean winter scene.—It's the breaking-up of
the ice-bound stream of Time.*

—Herman Melville, *Moby-Dick*

NORTHWEST PASSAGE

Three years after setting out on the trail of the toys, a few weeks before reaching its end, I find myself shoeless and prone on the red helipad of a Canadian Coast Guard icebreaker, trying to wriggle my feet into the rubber booties of a yellow survival suit. It's a sunny afternoon in early July, and the icebreaker is tied up at the docks in Dartmouth, Nova Scotia, across the harbor from Halifax. Spread out beneath me, the survival suit looks like the neoprene hide of a yellow giant. It's lined with buoyant foam and some sort of silvery, space-age fabric—quilted titanium, perhaps. Also on the helipad struggling to wriggle into survival suits, with varying degrees of success, are thirteen scientists, a few dozen Canadian coasties, the three members of a television news team from Australia, a Swiss reporter for a German wire service, and a Canadian photojournalist whom the Australians have taken to calling "the Snapper." Tomorrow, aboard this big red icebreaker, the CCGS *Louis S. St-Laurent*, or the *Louis* for short, we will embark on a grand, Arctic expedition.

As the *Louis* makes its annual run through the Northwest Passage, the scientists aboard will search for clues to the planet's history and future—clues written in sediment cores pulled from the seafloor, in water samples collected in a conductivity-temperature-depth rosette, in the migrations of copepods and currents and pollutants. The Australians will shoot footage. The Swiss reporter will file dispatches. The Snapper will snap money shots of polar bears. And I, at long last, will chase the toys into, and I hope out of, the labyrinth of ice.

Or into and out of whatever's left of it. Last year's summertime melt broke all records, exceeding the fears of even the gloomiest climatologists. In satellite photos taken on September 16, 2007, the day before I boarded the *Knorr*, the polar ice cap doesn't look like a cap at all. It looks like the white half of a yin-yang symbol, a swirl of ice opposed by a swirl of open water. The open water stretched from the coast of Siberia to within three hundred miles of the North Pole. No icebreaker in history has attempted what the officers of the *Louis* are attempting this summer—to transit the Northwest Passage in early July. In years past, the sclerotic, ice-clogged channels of the Canadian Arctic Archipelago remained impassable until late July or early August. Not even icebreakers could smash their way through. But at the height of last year's summertime melt, the European Space Agency announced that for the first time since satellites began monitoring the Arctic, the Northwest Passage was almost totally ice-free. Six yachts had sailed right through it, completing in a matter of weeks a voyage that a hundred years ago had taken Roald Amundsen three years. Canada's Ice Service is forecasting that this summer's big melt, like last summer's, will begin ahead of schedule—two to three weeks ahead of schedule, to be precise, and so the *Louis* will begin its annual Arctic voyage ahead of schedule too. "It's going to be hot up there," I overheard a deckhand say this morning. "Last time we were up there it was 24 degrees colder here in Halifax than it was up there"—24 degrees centigrade, that is.

The air up there may be warm; the water, I've been assured, will be plenty cold—below zero in some places, since the freezing point of seawater is two degrees lower than that of fresh. Hence the survival suits. Supervising the safety drill is First Officer Cathy Lacombe. A French Canadian, when speaking English, Lacombe tends to aspirate words beginning in vowels. "In Harctic water you're not going to last two minutes," she informed us when we mustered on the hangar deck. "It's going to cut your hair and you are going to die." She held up her own survival suit. "In this you can last four howers."

Listening closely to Lacombe's speech was a safety inspector dispatched on short notice. "Whatever you do," he added when she'd finished, "if you're wearing a hooded sweatshirt, get the damn thing off, because that hood is like a wick. It will suck the cold water in, and, again, *you'll die.*"

These aren't the only warnings we've received today. We've been told not to visit the crow's nest without permission because in rough seas we could tumble to our deaths. We've been told not to squeeze, Indiana Jones–style, through a watertight door that has begun to close; do so and the watertight door could amputate a limb, or slice you in half.

THE LAST CHASE, PART TWO

Once a watertight door has begun to close there's no shutting it off. We've been told that "it's not a good idea to get drunk on a ship."

Even if you're sober, even if the sun is shining and you're on the helipad of an icebreaker still tied up at the docks and yachts are sailing picturesquely around on Halifax Harbor, putting on a survival suit, I now know, can reduce you to a toddlerlike state of haplessness. While I tug at my suit's recalcitrant zipper with fingers gloved in spongy rubber, across the helipad, trying to extrude his right hand through the elastic cuff of a watertight sleeve, Gerd Braune, the Swiss reporter for the German wire service—balding, bespectacled, mustachioed, looking clerical—is muttering under his breath what I assume are German profanities. Other people are stumbling about like clumsy contortionists, tipsy Houdinis, yellow sleeves flapping. In the manner of a preschool teacher readying her wards for an outing in the snow, Lacombe assists us one by one, the Coast Guard inspector following after her, scribbling observations on a clipboard and checking his watch.

Lacombe, though she has the high voice and haircut of a choirboy, is a big woman, almost as tall as I am, and considerably wider and stronger. When she reaches me, she yanks my suit's zipper to my chin, pulls the yellow hood over my head, and clasps the Velcro face mask across my mouth like a gag. Only my eyes and nose are left exposed to the warm, midsummer Nova Scotian afternoon. I feel like a cosmonaut, or astronaut, or aquanaut—some sort of naut. A cryonaut, I suppose you'd call someone equipped for a dip in the ice. Mummified in yellow neoprene, poaching in my own body heat, waiting to be inspected by the inspector with the clipboard, I find myself thinking that, overboard in a survival suit, if one wished, one could reenact the Arctic journey of the toys, or at least four hours of it. "When seamen fall overboard" in Arctic waters, Melville writes, "they are sometimes found, months afterwards, perpendicularly frozen into the hearts of fields of ice, as a fly is found glued in amber."

The first leg of our voyage will take us to Resolute (population 229). There, the Snapper, the Australians, and Gerd Braune will all disembark, and several new supernumeraries will come aboard. Among them will be this expedition's architect, a visionary oceanographer named

Eddy Carmack. At Woods Hole, whenever I asked about the mystery I
was trying to solve, those who didn't summarily dismiss the journey of
the toys as "folk science" and me as a fool in possession of an errand,
John Toole among them, almost always said the same thing: the person
you should really talk to is Eddy Carmack. This turns out to have been
good advice.

Carmack, I was pleased to discover, is a believer in driftology. Last
January, the day after the Hanjin *Ottawa* tied up in Seattle, I paid a visit
to his office at the Institute of Ocean Sciences in Sidney, British Columbia. A fan of *Beachcombers' Alert!* Carmack had known Ebbesmeyer since
his graduate school days at the University of Washington, where the
young Dr. E. was Carmack's teaching assistant in an advanced physical
oceanography course. "I just find it fascinating," he said of Ebbesmeyer's
flotsam studies. "High-tech aside, this is telling us stuff million-dollar
instruments can't tell us—where stuff really goes."

Although Carmack has spent much of his career studying the ocean
the high-tech way, with conductivity-temperature-depth rosettes, geo-
chemical tracers, and expendable bathythermographs, he's also, every
year since the year 2000, studied it the old-fashioned way, by setting
bottles adrift. The Drift Bottle Project, he calls this ongoing Lagrangian experiment. With the help of sailors and scientists and Nova Scotian
schoolchildren, he's scattered more than four thousand bottles in icy
water. Replies have come in from Alaska and Nunavut but also from exotic destinations—Russia, Brazil, England, France, Norway. "Norway,
the Faroes, Orkney—those places are bottle magnets for bottles dropped
in the eastern Arctic or the Irminger Sea," Carmack told me.

It's thanks to Carmack and to the Drift Bottle Project that I've been
offered a cabin aboard the *Louis*. Officially speaking, I'm not a member
of the press corps. Officially speaking I'm an unpaid research assistant,
a volunteer bottle tosser. We've all been given a list titled "Crew on
Board" and beside my name appear the words "Scientific Staff," words
that I would like to photocopy and send—triumphantly or perhaps vin-
dictively or perhaps, come to think of it, pathetically—to my eleventh-
grade chemistry teacher, Ms. H——, who snuffed out, as if they were
the blue flames of so many Bunsen burners, the fanciful, marine biological dreams I'd once entertained. In fairness to her, my idea of a marine
biologist was a romantic one, influenced far more by Doc from John

Steinbeck's *Cannery Row* than by the Krebs cycle or the table of the elements or the melting point of magnesium. As it so happens, Doc, whom Steinbeck based on his friend Ed Ricketts—marine biologist, pioneering ecologist, founding father of fish-boat science—also happens to be one of Eddy Carmack's heroes.

At three points along our route (I've been furnished with a map, one far more detailed than the one Curtis Ebbesmeyer gave me), I am to fetch from the *Louis*'s wet lab one of three cardboard boxes. Inside each of these boxes are forty-eight corked beer bottles, all carefully numbered. Inside each bottle is a form letter addressed to beachcombers. "This bottle you have just found was dropped from the CCGS *Louis S. St-Laurent*, a Canadian Coast Guard icebreaker that travels the Arctic Ocean from Dartmouth, Nova Scotia," the letter reads. "We have received reports from all around the world." This year a freckle-faced, frizzy-haired fifteen-year-old oceanography enthusiast named Bonita LeBlanc is assisting Carmack with the project. She collected the bottles from a Canadian brewer. She spent weeks corking and sealing them. Before sealing them she recruited Nova Scotian schoolchildren to illuminate the letters in crayon and, if they wished, to add their own messages—messages like, *"Bonjour, je m'appelle Mélanie Maillet. J'ai 12 ans et je joue au flûte, au piano, et le violon."* (Personal touches, especially those of schoolchildren, increase the response rate, Carmack has found.) Unfortunately for Bonita LeBlanc, minors aren't permitted to travel aboard Coast Guard ships—unfortunate for her, good for me. I'm her substitute, her proxy. From the stern of the *Louis*, I am to lob the bottles overboard and record assorted data—latitude, longitude, bottle numbers, time—in a logbook. It should be an easy job, but after my lackluster performance on the *Knorr*, I'm already anticipating ways in which I might mess it up—sleep through my alarm, send a box of bottles tumbling down a flight of slippery stairs or a bottle hurtling into a bulwark.

In return for my bottle-tossing services, I get to remain aboard the *Louis* for the voyage's second leg, which will take us from Resolute even deeper into the labyrinth, to the Canadian Arctic's sanctum sanctorum—a town called Cambridge Bay. A few weeks ago, late one night in Manhattan, I looked up Cambridge Bay in my *Atlas of the World*. There it was, 2,000 miles due north of Denver, Colorado, 1,444 nautical miles

south of the North Pole, 3,139 nautical miles from the site of the toy
spill, a little dot on the south coast of Victoria Island, at the very heart
of the Northwest Passage.

Much as, seated in a taxiing jumbo jet, you can sense the precise moment when the wheels lift from the tarmac and the big steel machine
takes improbably to the air, so too you can sense, standing on the
bridge of an icebreaker—or for that matter at the taffrail of a ferry, or
on the deck of a research vessel, or in the copilot's seat of a homebuilt
cabin cruiser—that moment when, loosed from the bollards, a boat or
ship goes adrift. Even in calm seas, you can feel it, the sensation of
float.

Under hazy skies—the provisions all stowed, the safety drills all
conducted, the gangway raised—the CCGS *Louis S. St-Laurent* de-
parts. Up on the bridge, officers stand before levers and buttons and instruments and computer screens. The quartermaster sits at the helm,
beside the gyro, a compass encased in a glass sphere. I stand before the
windows, a paleochemist named Robie Macdonald to my left, and to
my right, Captain Marc Rothwell, looking appropriately nautical in a
navy blue sweater with gold epaulets. Neale Maude, the Australian
cameraman, is shooting footage, panning this way and that. Over
Maude's shoulder, his soundman, Daniel O'Connor, a lanky guy with a
splendidly Victorian mustache, is holding up what looks like a feather
duster on a stick, a gray, furry microphone.

Captain Rothwell radios the engine room and commands the engineers to test the bubblers. The bubblers are nozzles spaced along one-
third of the *Louis*'s hull. Icebreakers were invented about 150 years ago
but only perfected in the last century. Mariners aboard the earliest models discovered that broken ice adheres to the cold steel of a ship like glue.
Bubblers, Captain Rothwell tells me, "put a little air under the ice." The
air, like grease, makes the ice slip past.

Over the radio, from the engine room, four decks down, crackles the
voice of an engineer: "Bubbles, roger, bubbles." A moment later, the
bubblers fire up, screaming. From the bridge they sound like a choir of
damned souls.

"And that's the end of the first period," one mariner says.

"The Canadiens are winning!" another adds. (The Canadiens, evidently, are an ice-hockey team.)

The *Louis*'s helicopter, parked until now onshore, comes flying out, and Neil Maude swivels in its direction. The pilot, Chris Swannell, circles the ship then nimbly sets his pretty red mechanical dragonfly onto the helipad. Once word arrives from the hangar deck that the helicopter has been safely stowed inside its hangar, away we steam. The industrial districts of Halifax pass by.

By nightfall, we are out on the high Atlantic, making sixteen knots, and by first light, we've reached the subpolar front. There, where the warm water of the Gulf Stream meets the cold air flowing out of the Arctic, thick fog banks form. It's like steaming through a cloud. Our speed drops to 13.9 knots. Visibility drops almost to zero. Every three minutes, the foghorn of the *Louis* bellows out its two-tone song.

Late the following morning, the fog clears and there, to port, is the pale, mostly barren Labrador coast. Up on the bridge, I talk to Dave Fifield, the expedition's ornithologist, who is conducting a census of seabirds, one long-term aim of which is to assess the impact of global warming on populations and migration. To take a census of seabirds, what you do is station yourself in front of the thick windows, a pair of binoculars at hand, a pair of hiking boots on your feet, a pair of jeans on your legs, a pair of glasses on your face, a pair of ornithological field guides perched on the windowsill beside you, and spend all your waking hours gazing out at the ocean. And most of the time all you see are the same gray waves. The same encircling horizon. The same wedge of bow, pitching up and down, ever onward. Then, every so often, you spy a seabird and scribble something down in your logbook.

Considering myself an amateur bird-watcher of sorts, having gone bird-watching in the annals of folklore and myth, I borrow the second mate's binoculars, stored in a little box under the window, and give it a try. In the Strait of Belle Isle, with both Newfoundland and Labrador close by, seabirds are not uncommon. Before long, one appears to starboard, swooping down, skimming the waves, foraging for plankton and squid. Fifield's trained eyes spot it first. It has a snow-white underbelly, a dove-gray back. If forced to guess, I'd call it a gull. "Fulmar," Fifield says, and notes it in his log. A northern fulmar, to be exact.[33]

"The thing to distinguish fulmars from gulls is the wing stroke," he explains. "The gull will flap its wings slowly and continuously. A fulmar will hold its wings straight out and go *flap, flap, flap—glide; flap, flap, flap—glide*." As he speaks, the fulmar demonstrates the accuracy of this description. There it goes, three quarter notes of flapping and then, wings locked, a long, whole note of glide, the waves sparkling beneath it. Although I have no use for such ornithological information, I admire it. I'd like to be able to read all the world so closely; to distinguish fulmars from gulls not by their coloring but by the subtleties of their aerodynamical styles.

In addition to fulmars and gulls, we spy gannets, and black-legged kittiwakes, and storm petrels. But no ducks. No plastic ducks. No real ones either. Ducks do frequent Arctic and subarctic waters. In fact many of the earliest and bravest Arctic explorers were fortune-seeking prospectors who went in search not of silver and gold but of ducks, eider ducks. Fowlers, these feather hunters were called. Female eider ducks line their nests with breast feathers—known to those shopping for an excellent parka or an excellent quilt as eiderdown. The plumage of eider ducks, whether newborn or juvenile, male or female, is never yellow. The females are brown. The striking black-and-white patterning of the males makes them recognizable from afar. Like the northern fulmar, the eider duck nests on Arctic cliffs. Eider ducks are by nature docile and defenseless birds. Hence their preference for cliffs.

To harvest eiderdown, you had to rappel down on a braided seal-hide rope, coax the mother duck from her nest, and then, dangling hundreds of feet above the icy, rocky surf, plunder her feathers, pocketing a few of her pale green eggs for tomorrow's breakfast, being sure to leave at least one, so that she would pluck more feathers from her breast, which you could come back to harvest later. Then you'd wad the harvested feathers into balls and lob these down to a boat pitching around in the rocky shallows below. Fowling was a perilous and often fatal form of avian husbandry that speaks both to the magically insulating properties of eiderdown, properties the manufacturers of synthetic down have attempted to simulate, and to the desperation and courage of fowlers, of whom one nineteenth-century journalist wrote, "We who have been brought up in comparative ease and luxury can scarcely picture to ourselves a more wretched lot than that of these poor islanders, compelled to undergo such toils, and expose themselves to so great dangers, for acquiring the mere necessaries of life."

"Minke whale," Dave Fifield says, pointing to port. As soon as I spot the black fin, it disappears. Ten seconds later it surfaces again, this time directly ahead of us, at twelve o'clock. Then ten seconds later it's to starboard, at two o'clock, headed south. Then it's gone. Then we are emerging from the Strait of Belle Isle into the frigid waters of the Labrador Sea, where for all I know an Irminger Ring is churning secretly beneath us, a yellow float or a sun-bleached duck caught in its watery coils. The next day, after passing through a gale-force storm, we're farther north than I've ever been.

THE UNKNOWN NORTH

In 2008, a British cell phone company launched an ad campaign inspired by the legend of the rubber ducks lost at sea. There were print ads, billboards, television spots, all of which prominently featured ducks—yellow, of the classic sort, with clownish lips and cherubic cheeks and that innocent look of wonderment in their eyes—drifting merrily along through a digitally enhanced landscape of glaciers, icebergs, dark-blue water, sky-blue skies.

It's a compelling image, at once ironic and enchanting. After many lucubratory hours spent in spellbound contemplation, I am prepared to hazard an interpretation of the image's magic: The rubber duck is an icon of childhood, and of hygiene, and of domesticity, whereas the Arctic—or at least the Arctic of the mind, the Arctic as those of us who've spent our lives in temperate latitudes tend to imagine it—is everything that the rubber duck is not, the ultimate Ultima Thule, the most otherworldly place that is in fact of this world. The place where, before the space age began, virile men went to test their manhood, or, failing the test, to die.

From antiquity to modernity, what lay hidden within the ice was anybody's guess, and plenty of people guessed. Plato guessed that at the North Pole a tunnel conducted water to and from the earth's watery yolk—and as recently as 1909, his theory claimed adherents. The ancient Finns guessed that the Arctic was a parallel universe and that every one of us had a shadow self who dwelled beneath the ice. I like this theory and kind of wish it were true. More accurately, if less hauntingly, an ancient Finnish epic described the Arctic as the "dark Northland, the man-eating, the fellow-drowning place." In Greek myth, the Arctic was both the Cimmerian hell of perpetual darkness and the Hyperborean paradise of perpetual youth, a sunlit realm where, according to Pindar, dancing girls forever swirled around to "the lyre's loud chords and the cries of flutes." According to the biblical prophet Jeremiah, "Evil is brought from the north over all the inhabitants of the Earth," and in the last canto of *Inferno*, Dante's pilgrim visits a circle of hell where the damned are, as the historian Peter Davidson puts it in *The Idea of North*, "frozen into the ice of their own selfishness."

In short, the Arctic has always been, in Davidson's words, "a place of extremes and ambiguities." I think this is still true today, at least for most of us. Even in the narratives penned by explorers who actually traveled there and saw the place with their own eyes, you can detect traces of the old contradictory nightmares and dreams. The "unknown North," as explorers often call it, is at once a "desert of ice," a "howling waste," "a frozen hell" and a sort of heavenly wonderland—"majestic," "glittering," "sublime." The ice pack is "hydra-like" but the icebergs are "angelic" and the aurora borealis "celestial." Adventurers still go to the Arctic seeking thrills, and find them; tourists on icebreaking cruise ships equipped with saunas and movie theaters go there seeking wonder and beauty and strangeness; scientists, meanwhile, go there seeking data with which to reduce their uncertainties to certainties, and find signs of fragility—thawing permafrost, melting ice, native species losing ground to invasive ones.

Before we went to the moon, the Arctic was the moon. Before we went to the moon, when we went to the Arctic, hygiene and domesticity were the least of our concerns. Who ever heard of an Arctic explorer taking a bubble bath? Arctic explorers had more pressing things to worry about than the odor of their undergarments—the odor of their gangrenous extremities, for instance. The Arctic is the symbolic if not the geographic antipode of the bathtubs of America, and the yellowness of the duck the antithesis of the whiteness of the polar bear, which is to the Arctic what the rubber duck is to the bathtubs of America—totem, emblem, mascot.

Melville himself gave some thought to the whiteness of *Ursus maritimus*: "With reference to the Polar bear, it may possibly be urged . . . that it is not the whiteness, separately regarded, which heightens the intolerable hideousness of that brute; for, analyzed, that heightened hideousness, it might be said, only arises from the circumstance, that the irresponsible ferociousness of the creature stands invested in the fleece of celestial innocence and love; and hence, by bringing together two such opposite emotions in our minds, the Polar bear frightens us with so unnatural a contrast. But even assuming all this to be true; yet, were it not for the whiteness you would not have that intensified terror."

With reference to the rubber duck, the emotions are not opposite. The only unnatural contrast is between the birdlike, childlike form of the thing and the synthetic, petrochemistry of the thing. The yellowness of the duck reinforces the harmlessness and cuteness and happiness of the duck. In the symbolic Arctic, the imaginary Arctic, a yellow duck is an invasive species, a puerile, comical interloper in the whitest, most hostile wilderness of all.

THE REALM OF ICE

When we reach the Arctic Circle, 66.56083°N, First Officer Cathy Lacombe pipes a message over the PA system: all Arctic Circle initiates are to report, pronto, to the hangar deck. For two days now, the officers and crew of the *Louis* have been subjecting those of us who've spent our lives in temperate latitudes to an elaborate hazing ritual. We've been made to wear, on lanyards, red ID cards that read ARCTIC CIRCLE INITIATE. We've been made to carry raw eggs around, under strict orders to prevent their breakage; if the egg breaks, we've been given to understand, there will be consequences. I swaddled mine in tissue and packed it into my thermal coffee mug. We've been ordered to write poems about the Arctic.

Now, after we've mustered on the hangar deck, Cathy Lacombe conducts us initiates into a mechanic's workshop, a windowless room with a watertight door, from which, blindfolded, we're led out, one by one, to the starboard deck. The idea behind the ritual is to sustain the suspense, to make those of us left behind in the workshop tremble in fear of what

awaits us. It all seems a bit silly, but a young graduate student named
Marie-Ève Randlett is visibly in distress. And she's not alone. Mine is
among the last names to be called out.

Cathy Lacombe ties on the blindfold and leads me through the watertight door. On the starboard deck, I'm commanded by a male voice
I can't recognize to drop onto all fours and crawl. Obedient by nature, I
obey. As I crawl Canadian coasties pour buckets of icy seawater and
kitchen slop down on me, kitchen slop that smells—and tastes, I can
attest—like vomit. Then, suddenly, my blindfold is removed and I find
myself, still on all fours, at the feet of the boatswain and a steward, the
king and queen of the Arctic, so I'm told. The boatswain—Bosun Bob,
everyone calls him—has a Rasputin beard. He commands me to produce my poem. I unfold the crumpled paper from my soggy pocket and
begin to read: *It was kind of like forgetting / the way the earthly world diminished / in our wake, then disappeared. . . .* Before I've reached the
fourth line, Bosun Bob says, "Yeah, yeah, yeah, enough," and commands me to produce my egg. My thermal coffee mug, mercifully, has
remained hooked to a belt loop. I unscrew the lid and unswaddle the
fragile cargo, signed in felt pen by Captain Rothwell, then proffer it to
Bosun Bob, who proceeds to shatter it on my head. For some reason I
find myself thinking, yolk and egg white and kitchen slop dripping
down my cheeks, *I deserve this.* Then, wiping egg and slop from my hair,
I stumble belowdecks for a hot shower.

This is not what I imagined crossing the Arctic Circle would be like. I
imagined that crossing the Arctic Circle would be like waking from a
dream, one I'd been dreaming ever since the night I first traced the
route of the toys in my *Atlas of the World.* How astonishing it must have
been for those explorers and sealers and fowlers, let alone those first migrating Vikings or Inuit, who if they'd heard of icebergs at all had only
heard of them. Here I was on the *Louis,* showered, safe and warm, in
clothes—polyester underwear, a synthetic eiderdown jacket, an ecological windbreaker spun from old plastic bottles—that no longer smelled
of kitchen slop, and out there, in the distance, was an iceberg, so distant
that it shimmered on the overcast horizon very much like a pixelated
image on a screen. I felt none of the feelings of terror and awe that ice-

bergs elicited from my nineteenth-century predecessors. What I saw felt no more real, if anything less, than the icebergs I'd previously glimpsed on television and in magazines.

I wondered again about the limitations of sight. The wind stinging my ears and the cold, rusty rail under my hands, its paint chipping off, both felt more real than the white spectacle in the distance. So did the prospect of lunch, three decks down, in the main mess, a big room the size of a high school cafeteria, where there was a metal track that you slid your plastic tray along, and your tray would still be warm and steamy from the Hobart dishwasher, and you would look through a hygienic visor of glass at the smorgasbord laid out in steel trays, and to be heard over the engine's throb and the lunchtime chatter, you had to shout your order at a steward, who, after asking you to repeat your order more loudly, would scoop onto your steamed white plate mashed potatoes or beef stew, or buttered carrots. That felt real. Looking at the iceberg on the horizon felt like looking at the idea of an iceberg.

I was reminded of the afternoon the previous April when I took my son for a free ride on the Staten Island Ferry, which travels past the Statue of Liberty—not to it, but past it. Having lived more than a decade in New York, I'd never gone to see the famous bronze monument, and when I finally did, it looked small and unimpressive, the size of a souvenir replica. It was fun being out on the harbor, though. We'd passed close by a big green container ship along the hull of which, in white block letters, could be read the name of, I later learned, a Swedish shipping line, WALLENIUS WILHELMSEN. I told Bruno that I'd crossed the ocean on a ship like that. But what I remember most from that afternoon were the moments that you'd think would be forgettable. That the ferry windows were so scratched and dusty the sun made them glow. That I'd brought along a banana for Bruno. That he ate it while sitting in my lap. That when he was done I returned the peel to its plastic ziplock bag and tucked peel and bag into the outer pouch of my backpack next to the spare diaper. That it felt good to have my toddler son in my arms, on my lap. That on the return trip, from Staten Island to Manhattan, Bruno and I stood outside, up at the ferry's crowded bow, I holding him in my arms, the headwind in our faces, while seagulls coasted on an updraft. That he opened his mouth wide and shouted, "Papa, I'm eating the wind!" That I opened my mouth too. As we approached the docks, he said, of Manhattan, "The buildings are floating!" I saw what he meant. It was hard to tell whether the prow of the island was moving toward us or we toward it.

After disembarking, we'd walked home from the Battery, up the promenade, along the Hudson River, where only yachts and ferryboats plied the olive-green water. The surf of commerce no longer washed loudly over these shores, only the surf of leisure. This wasn't Melville's Manhattan. By Battery Park City we came upon a string of decorative riprap, granite boulders, sparkly with quartz, piled along the promenade's edge. For a half hour I let Bruno clamber up and over them, a towheaded colossus in a red hoodie striding over a miniature Himalayan range, a papery fringe of diaper sticking out from the waistband of his pants. I hung close, shadowing him, ready to catch him if he slipped. When he attained a summit he would hold his arms up and flash a big look-at-me grin, and when the descent was too steep I'd hoist him down.

It was getting late. To make it home by dinner, I ended up having to jog two miles, pushing his stroller along the bike path, ten-speeds whooshing by us, bells chiming, and our long shadows stretching out toward the West Side Highway. When I paused to catch my breath, Bruno would shout to me, "Faster!" and winded as I was, wearing the wrong shoes, I granted him his wish.

Such are the disappointments and the pleasures of travel, I thought now, as the *Louis* steamed north, stemming the Labrador Current.

Then, in Davis Strait, the icebergs doubled, tripled, multiplied. The first time we drew close to a big one, we all hurried out onto the decks with our cameras. Now I understood the rhapsodies I'd read in the journals of explorers. Now we'd attained the proper proximity, the proper scale. The iceberg was as close as the container ship Bruno and I had passed in New York Harbor, and as colossal. Its walls were palisades of ice. Its summit towered over the *Louis*'s bridge. You could imagine how small you'd feel walking around atop it. You could almost imagine what it would taste like. This was our first close encounter. There were more to come.

In Baffin Bay we entered an iceberg armada. They were everywhere— north, south, east, west, northwest, north by northwest. Supernumeraries and off-duty coasties alike spent most of the day out on the bow or up on the bridge, snapping photographs and exchanging exclamations of awe. Paleochemist Robie Macdonald, whose sense of humor tended toward the corny, told me that scientists had proposed an international standard unit to quantify the beauty of natural wonders such as these—a millihelen. Get it? I didn't. "One millihelen is equivalent to a face beautiful enough to launch one ship."

Not even those photographs taken by professionals like the Snapper, armed with zoom lenses the size of telescopes, do icebergs justice. Photographs fail to convey the grandeur of icebergs, but they also fail to convey how mutable they are. An iceberg that looks like a mesa in the distance as you approach transforms into something architectural, with melt-carved towers and wind-sculpted outcroppings suggestive of angels—as European explorers noted—or birds. Explorers, in their journals, grasping for comparisons with which to familiarize the strange, likened icebergs to cathedrals as well as angels. But icebergs lack the symmetries and patterns of a church. They exhibit form, but organic form, form sculpted by the subtle force of the coincident, form verging on the chaotic. Every change in angle is a revelation. The light drapes differently. The shapes shift. The colors turn from white to turquoise to blue. In some there were grottoes or canyons or isthmuses terminating in a peak that seemed about to break off. From the big ones, cataracts of meltwater gushed into the sea. It occurred to me, admiring those waterfalls, that before my eyes the past was dissolving into the present. Those melting molecules of H_2O now gushing from an iceberg, joining the currents of Baffin Bay, hundreds or even thousands or possibly tens of thousands of years ago fell as snow on the mountains of Ellesmere or Greenland. Were you to drink a glass of that meltwater, you might well be degusting the climate of the Iron Age. You would be quaffing centuries.[34]

I found myself wishing that Eddy Carmack, who knows this place as well as or better than anyone, were here to help me solve the riddles written in water and ice; to help me see the microscale, and the mesoscale, and the megascale. Dante had imagined the innermost circle of Hell as an Arctic landscape, and clearly this place wasn't as hellish as benighted Europeans formerly believed, but I did find myself thinking that, like Dante's pilgrim, I could use a guide, a Virgil, like those I'd met on previous trips—Ebbesmeyer, Pallister, Moore, Henry Tong, Willa France, Amy Bower.

In Carmack's absence I relied on those scientists who were aboard, especially Robie Macdonald, who taught me this: there's another way that icebergs are time capsules; frozen into the ice are sediments scoured from glacial moraines. As the icebergs melt, those sediments sink to the seafloor, where they accumulate in layers, mingling with sediment deposited by sea ice, or terrestrial silt delivered to the ocean by rivers, swept out by deep currents from the continental shelf. These layers of sediment are one reason why we're here.

Yesterday, in Davis Strait, the science began. I chucked my first batch of forty-eight bottles off the *Louis's* stern. Amidships, a team of oceanographers collected water samples, which they would test for temperature, salinity, and certain meaningful isotopes—chemical tracers or signatures that would reveal the provenance of the water and its age. But the main action was up on the bow where Macdonald and his Francophone partner, Charles Gobeil, were wrestling a stainless steel contraption over the starboard rail. This was a box corer, a primitive device that looked like something John Swallow might have soldered together out of scrap.

A cold wind was blowing, and the deck was slick with spray, and the bow was crowded with deckhands and scientists all dressed in orange foul-weather gear and blue or yellow hard hats. Neale Maude, shouldering his big camera, was angling for close-ups, and over Maude's shoulder protruded his soundman's fuzzy gray microphone. A crane hoisted the box corer over the starboard rail and dropped it, with a big splash, into the dark water. As it sank, a float on its cable turned from yellow to green and then, like a dimming lightbulb, went out. For a long time, the cable unspooled from a winch. More than 2,500 feet down, the contraption hit bottom, its boxy stainless steel maw sank one and a half feet into the primordial mud, the cable fell slack. Bosun Bob did a chin-up on the cable, checking its slackness. He flashed a crane signal. The winch operator began to reel the box corer back in; 2,500 feet down, a scoop clamped the box closed. Many minutes later, the yellow float appeared, turning from green to yellow as it surfaced, and the box corer followed it, streaming water and dripping mud.

"Bring it down!" shouted Bosun Bob, as the box corer spun around above us. "More slack, please. Down slow! Let it down easy! Hold it!

There." The crowd on the deck closed in, possessed of the same impulse to discover.

Macdonald and Gobeil pried loose the box of sediment and wrangled it to their prefabricated lab, a kind of steel hut resembling a shipping container, bolted beside the port-side rail. Therein, on a stainless steel table, a surgically meticulous excavation began. Gobeil had brought with him Marie-Ève Randlett and another graduate student, Danielle Dubien, both in their early twenties, both French-speaking, both ponytailed, both dressed in brand-new Day-Glo orange jumpsuits and brandnew boots. For the next few hours, with instruments resembling scalpels or palette knives, Randlett and Dubien scraped away layers of sediment centimeter by centimeter, archiving each layer in labeled ziplock bags. Every fifty centimeters represented about two hundred and fifty years.

Slice, slice, slice—there, into baggies, went the twentieth century. Slice, slice, slice—there went the nineteenth century. Way out in the Arctic Basin, farther still from the continental shelf, a single centimeter of mud constitutes a millennium's worth of sedimentation. Out in the Arctic Basin, a slice thin as a sheet of cardboard would transport you back to the Norman conquest. Two centimeters would transport you back to the birth of Christ. A third, to the Babylonian Empire. A core as deep as the one we'd collected from Davis Strait would transport you back to the Pleistocene.

Recorded in this mud were "biomarkers," most of which were compounds of carbon. Most evenings, I ended up talking to Macdonald, who was happy to tutor me in organic geochemistry. After dinner, scientists and sailors alike would gather in the *Louis*'s main lounge, a room as wide almost as the ship. The walls were paneled in wood. The big picture windows looked out over the bow, and over the bulwarks, toward the far horizon where icebergs drift. There was an upright piano in the main lounge, and a drum kit, and wooden card tables, and a bar wellstocked with spirits and beer. Although there was a two-drink limit, minutes after the bar opened, the place had the loud, convivial feel of a saloon. You almost forgot that you were on an icebreaker, steaming north into the Arctic. That it was nine o'clock and out on the decks the sun was so bright it looked like late afternoon.

"Where does the carbon come from? That's perhaps the most important question," Macdonald told me one evening, while swirling a thimble of Scotch. I was inclined to attribute his bardically long mane of white hair to a romantic disposition until he confessed that for weeks he'd been meaning to visit the barbershop. He had bags under his eyes, the left one baggier than the right, and wore on his wrist a chunky, black, heavily instrumented waterproof watch. When he wasn't out on the deck in his foul-weather gear, he walked around the ship in sandals and socks.

"We can look at stories in these sediments," he said, stories written in chemical elements. Manganese, iron, sulfur, phosphorous, and radionuclides all play supporting roles in these stories, but the protagonist is carbon. There are two kinds of carbon, organic and inorganic, Macdonald said. The ratio between them reveals a great deal about the ancient climate of the Arctic, and about how swiftly the Arctic has changed during the industrialized heyday of fossil fuels. Photosynthesizing flora, both terrestrial plants and phytoplankton, convert inorganic carbon—CO_2—into organic, carbon-based compounds, CHO, for instance. "Chloroplasts"— photosynthesizing cells—"are the little engines that run the biosphere," Macdonald said.

Petroleum consists mostly of organic carbon, as does a polyethylene duck, since polyethylene is a by-product of petroleum. A polyethylene duck is in fact made of carbon compounds assembled by plankton in primordial seas. When planktonic plants and animals died, they sedimented to the ocean floor, and were eventually subsumed into the earth's mantle. There, under great pressure and great heat, over the course of eons, the planet cooked it into that primordial ooze we call oil.

To turn oil into ethylene you need one of those industrial kitchens known as a refinery, as well as a chemical plant. Here's a recipe, albeit a useless one: In something called a "crude tower," through a process known as fractionation, skim away the ethane, the most ethereal part of crude oil, too light to use as gasoline. Ethane looks like this:

$$H \quad H$$
$$| \quad |$$
$$H - C - C - H$$
$$| \quad |$$
$$H \quad H$$

Pump your ethane to your chemical plant. Now it's time for steam cracking: Dilute your ethane with steam. Briefly heat it to over 850 degrees Celsius. Eventually you'll end up with an ethylene molecule that looks like this:

```
H   H
|   |
C = C
|   |
H   H
```

Take a bunch of ethylene molecules. Cook them. Add a polymerization catalyst, usually a chemical derived from titanium, and you get polyethylene.

```
H   H   H   H
|   |   |   |
C — C — C — C —
|   |   |   |
H   H   H   H
```

String enough of these molecules together and you get plastic resin. Ship the resin from the refineries of the Gulf of Mexico to Los Angeles, mill it into nurdles, ship them to Guangzhou, add some yellow dye, throw them into the hopper of an extrusion blow-molder, melt them into parison, extrude the parison through a steel mold milled by Henry Tong's late father, hire a teenage immigrant from the Chinese interior to press a button that sends a blast of air into the mold, and primordial sunlight becomes a yellow duck. And a yellow duck is a delightful toy. But in alchemizing oil into objects and energy, we are, as Macdonald put it to me one day over lunch in the main mess, "spending down our capital," "living on borrowed time."

Out on the bow of the *Louis*, in the little prefabricated lab, Marie-Ève Randlett and Danielle Dubien were making progress. The place resembled a potter's studio. The steel table was splattered with gray goo. Atop the table, enclosed on three sides by the steel box, was the core of sediment. In just an hour Gobeil's graduate students had already bagged a few centuries of mud. Randlett showed me a weird prickly ball she'd exhumed—a dead sea urchin. Ascending from the high pressures of the benthic zone to the low pressures of the atmosphere, the poor thing had exploded, to Robie Macdonald's dismay. "I don't like to see anything die," he'd tell me that night in the main lounge.

Here are the sorts of stories that can be read in a box of Arctic mud: A low level of organic carbon indicates a low level of photosynthesis, which in turn indicates a frigid, polar climate hostile to phytoplankton and plants; a high level of organic carbon indicates a high level of photosynthesis. Fifty-five million years ago the Arctic was ice-free. Trees grew along its shores. Crocodiles transited the Northwest Passage. Then the planet wobbled on its axis and a new ice age—the Pleistocene—began. The Arctic Ocean chilled. Ice formed on its surface—not the way ice forms on the surface of freshwater, as a skin; when the surface of saltwater first reaches the freezing point, it thickens into a kind of crystalline soup many inches deep. This soup has a savory name: frazil ice. Frazil ice turns into grease ice—so called because, in the memorable words of the naturalist E. C. Pielou, "as the swell passes beneath it" and as the wind blows over it, grease ice "forms ripples like those on cooling fat in a tilted frypan." Grease ice can bear, and has borne, the weight of a man wearing snowshoes. Frazil ice can't. As frazil turns to grease it congeals around flotsam.

During the Pleistocene, snow began to fall on the mountains of Greenland, Svalbard, and Ellesmere. From that snow, glaciers formed. The glaciers began to flow down the sides of the mountains, scouring the earth as they went. By the end of the Pleistocene, ice sheets stretched all the way to Chicago. All the way to Manhattan. The crocodiles, by this point, had moved south, and so had the Arctic trees. Biomarkers in cores of mud and ice record every one of these changes.

But these changes happened over millions of years. The current pace of climatological change is unprecedented, and natural causes—solar flares, for instance—will not account for it. What happened over the course of millennia is now happening over the course of decades. "Basically what we're doing is turning the polar ocean back into a temperate one," Macdonald said. Whether this was a good or bad thing he at first hesitated to say. "Good, bad—it depends which species we're talking about. Good for some, bad for others. Good for invasive species like the spruce beetle. Bad for the polar bear." After a second Scotch he was less equivocal. "Basically we're going to hell in an incremental handbasket," he said. Like most of the scientists I met, he was quick to insist that uncertainties remained—gaps in the data record, puzzles yet to solve, which is why we were here. Nevertheless he also believed that we didn't have time to wait for all the results to come in. "We've never conducted this experiment, and we only get to conduct it once; we don't have a duplicate planet."

In the middle of Baffin Bay we encountered a diffuse flock of bergs and floes stranded at the heart of the convergent currents. The sea state had fallen to Beaufort force 0, "Calm. Flat. Smoke rises vertically." No wind. No breezes. No waves. Not even capillary waves. The surface of the sea was silken, and the bow wrinkled it without breaking it. It seemed as though we were steaming across a reflecting pool that stretched to the horizon. "The sweet water," Captain Rothwell called it. A captain of a ship in the age of sail would have called it purgatory, or hell. It occurred to me that the conditions here resembled those Charlie Moore had described encountering in the Garbage Patch. Here, as there, a high-pressure system dominated. If there had been anything other than ice floating in the water, you would have seen it. It was beautiful, eerily beautiful, that glassy stillness, the bergy bits and floes motionless atop their own reflections—until the waves radiating from our wake made them rise and fall.

WEYPRECHT'S DREAM

Before making Cambridge Bay, I don't really expect to find a duck frozen into a floe, or a beaver stranded on some barren shore, or a turtle perched on some Inuit windowsill. But you never know. In the annals of drift, stranger things have happened. Consider, for instance, the strange thing that happened on March 2, 1885, to a certain C. Brainard, whose first name seems to have been lost to posterity, along with most of his biography. On March 2, 1885, Brainard was working as a "leveler" for the Mississippi River Commission, a federal entity charged with a task now carried out by the Army Corps of Engineers, that of surveying and improving the Mississippi River. A leveler's job was to operate a transit, which in the nineteenth century would have comprised a telescope and spirit level mounted on a tripod. On that March afternoon, three miles south of Point Pleasant, Missouri, two hundred miles south of St. Louis, three feet from the river's edge, at the high-water line, Brainard noticed something—not a rubber duck (in 1885 the only rubber ducks around were hunting decoys) but a sheet of parchment covered in script.

We don't know for certain the exact sentences on which Brainard's eyes first fell. Perhaps these: "Our only foods now consists of 3 ounces of shrimps daily per man. Lichens and saxifrage and reindeer moss are eaten in the stew by those who like it." Or these: "The evening dinner consisted of sealskin soles entirely, no shrimps being on hand, and the stew was enjoyed by all and gave great satisfaction. I had a good stool in the evening." Or these: "Although Henry has told before his death that I had eaten a lot of sealskin, yet, although I am a dying man, I deny the assertion he made against me. I only eat my own boots and a part of an old pair of pants which I received from Lieut. Kislingbury." Or these: "Poor Gardiner died at 11 A.M. from inflammation of the bowels and starvation; he will be buried in the ice foot, as it is seen that the rest of the bodies are uncovering with every light wind, and are thus laid bare to animals."

Onward C. Brainard trudges, urgently now, following the Mississippi's high-water line, chasing pages, shuffling them into chronological order, pausing to read, hoping to learn how this story will end. He will never record for posterity the thoughts now passing through his mind but I can imagine those that would have passed through mine: *Who was Lieut. Kislingbury? And Henry? Did they also die of inflammation of the bowels and starvation? And what the hell is saxifrage? And reindeer moss? Who wrote this? Ate his own boots! And Kislingbury's pants too!* (It is remarkable, reading histories of Arctic exploration, how often explorers were reduced to a sartorial diet. In this one respect the woolen and leathery wardrobes of Victorians were clearly superior to a neoprene survival suit, a potage of which would more likely poison than sustain.) Of all the thoughts that would have passed through my mind had I been in C. Brainard's muddy, edible shoes I suspect the most pressing one would have been this: *How in the hell did these pages turn up here? On the banks of the Mississippi, 2,313 miles south of the Arctic Circle?*

After three hundred feet, the trail of pages ends. The story they tell does not. The latest entry Brainard finds, dated June 17, breaks off abruptly, after this sentence: "Only one meal is cooked a day now, as Fredericks is getting so weak; yet it is remarkable how he keeps up at all on this food with the work which he does."

I picture Brainard looking up from this last sentence and gazing out in wonderment at the muddy, mile-wide river, on which trading scows and rafts and perhaps a paddle-wheel steamboat travel. Had I been in his shoes, I think I might well have set out for the Arctic that same day. Level-headed leveler that he was, Brainard did the far more sensible thing. He turned the papers over to his boss, thanks to whom we know the answer to this particular riddle. The pages were written by Private Roderick R. Schneider, Company A, First Artillery, United States Army, a member of the 1882 American expedition to Ellesmere Island led by Lieutenant Adolphus Greely.

The Greely expedition, after spending two years collecting scientific data and planting the American flag at 83.20°N, at the time the highest latitude ever attained, concluded in the usual way of disastrous nineteenth-century expeditions to the Arctic—gothically, in scurvy, frostbite, gangrene, starvation, drowning, shipwreck, dog meat, bear meat, pemmican, execution by firing squad, and, possibly, cannibalism. Caught in the ice, the relief vessel sent to fetch Greely and his men in the summer of 1883 had sunk into the depths of Baffin Bay, thereby condemning the explorers on Ellesmere Island to a third, unanticipated winter in the Arctic.

On foot and by boat, they retreated south, searching for caches of food and supplies, making it as far as Cape Sabine, where they set up camp on the bleak shores of Smith Sound. It was there that the events Private Schneider recorded in his logbook transpired. Out of twenty-five men, seven survived. Schneider, whose main duty had been to look after the sled dogs, all long dead, some eaten, held on until June 18, 1884, before he too succumbed to the effects of starvation. Just four days later a relief vessel arrived, and one of its sailors, after a copy of Schneider's diary had been made, purloined the original. A year later, as this sailor was steaming up the Mississippi on a riverboat, a thief purloined his suitcase and cast the purloined diary upon the muddy currents, which, on March 2, 1885, delivered a dozen of its pages into C. Brainard's lucky hands.

In the annals of drift, things get stranger still. On June 18, 1884, while Private Schneider, whose diary a year later C. Brainard would discover on the banks of the Mississippi, was expiring on Cape Sabine, 500 miles to the southeast, Eskimo fishermen from Greenland noticed some strange flotsam—or was it jetsam?—stranded on a floe.

History records neither the first initials nor the last names of these keen-eyed Inuit, which is a shame, for their discovery would prove far more consequential than C. Brainard's, and far more relevant to my own investigations. On that floe were some fifty-eight items (clothes, gear, papers)—the relics, evidently, of a shipwreck. In the journal of the Danish Geographical Society, Greenland's colonial director later cataloged them. They included, notably, a list of provisions hand-signed by Lieutenant George De Long, captain of the USS *Jeannette*; a list of the *Jeannette*'s boats; a pair of oilskin breeches in which a certain Louis Noros had written his name; and the peak of a cap inscribed by one F. C. Nindermann. [35]

In 1879, in the mistaken belief that at the top of the planet there lay an open sea, the officers and crew of the *Jeannette*, a 142-foot steam yacht, had sailed through the Bering Strait, plowed boldly into the ice, expecting to break through it, and promptly, off the coast of Wrangell Island, found themselves beset—a fate that at the time few ships had been known to survive. The steamer lasted surprisingly long, drifting about for twenty-one months, before the ice crushed it. To the Norwegian scientist-explorer Fridtjof Nansen, the relics of the *Jeannette* discovered three years later off the coast of Greenland were not merely historical curiosities but data, data that seemed conclusively to prove the existence of uncharted, transarctic currents.

At the time, many geographers were reluctant to accept Nansen's conclusions. Some questioned the provenance of the relics. Wasn't it far more likely, they speculated, that they'd come from the sunken relief vessel sent to rescue the Greely expedition? Furthermore, the *Jeannette* had made it only to within 884 miles of the pole. Who could tell what lay beyond? Some obstinately or wishfully or devoutly clung to the theory of an open polar sea. Those fond of symmetries believed that at the

North Pole as at the South there must exist a continent, or perhaps an undiscovered archipelago.

"It is doubtful if any hydrographer would treat seriously [Nansen's] theory of polar currents," wrote none other than Lieutenant Adolphus Greely, namesake and leader of the disastrous expedition to Ellesmere Island that had claimed the life of Private Schneider. Seeking vindication, Nansen proposed to do what at the time seemed suicidal: in a smaller, better, more iceworthy ship, he would reenact the voyage of the *Jeannette*—or at least the first twenty months of it. He named this vessel the *Fram*, Norwegian for "forward," because once he and his crew found themselves beset off the northeast coast of Siberia, Nansen foresaw, there would be no turning back.

The voyage of the *Fram* is one of the most heroic and least gothic in the history of Arctic exploration. Everyone survived, and everything went almost perfectly according to Nansen's carefully preconceived plan, and when it didn't, good fortune seemed almost miraculously to intervene.[36] The *Fram* entered the ice pack in 1893. Three years and fifteen hundred miles later, passing through the strait that now bears that vessel's name, it emerged into the open waters east of Greenland.

When you go beachcombing in the annals of drift, you tend to notice coincidences, coincidences that seem to indicate the presence of subtle currents or eddies, currents or eddies that flow through time as well as through oceans. Consider this: In charting the likely transarctic route the castaway toys would take, Ebbesmeyer examined the precedents set by both the *Jeannette* and the *Fram*. Or consider this: The Greely expedition took place under the auspices of the first International Polar Year, or IPY. This expedition, the one I've joined, the one Eddy Carmack conceived, is taking place under the auspices of the fourth International Polar Year, in honor of which the *Louis S. St-Laurent* recently received a new paint job. Adorning the starboard side of the *Louis's* red hull is the new IPY logo, rendered in what might be called the United Nations style. A blue figure reminiscent of those that appear on the doors of men's rooms, assuming the splayed posture of Leonardo's famous portrait of man, stretches his limbs to the four corners of an abstract planet crisscrossed with longitudinal and latitudinal lines. Orbiting this planet is an alphanumerical caption: INTERNATIONAL POLAR YEAR 2007–2008.

The series of improbable events that would eventually deliver the pages of Private Schneider's diary to the banks of the Mississippi and me to the shores of the Northwest Passage began in Vienna in 1875, when, addressing a meeting of the Austro-Hungarian Academy of Sciences, Lieutenant Karl Weyprecht (naval officer, scientist, dreamer) laid forth a scheme as inspirational as it was implausible. For more than a century, ever since Captain Cook sailed to 70 degrees south and then to 70 degrees north, European explorers had been exploring the earth's two poles, or trying to. But to what end?

Yes, they'd charted previously uncharted coasts, and planted flags on previously unclaimed land, and pushed the point of farthest north farther and farther north, and that of farthest south farther and farther south. Yes, they'd performed heroic feats of derring-do, attaining glory and fame, but they'd also performed many disastrous feats of gothic folly ending in a diet of sled dog and lichen and boiled boot. But from that data what had been learned? Not much, Weyprecht believed. "Immense sums were being spent and much hardship endured for the privilege of placing names in different languages on ice-covered promontories," he once wrote, but "the increase in human knowledge played a very secondary role."

He knew whereof he spoke: From 1872 to 1874, Weyprecht and another naval officer, Julius von Payer, had led an Austro-Hungarian expedition to the Arctic. The objective was to reach the North Pole by ship, or, falling short, to transit the Northeast Passage, from Norway to the Bering Sea. Payer, Weyprecht, and the men under their joint command made it as far as Novaya Zemlya, an island due north of the Ural Mountains, before the ice closed in on their three-masted schooner. The Arctic currents carried them, fortuitously, to an archipelago of ice-covered promontories on which Weyprecht and Payer bestowed the name Franz Josef Land, in honor of their emperor. From there, all had made it home alive. Weyprecht was a hero. He'd earned his place in history books. Nevertheless, he considered the expedition a failure.

The two "forces of Nature" Weyprecht wished most urgently to illuminate were terrestrial magnetism and the aurora borealis. What was needed, Weyprecht told the scientists assembled in Vienna, was a coordinated, synchronous series of expeditions that would set up a ring of research stations around the poles and, using standardized instruments, carry out meteorological observations for at least one year. No single nation could accomplish a project of this scale. For the plan to work, the scientists and governments and militaries of many nations would have to set aside their rivalries and join in common cause.

Six years later, in 1881, just forty-two years old, a man of his time, Weyprecht died of tuberculosis. By then his plan had gained prominent adherents, and a year after Weyprecht's death, scientists from eleven nations put it into action. This first International Polar Year, like subsequent International Polar Years, would last longer than a year. Germany established an Arctic research station on Baffin Island, and another at the Antarctic island of South Georgia. The Austro-Hungarian empire established one on Jan Mayen Island, off the east coast of Greenland, Sweden another on Svalbard. The Dutch sent a ship into the Kara Sea. The French, Danish, Norwegians, Finns, and Brits also participated, as did the Russians. In all, fourteen polar expeditions took place between 1882 and 1884. The one that was the most ambitious turned out also to be the most disastrous and, historically, the most notorious—the Greely expedition to Ellesmere Island.

The scientists of the first IPY didn't manage to solve the mysteries Weyprecht set out to investigate, but they did manage to collect meteorological data thanks to which we now know just how much the Arctic has warmed in the past century—up to 4 degrees centigrade on average, 5 degrees on land, two to three times faster than the rest of the planet. Climatologists predicted this accelerated Arctic warming years ago. Its primary cause? The ice-albedo effect. Albedo is the amount of sunlight that the planet reflects back out into space. As anyone who's walked barefoot across a parking lot on a hot summer's day knows, white surfaces reflect the sun's rays; black surfaces absorb them. In the Arctic, when the ice retreats and the snow melts, earth and ocean absorb more heat, thereby melting more ice and snow, thereby absorbing more heat.

None of this was on the mind of Weyprecht when he organized that first International Polar Year, nor was it on the minds, fifty years later, in 1932, of the scientists who staged the second IPY. The organizers of the second IPY decided to reenact Weyprecht's dream on a more ambitious scale. The phenomenon they hoped to explain was the jet stream. Twenty-five years later, in 1957, yet another generation of scientists staged yet another synchronized scientific assault on the poles. Armed with technological instruments many of which had been developed during the Second World War, they confirmed the existence of mid-ocean ridges, measured for the first time the mass of Antarctica's ice, sent the first weather satellites into space, and collected some of the first evidence that man-made greenhouse gases were influencing the climate.[37] By then, the theory of man-made global warming had been around for almost half a century, though few scientists took it seriously. Most physicists and oceanographers and meteorologists were confident that the oceans would absorb all but an irrelevant fraction of the CO_2 we were adding to the atmosphere.

But in 1957, a group of scientists led by Roger Revelle, an oceanographer at the Scripps Oceanographic Institute, found that the ocean removed CO_2 from the air far more slowly than expected. Also in 1957, another beneficiary of Weyprecht's dream, the climatologist Charles Keeling, set up a pair of observatories, one on Hawaii's Mauna Loa, another on Antarctica, and for the first time accurately measured atmospheric levels of carbon dioxide. In the following decades, his results would vindicate Revelle: atmospheric CO_2 levels were rising dramatically.

After 1957 scientists and the rest of the world seemed to lose interest in the poles, which by then seemed to have given up their secrets. We now knew that there was no open sea at the North Pole, only ice. We now knew that the Northwest Passage wasn't a commercially viable shipping lane, so why study it? We had new frontiers to explore. Nuclear waste was our most pressing environmental threat. The moon was the new Arctic; astronauts, the new explorers. It was only later, in the 1980s and 1990s, after the space race had ended, after scientists had begun to investigate the fourth dimension, after the evidence for global warming began to mount, that we once again turned our attention toward the poles. And what we saw was hard to reconcile with what we'd seen before.

Where before we'd seen permanence, we now saw evanescence—and it was on this theme, the theme of change, that the scientists participating in the fourth International Polar Year would play their variations.

The fourth IPY would be the most ambitious one yet, lasting longer than its predecessors—for two years, not one, from March 2007 to March 2009. By the time it was over, 10,000 scientists from more than sixty nations had investigated everything from "ocean–atmosphere–sea ice–snow pack interactions" to "the biodiversity of Arctic spiders" to the threat to Inuit oral traditions posed by hip-hop and rap.

Of the twelve Arctic expeditions taking place this summer, this one, organized by Eddy Carmack, was among the most ambitious, involving around sixty scientists and not one icebreaker but two. As we entered the Northwest Passage from the east, the *Louis*'s sister ship—the *Sir Wilfrid Laurier*, based in Vancouver, carrying yet another company of scientists—would enter it from the west. The shared goal of the scientists aboard both icebreakers was also the goal of Carmack's life work: to form a high-resolution, megascale picture of how the North Pacific, North Atlantic, and Arctic oceans interact. Underlying this goal was an insight lost on the explorers of the past: far from being an otherworldly place, fortified by ice, mystified by myth and mist, the Arctic is very much of this world, connected to temperate oceans and to Manhattan, Hong Kong, Hilo, Guangzhou, Sitka, Kennebunkport—and to Helsinki and to Dar Es Salaam and to Tasmania and even to Antarctica by currents and winds.

ICE PICKING

Around dawn, when I wake in my little cabin belowdecks, it sounds as though we're sailing through riprap, through boulders of the sort my son would like to climb. Skipping breakfast, I ascend to the bridge. Where the day before we looked out onto a glassy ocean, we now look out onto a white labyrinth—ice puddled with blue pools of melt, fissured with cracks and leads, which are canal-like channels that open up between plates of ice. "We call it an icebreaker," says the quartermaster, Dale Hiltz, seated at the helm, "but really it's an ice-avoider." You can tell that this is a line he's delivered before. At the helm, Hiltz, a big man, looks like John Goodman seated in an easy chair, or a king played by John Goodman presiding from his throne. Or picture Captain Kirk at the helm of the *Enterprise* but imagine that instead of William Shatner,

Goodman had played Kirk. That's what Hiltz at the helm of the *Louis* is like. With Captain Rothwell stationed by the windows, looking anxiously on, Hiltz seeks the path of least resistance, following leads, or "puddle jumping" between patches of open water. When the leads close up and there are no puddles, Hiltz seeks out the "rotten" ice, ice so old and thin and waterlogged that the *Louis* can easily steam through it, bubblers bubbling away.

When we come to a dead end, when the ice at the terminus of a lead is hard and thick, the breaking begins. Hiltz pilots us straight at it, so that the bow of the *Louis* hauls out like a kayak onto a beach. An officer at the controls cranks up the RPMs. The three propellers churn in our slushy wake, pitching the ship forward, bringing its weight and the thrust of its engines down onto the ice, snapping it like a wafer. On the rare occasion that the ice fails to snap on the first attempt, the officers perform a maneuver they call backing and filling. We reverse a hundred yards, and the water churns forward, sending out a white-and-turquoise lacework of eddies and foam. The officer at the controls shouts, "Here we go!" The ship charges forward under full steam, gaining speed, hitting the ice at eight or nine knots. Inevitably the ice gives way. It isn't elegant, this diesel-powered battering, nor is it fast, but it works.

Near the entrance to the Northwest Passage, we emerge once again into open water and, assisted by a following current, make up for lost time. Speaking of time, I've begun to lose my sense of it. We're now in the latitudes of perpetual daylight, and the schedule of fieldwork requires those of us on the scientific team to keep erratic hours, catnapping by day, working by night, taking meal breaks and cocktail breaks when we can.

A week out of Halifax, I wake up at 3 A.M. to fetch my last batch of bottles. Outside it feels like midday. Off the port bow there floats an iceberg, a white cake-shaped island on the dark water. I hardly give it a glance. How quickly wonders degrade into the ordinary. A fulmar swoops past, and the sunlight reflecting from the ship's red hull turns its white belly feathers pink. How quickly the ordinary becomes wondrous. It's cold out, near freezing. So much for the balmy forecast I overheard back in Halifax. If I didn't know the difference between climate and weather I might well be inclined to dismiss global warming as conspiratorial bunk.

As I carry my box of bottles to the taffrail, Dale Hiltz, bleary-eyed, soon to begin his next watch, notices my cargo, hitches his trousers, lumbers after me, and asks, sheepishly, "Can I throw some?"

"By all means," I say. "As many as you like." Hiltz's face lights up. He has been working on this ship for thirty years, and yet the prospect of bottle throwing has elicited from him boyish delight. A graduate student working with the marine biologist Glenn Cooper also requests permission to join us, as does Glenn Cooper, as does the chief scientist, Jane Eert. Our little merry band defiles to aft, along the starboard rail, down the flight of steel stairs, I leading the way, my box of bottles clinking cheerfully.

There's something irresistible about throwing bottles into the ocean. You take the bottle by the neck and send it flying tomahawk-style, and as it flies, end over end, there's a faint whistle, and it catches the light and describes an exuberant arc through the sky, an arc that ends in a sad little splash. Cooper, the marine biologist, starts calling out "Launch!" as if preparing to shoot clay pigeons. Bottles fly—to port, astern, to starboard—tomahawking through the air, plopping into our frothy wake. Watching them drift away, it's hard not to dream of distant shores. Just two weeks ago one of Eddy Carmack's bottles, tossed into Baffin Bay last summer, was discovered on an uninhabited island south of Iceland. By tourists on horseback. Tourists who then sent Bonita LeBlanc a reply that read, "Hello! We found this drift bottle last Saturday, the 14th of JUN 2008, off the southwest coast of Snaefellsnes." Evidently you can reach the southwest coast of Snaefellsnes only at low tide, hurrying out on horseback over the mud. Trying to picture that scene makes me want to saddle up and go there, right now.

On the far horizon a white line appears, a band of snow-white radiance. Heavy pack ice ahead, Captain Rothwell assumes, and since we're running late, due in Resolute by tomorrow morning, he decides to dispatch the ice observer, or the "ice pick," as everyone calls her. The ice pick is Erin Clark, a small twenty-eight-year-old Toronto native who wears her dark brown hair in a bob the pointy tips of which she is forever tucking behind her ears. Though an officer, one of only two female officers on the *Louis* and one of the youngest, she favors the company of the crew, hirsute oilers and deckhands most of whom are twice her age, spending much of her free time in the smoking lounge, puffing on Player's Lights, a habit she means to give up one of these summers, but life on ship—the long hours of tedium, the occasional sleep-deprived bursts of action—lends itself to smoking. Her baggy uniform is at least one size too big, and she complements it with a pair of steel-toed shoes. Her duties are twofold. She helps the officers of the *Louis* navigate the channels of the Northwest Passage, and she e-mails reports to the Canadian Ice Service. Satellites can show you where the ice is, but not what it's like, not in real time. They can't show you whether it's rotten or thick, young or old, where the leads and pressure ridges are. Clark reads ice the way nephologists read clouds, or psychics, palms.

She conducts her surveys via helicopter, equipped with a specially designed tablet computer on the screen of which she taps out observations with a stylus, using a color-coded system of alphanumerical glyphs to indicate the varieties and qualities and quantities of ice. She is part cartographer, part meteorologist, a mapmaker mapping mutable terrain. The varieties of ice, like those of clouds, have excellent names—nilas ice (halfway between frazil and grease), pancake ice (round floes that look very much like pancakes on a griddle), fast ice (frozen to the land).

There happens to be a spare seat in the helicopter, and it's mine if I want it. I do want it. Yes, please. Very much. Thank you. I want it because I'd like to see the ice through Erin Clark's eyes, and because from the airborne vantage of a helicopter I'll be better able to search for flotsam, and because this is the sort of moment I've been dreaming of ever since I looked up the Arctic in my *Atlas of the World*, but also because, as I first learned during the airlift at Gore Point, riding in helicopters is fun. It is in my opinion a grave injustice that Igor Sikorsky is not as famous as Orville and Wilbur. The helicopter, in my opinion, is superior to the airplane because it more closely approximates my recurrent dreams—in which I swoop over the water, and soar into the sky, with the aeronautical grace of a fulmar.

Chris Swannell fires up the rotors: *Whoosh, whoosh, whoosh*, the blades turning from a ceiling fan into a loud discus of blur. Wearing insulated jumpsuits and life vests, escorted by Chief Officer Stephane Legault, Erin and I rush from the hangar onto the helipad, assuming that hunched posture that helicopter passengers in movies always assume, as if the rotors might lop off their heads, which, as the rotors whoosh inches above your skull, seems a distinct possibility. Clark climbs into the copilot's seat, I climb into the backseat.

Before we return, dinner will be over, and so the kindly officers on the bridge would like to take our orders. The menu is simple, beef or pork. "Pork," Erin Clark says.

"Pork," says I.

Liftoff.

To the north rise the snow-swept cliffs of Devon Island, a place so remote, so lifeless, that NASA has set up a training camp there the purpose of which is to give would-be astronauts a taste of what it might be like to live on Mars. The dark but sparkly water flashes below. The white line seen from the bridge was not, in fact, pack ice but a mirage. It disappears as soon as we're aloft. The Arctic is, climatologically, a desert, and as in any desert, mirages are not uncommon. Viewed from a distance, the peak of a mountain can turn into an hourglass, casting a reflection upward into the sky. An eerie blackness can spread across low clouds—a phenomenon known as a water sky; the blackness is a reflection cast up onto the clouds by a patch of open water.

Now another band of radiance appears on the horizon, and this time it's for real: ice straight ahead. "There's a narwhal, a whole school of narwhal," Swannell says. "There's one that's about to break the surface." At first I don't see them, but then Swannell banks, and I press my face against glass, and there they are, five or six of them, white, speckled whales that—from this height, three hundred feet—seem tiny as minnows. "They're nursing their young," says Swannell. "When they do that they turn on their sides and you can see the whites of their bellies."

We keep seeing them, narwhals as well as belugas, foraging at the ice's edge. Below us lie a million white polygons jumbled loosely together as if a sheet of ice had fallen from the sky and shattered, polygons outlined in dark water. When the wind blows from the west, as it's blowing now, Erin Clark explains, the edge of the pack loosens and frays, scattering east. As we fly, against the wind, the white puzzle below assembles itself, until it stretches, uniform and solid, from barren shore to barren shore. Above us, the sky is clear. Below us, you can see our helicopter-shaped shadow made nervous against the white fields below. The way through the maze is no longer obvious.

It's altogether possible that there's flotsam down there, but if there is, from this height, not even Clark's trained eyes would be able to spot it. I wish we could zoom low, but the lower you go, Chris Swannell tells me, the more slowly and cautiously you have to fly; the closer you are to the ground or to the water, the less time a pilot has to react, should something go wrong. Out here, you don't want to have to make an emergency landing. Our jumpsuits are warm, but they aren't watertight survival suits. Although helicopter 363 carries enough fuel to fly for two hours and twenty minutes, it never flies that long. What limits its range is the *Louis*'s radio reception.

"There's some old stuff in here," Swannell says to Clark, meaning old ice.

"Most of this stuff is first year," Clark says. "There was some multiyear back a ways."

"It comes from up there, doesn't it?" Through the bubble of the windshield, Swannell points north.

Clark nods, then says, "This isn't exciting. It's all pretty much the same."

Says Swannell, "So what are you telling me—you're not having a good time?"

Me, I'm having a good time. We're sixty-two miles from the ship, ninety miles from Resolute, off the coast of an island that NASA considers to be an approximation of Mars.

"Lots of seal holes," Swannell says, banking east, back toward the *Louis*. There are indeed lots of seal holes, black dots on the floes below, easy to see because of their roundness, an aberration of geometric perfection in a landscape that abjures geometric perfection. Seals, it seems, are the gophers of the Arctic. A lead appears, and flying over it, hundreds of feet below us, is a flock of white birds, too far down for an amateur bird-watcher like me to identify. Ahead on the horizon, just entering the scattered puzzle pieces at the edge of the pack, a red speck appears, a red speck that grows into the *Louis*.

RESOLUTE

Named for one of the ships sent in search of John Franklin, the ill-fated British explorer who sacrificed his life and those of his men on his fourth and final search for the Northwest Passage, Resolute is one of the coldest, northernmost settlements in the world, so cold and northern and bereft of edible game animals that the Inuit avoided the place until 1953, when the Canadian government, eager to establish the northern limits of Ottawa's sovereignty, compelled four Inuit families—twenty-three people, twenty-seven dogs—to move there. These unhappy pioneers, after a single hungry winter, regretted the move, but their requests to return south were denied. Since Resolute boasts an airstrip and a military base and a scientific research station, the Canadian government likes to call it "the gateway to the Arctic." The Inuit like to call it either Qausuittuq, meaning "the place with no dawn," or else Qarnartakuj, meaning "the place of the ruins."

To me it seems like the place of corrugated metal buildings painted in primary colors, or perhaps the place of many all-terrain vehicles, or the place of wolfish, xenophobic sled dogs on staked chains who, when roused from their slumber by a curious stranger, bristle with canine menace. Most of all it seems like the place of pebbles and dirt. The roads are unpaved and vehicles traveling along them kick up curling plumes of dust. There is, so far as I can see, no flora to speak of. The hills rising from behind the village are as denuded as dunes. On one of them, written in white, appears the word RESOLUTE, as if the landscape itself were a kind of giant map. This is not at all how I pictured an Inuit village in the Arctic. There's no snow on the ground. No igloos. There are, however, polar bear skins, draped over porch railings like beach towels drying in the sun.

Resolute's scientific research station, on the outskirts of town, in convenient proximity to the unpaved airstrip, is the northern base for something called the Polar Continental Shelf Project, the primary purpose of which is to provide logistical support to visiting scientists. By design, we've arrived just in time to attend festivities commemorating the project's fiftieth anniversary, festivities taking place in a blue building the size of an airplane hangar. The festivities include an assortment of speeches, a throat-singing performance by Inuit teenagers, and a free lunch to which all 229 residents of Resolute have been invited. Politicians from Ottawa have flown in for the occasion, as has a business executive from the Ontario-based cell phone company responsible for the BlackBerry.

It is a disorienting experience, after ten days of voyaging through Arctic waters, to find oneself seated in a stackable chair, in a hangarlike room crowded with media and mariners and Canadian politicians in blue jeans, and Inuit mothers, some of them teenage, dressed in a kind of garment called an *amautik*, meaning "coat with pouch," the heads of their babies poking up out of the hoods. It is even more disorienting to find oneself politely trying to remain awake for the duration of multiple PowerPoint presentations on, for instance, the "over predation of musk ox," or the ecological importance of the wetlands of Creswell Bay, or, most disorienting and tranquilizing of all, a speech delivered by the executive from BlackBerry, who has, it turns out, come all this way in order to bestow some playground equipment on the children of Resolute.

Captain Rothwell, dressed in his nautical finest, looking very much the part of a sea captain, the brass buttons of his navy blazer polished to a shine, also has a gift to bestow, a framed photograph of the *Louis*. To make it to Resolute in time for today's festivities, Rothwell had to cut a number of research stations from the scientific plan, much to the dismay of the scientists. The *Louis* was supposed to play a large role in today's festivities. Crossing the ice on snowmobiles, locals and visitors alike were supposed to come out for guided tours of Rothwell's big red ship, but this morning village elders informed us that the ice conditions were too dangerous. Robie Macdonald was pissed. "We rush out here and then they can't do their open house—typical."

After a half hour of speeches, the juvenile beneficiaries of the playground equipment have left the room, and, following their lead, I join them, outside. In the sunshine, on a piazza of dust, the kids joust at each other with little complimentary souvenir flags commemorating today's anniversary. Most wear free Polar Continental Shelf Project baseball caps, too big for their little heads, from which paper tags still dangle on plastic threads. Others, also wearing free ball caps, have gathered in the lunchroom, in anticipation of the feast to come. There on folding tables draped with red-and-white-checkered tablecloths, a catering staff has begun to set out the spread—urns of coffee, towers of stacked Styrofoam cups, hamburgers, hot dogs, potato chips, paper plates, the usual North American picnic.

Wandering through the crowd, searching in vain for my Arctic Virgil, Eddy Carmack, I notice an elfin young woman—rusty hair, pale complexion, big blue eyes, worrisomely thin—filming Inuit children with what looks to be an expensive video camera. Making small talk over hot dogs, I'm surprised to learn that rather than the television journalist I mistook her for she is in fact a Rhodes scholar. What the hell is a Rhodes scholar doing in Nunavut? Shouldn't she be in tutorials with an Oxford don?

The answer is complicated. For one thing, she's from here—well not from here, not from Resolute, but from the North. In Canada, apparently, the 60th parallel is a kind of Mason-Dixon line, dividing North from South. (In my head, learning this, I perform a kind of geographical somersault: how strange to think of Toronto and Edmonton as the South.) Her name, Erin Freeland-Ballantyne, suits her, suits her so well that if a novelist were to give such a name to an anthropological Arctic explorer with a Rhodes scholarship and an Irish complexion, it would seem way too heavy-handed, the first name and surname both. Freeland-Ballantyne grew up in Yellowknife, capital of Northwest Territories, made famous by a television show called *Ice Road Truckers*.

Like me, she's in Resolute because of Eddy Carmack, who's hired her to assist with "community outreach." She's supposed to help the scientists learn from the Inuit and the Inuit from the scientists, a summer job for which she is well qualified. A biogeographer, she's writing her Rhodes thesis about, she tells me, "the effect of oil and gas development on Dene and Inuit community health." She's also, at twenty-six, the mother of a two-year-old daughter whom, for the next few weeks, she's left back in Yellowknife, in the care of her husband, an unemployed poet who drives a Volkswagen microbus and lives in an apartment above a restaurant. She appears to be experiencing far less separation anxiety than I am. In the subsequent days I'll spend in her company, I'll hear Freeland-Ballantyne say things like, "Fresh caribou liver is good for your breast milk."

ARCTIC ATTACK

Two days out of Resolute, four days from Cambridge Bay, in the smoking lounge, where the yardlong penis bone of a walrus hung, trophy-style, above the wet bar, members of the *Louis*'s crew were drinking cans of Pepsi purchased from vending machines, and tapping their cigarettes against the crenellated edges of black plastic ashtrays, and watching, on the flat-screen television, a National Geographic documentary called *Hunter and Hunted: Arctic Attack*. Ice pick Erin Clark was there, as was Paul Devlin, the chief cook. As was Doug Murray, chief technician. As were off-duty oilers and able-bodied seamen. Through the two portholes of the smoking lounge could be seen blue skies and white ice.

No one in the smoking lounge was paying attention to the view through the portholes. Everyone was paying attention to the documentary, which recounted the tragic tale of a starving polar bear who happened upon a pair of delicious young women, Swedish graduate students, hiking on the Arctic island of Spitsbergen. They'd dressed appropriately, in brightly colored Polartec, and they'd brought along carabiners. They'd neglected, however, to bring along firearms. "Nearly three thousand polar bears live in the region," the baritone narrator intoned.

Paul Devlin, in his chef's whites, leaned forward, intently, on his elbows.

"They outnumber humans here, by two to one," the baritone narrator continued. "Despite these considerable odds, there have been very few fatal attacks since Spitsbergen was settled nearly one hundred years ago. The bears simply haven't been a problem"—an orange special effect burned across the screen, accompanied by the sound (*whooosh!*) of a flash fire, as if a polar bear had spontaneously combusted—"but the animal, approaching the hikers"—portentous pause—"is." On the sound track there now erupted what I took to be an ursine roar.

"I'm not going on the ice, fuck that," Doug Murray, the technician, said.

That same day, Captain Rothwell piped over the PA system the news that a polar bear, a real polar bear, could be seen to port. I rushed to the bridge just in time. There it was, sitting beside a seal hole. Looking so cuddly you almost wanted to give it a hug. Contrary to Melville's meditations, it wasn't really white but a kind of creamy off-white and dirty about the withers. The Snapper and the Australians, having flown home from Resolute, weren't on hand to capture the moment, but everyone with a camera was snapping away, an impulse that, even as I was brandishing my own digital point-and-shoot, struck me as curious. I'd seen a polar bear before, at much closer range, most recently at the Central Park Zoo, where through thick windows you could watch the animal swimming around in its tank, or from above, witness it devour a lamb shank. Why photograph it here? Was an animal viewed in its natural habitat somehow more real than one viewed swimming around in a landscaped aquarium the size of a swimming pool? Even if, in the photographs one took of it, it appeared as a yellowy dot scarcely distinguishable from its icy habitat?

The way we look at polar bears is indicative, I think, of a larger confusion, a larger and perhaps untreatable blurriness in our vision. It's as though the more pictures we take of the world the less clearly we see it, as if our megapixelated screens weren't windows but kaleidoscopes. Even in the perpetual darkness of the Arctic winter, Fridtjof Nansen was less confused about the meaning of polar bears than I was, standing on the bridge of the *Louis* at the height of the Arctic summer. In my cabin belowdecks I'd been reading both *Farthest North* and Barry Lopez's *Arctic Dreams*, which perhaps explains at least some of my confusion.

Polar bears appear almost as frequently in *Farthest North* as does the eponymous whale in *Moby-Dick*. They sneak onto the *Fram*. They eviscerate sled dogs. One bites a harpooner named Peter Henriksen on the hip before Henriksen delivers a roundhouse to its noggin with an oil-burning lamp. One bear Nansen encounters in the dwindling light of October he describes as "a beautiful white fellow rummaging among the flotsam on the beach," but as darkness falls he's less given to aesthetic appreciation and instead, understandably, to either hunger or fear. One bear he calls a "monster," another a "demon," another a "ghost." Nearly every bear the mariners of the *Fram* meet, including the one that tasted human flesh, ends up dropping to the ice, felled by a ball, then flayed by knives, then turned into dinner and blankets. Of the execution of a mother bear by firing squad and her two cubs by a pack of sled dogs, Nansen writes, "It was a glorious slaughter, and by no means unwelcome, for we had that very day eaten the last remains of our last bear in the shape of meatcakes for dinner. The two cubs made lovely Christmas pork."

Compare this with the chapter of *Arctic Dreams*, published less than a century after *Farthest North*, in which Barry Lopez accompanies a team of biologists hunting for *Ursus maritimus* on Lancaster Sound. From Lopez, we learn many wonderful ursine facts,[38] but his ultimate aim is to dispel the old myths. European explorers, he notes, eventually saw in the polar bear "a curious image of themselves"—"a vaguely noble creature, wandering in a desolate landscape, saddled with melancholy thoughts"—whereas Lopez seeks to see the bear itself, and succeeds more than any writer I've read.

And yet, he too ends with a curiously anthropomorphic image—an image, he writes, "of vulnerability." The biologists he's traveling with have shot a female with a tranquilizing dart. "As I sat there," Lopez writes, "my companions rolled the unconscious bear over on her back and I saw a trace of pink in the white fur between her legs. The lips of her vulva were swollen. Her genitalia were in size and shape like a woman's." Lopez, feeling like a prurient voyeur, looks away. How far we'd come since the Victorians. This anesthetized beast is no monster of God but a damsel in both estrus and distress. The polar bear, in other words, remains symbolical, and since the publication of *Arctic Dreams*, it has grown more symbolical still. I can't help wondering whether vulnerable images of polar bears—usually less graphic and more sentimental than the one Lopez gives us—have become as common and therefore potentially as obfuscating as the monstrous images with which, in my mind at least, they compete.

Even in that National Geographic documentary, the portraiture of the bear kept wavering confusingly between pity and fear. As the ice diminishes, the baritone narrator informed us, so does the bear's habitat. Less ice, fewer seals, more hungry bears eager to snack on Swedish graduate students. If that weren't bad enough, evidence suggests that one class of toxins drifting to the Arctic (flame retardant polybrominated diphenyls) are elevating the rates of polar bear hermaphroditism. Roll a tranquilized polar bear over now and it might well be difficult to determine its sex. Whatever the sex, a polar bear is now the totem of global warming, and photographs of them—stranded on an ice floe, or swimming in open water—are in great demand, which is why the Snapper joined us on the first leg of the voyage. He shot pictures of Macdonald and Gobeil deploying their box corer, and of scientists tossing bottles from the stern (he asked me to step out of the frame). But the big money, he said, was in bears.

On the bridge of the *Louis*, somewhere in Peel Sound, we were all still watching the bear beside the hole. Its patience was so great that it resembled somnolence. I swear to both God and the monsters thereof, that as we watched, a seal popped up to catch a breath, and as it did the until now statuesque bear sprang forth, catlike, extending its fatal paw. With one terrible and yet somewhat leisurely swipe it snared the seal by the neck, punctured the jugular with one terrible bite, and then, limp carcass hanging from its jaws, trailing blood, lumbered off, making an exit that Nansen describes well, assuming "an easy shambling gait, without deigning to pay any further attention to such a trifle as a ship." Then it disappeared behind a pressure ridge to enjoy its meal in private.

CARMACK'S DREAM

Eddy Carmack, sixty-seven, does not resemble the Arctic explorers whose black-and-white portraits I've seen in history books—those pipe-smoking Victorians and Edwardians wearing layers of wool and layers of beard. Beardless, since boarding the *Louis*, Carmack has worn, as if it were a kind of uniform, a navy-blue fleece cardigan vest decorated with the logo of the Canada's Three Oceans Project. He has short brown hair, little round glasses, a thin, scholarly face, almost no visible jawbone to speak of (a trait he and I share, one that makes his neck and mine resemble that of an iguana), and a nervous habit of smiling for no apparent reason. Talking to him, I'm often left wondering why he's smiling and what he's thinking. When he smiles, the smile lines in his cheeks crinkle almost to his ears.

When I visited him at his office in British Columbia, back in January, after disembarking from the *Ottawa*, I asked him the same questions I'd asked John Toole, Amy Bower, and just about every other oceanographer I'd met: If flotsam can tell us where stuff really goes, as Carmack believed, then where, upon entering Bering Strait, had the ducks really gone? And why in the summer of 2003, despite Ebbesmeyer's predictions, hadn't the beachcombers of New England found them?

"You had one finding," Carmack said.

"One *sighting*," I said. "I interviewed them, the people in Maine who reported the sighting. Put it this way, I don't think their testimony would have held up in court." (How long ago and far away that drizzly morning in Maine seemed! Had a duck made an appearance in Kennebunkport or hadn't it? Science or no science? Proof or no proof? Was this a children's fable after all?)

In his office, Carmack had smiled, inscrutably, knitting his fingers, but said nothing. The window above his desk looked out onto fir trees. Affixed to the glass was the translucent likeness of a cardinal, and on one of his bookshelves was a framed photo of his own feet, in brown sandals, propped up on the prow of a red kayak afloat on some tropical lagoon.

To fill the silence, I'd continued: "So from what I understand, this question turns out to be a fairly difficult one to answer. Given that the spill happened in 1992, you now have fifteen years of climate to be thinking about, and from what I've read the climate of the Arctic has changed a lot since 1992."

From behind his little spectacles, Carmack's eyes seemed to twinkle a little. It was then that he'd told me about the Drift Bottle Project, fetching a beer bottle from a bookshelf, saying, "So here is the Canadian version of the rubber duckie." Then he shared with me one of his preliminary findings: "The Arctic is moving faster."

Down the hall, on a colleague's computer, he pulled up satellite footage of the polar ice cap. The footage began in September 2007, the same month that I'd voyaged to the Labrador Sea. Watching this grainy, pixelated, black-and-white, yin-yang, stop-motion montage of ice swirling and expanding and contracting inside the Arctic Basin, I was reminded, weirdly, of watching a fetus on a sonogram screen. "This shows you what an ice catastrophe last year was," Carmack said. "I mean it broke all records. Smashed them. It was hell up there."

Now, months later, one evening, before the bar opened for business, in the *Louis*'s main lounge, in heavy ice, the bubblers screaming, somewhere in Franklin Strait, Carmack delivered a lecture in which he expounded and expanded on the preliminary findings he'd shared with me seven months before. His was among the last in a series of lectures. When he boarded the *Louis* in Resolute, he'd brought with him, in addition to Erin Freeland-Ballantyne, a number of supernumeraries whom he referred to as VIPs—luminaries of oceanography, all dressed, like him, in matching fleece cardigan vests onto which was embroidered the logo of this expedition—the C30 project, Carmack had called it, for Canada's Three Oceans. His idea for the second leg of our voyage was to turn the *Louis* into a kind of icebreaking, traveling oceanographic lyceum.

During the lectures the VIPs delivered, I learned many interesting facts—for instance, that in the "microbiome" of the human body, only a portion of our cells, genetically speaking, are human in origin. The rest are bacterial. (While learning this, I found myself looking down, examining my midriff, into which, in the main mess, I'd recently deposited some potatoes, carrots, and buttery cod. The cloth between the buttons of my quick-dry adventure shirt was puckering over the waistband of my quick-dry adventure pants in an unflattering way, and I tried to smooth the puckers flat.) I learned that herring are shrinking but also that eighty-two "jumbo squids" had recently been caught in Canadian waters, and that these squids were members of an invasive, southern species. (To my mind this didn't sound so bad—less herring, more calamari.) I learned that approximately every ten years the prevailing winds of the Arctic shift, a phenomenon known as the Arctic Oscillation, and that this wind regime plays a role in the movement of ice.

In the main lounge of the *Louis,* Carmack stood before a screen onto which a computer projected various slides. The computer also projected slides onto Carmack. As he paced back and forth over the carpeting, rocking in his mocassins, knitting and unknitting his fingers, smiling his sphynxlike smile, maps and diagrams played across his plaid shirt and jeans and cardigan vest, and made the lenses of his little spectacles shine.

Around five million years ago, Carmack told us, the Isthmus of Panama, without the efforts of Teddy Roosevelt, had been a strait, conducting equatorial currents freely between the Pacific and the Atlantic. Only after the Isthmus of Panama closed, and deflected the currents in both the Pacific and the Atlantic, did the oceans come to resemble the oceans of today, "with high-pressure centers in the subtropics and low-pressure centers in the subarctic, and clockwise currents traveling around the subtropical gyres and counterclockwise gyres flowing around the high latitudes north of the subtropical front." Over the past five million years, two distinctly different kinds of oceans had formed.

There is on the one hand the ocean most of us have heard of and thought about. The one we like to go bathing in, on the beaches of Coney Island or Costa Rica or the Côte d'Azur. This is the "thermally stratified ocean," where variations in temperature and salinity and therefore in density make layers of water mingle and move. Then, on the other hand, there's the northern ocean, which is "permanently stratified by the accumulation of freshwater." Much of the freshwater—sixteen thousand cubic kilometers of it or so per year—comes from the comparatively fresh currents that flow into the Arctic from the Pacific through the Bering Strait. Another thirty to forty thousand cubic meters or so come from the northerly rivers of Canada and Siberia and Europe. It was thanks to the freshwater as much as to the low temperatures (1.7 degrees Celsius, last I checked) that right now, in the middle of July, through the closed curtains of the *Louis*'s main lounge, the creaky thunder of crumbling ice—six feet thick and stretching from shore to shore and riddled with seal holes and stalked by bears—could be heard.

The northern ocean, Carmack emphasized, is connected to the "global ocean" by subtle currents and winds. In fact, the climate as we know it depends on those currents and winds, which transport excess heat away from the equator, to be released back into space at the poles. And the subpolar front, the boundary between "these two counter-rotating gyres"—the boundary along which sixteen years ago twelve containers tumbled overboard, along which bath toys had traveled to Sitka—"acts as a bit of a wave guide for storm tracks." The weather in Myanmar as in Manhattan is the consequence, in other words, of the Isthmus of Panama and of those currents circling and spiraling through time and space, currents that with alarming swiftness, as the Arctic warms, are changing in measurable if not always visible ways.

"You look out now," Carmack said in the main lounge, directing his gaze toward the closed curtains, his computer projecting a cartoon planet Earth onto his shirt. "Well, you know what kind of ice we've had in the last couple of days." Since Resolute, far from ice-free, the Northwest Passage had seemed more like the Northwest Impasse. One might as soon attempt to transit the Gowanus Expressway in an icebreaker as the frozen thoroughfare in which, the previous evening, we'd found ourselves trapped.

While I was chatting about winged copepods with Glenn Cooper over beers, a strange look had passed over the biologist's face. He'd paused midsentence and was gazing, rapt, out the windows of the main lounge. I followed his gaze, expecting to see I'm not sure what—perhaps a polar bear feasting on a paleochemist. "We're not moving," Cooper had said. Yes, now that he mentioned it, I noticed it, too. I was experiencing the sensation of beer but not the sensation of drift—nary one of the six degrees of freedom. The *Louis*, weighing 11,345 tons, outfitted with three propellers each one of which can exert eighty tons of thrust, was frozen in, beset like the *Jeannette*, stuck. Out by the starboard rail a couple of graduate students were joking about having to get out and push. We were in Peel Sound, not quite halfway between Resolute and Cambridge Bay.

Up on the bridge, Quartermaster Dale Hiltz was at the helm and Captain Rothwell was standing nervously by, dressed casually, in sneakers and a sweatshirt, looking out the windows and rubbing his mustache. He was supposed to be off-duty, asleep by seven, in his cabin one deck down. His second-in-command, Stephane Legault, also dressed casually, in sandals and shorts, was supposed to be in the main lounge enjoying a beer. Cathy Lacombe, in uniform, was supposed to be the commanding officer. Having encountered thick ice, Hiltz and Lacombe had performed the usual maneuver. They'd backed. They'd filled. They'd charged. The bow of the *Louis* had hauled out onto the ice. And the ice had refused to break, and the *Louis* to reverse, and now we were parked there, like a beached whale, propellers churning. The late evening sun, however, still shone brightly as a midday sun, making the ten thousand pools of melt that speckled and marbled and gilded and puddled the ice flash, as if some Nordic or Inuit god had dropped a handful of change. The watery puddles and the white hummocks formed a pleasing pattern—not a perfect pattern, but a discernible if erratic one, a kind of messy paisley. If you've ever noticed the rills waves make on packed sand, you can imagine those that the wind and puddles make on thawing land-fast ice.

To Hiltz, in thirty years of icebreaking, this had happened only once before. "Come on, girl, patience!" he bellowed as the propellers churned. Robie Macdonald, his white hair resembling the nest of an eider duck, had never seen it happen. "It's all one piece!" he said of the ice, meaning that there were no cracks and that Peel Sound was frozen fast from shore to shore. Captain Rothwell said, "It's like an ice grip!" Then he said, "If it's like this the rest of the way it won't be good." And Macdonald, thinking of the years he'd spent planning his scientific fieldwork, said, "No, it won't be good."

When the captain of your ship begins speaking in perplexed exclamations and gloomy prognostications, it is hard, I've found, to keep the hysterical flights of fancy at bay. With a full tank the helicopter had enough gas, I knew, to travel 220 miles, which hardly seemed enough, considering how many of us there were aboard. Should we break out the survival suits? Would we spend weeks here in Peel Sound, eating our shoes and performing theatricals? Probably not. When we set out from Nova Scotia, the larder of the *Louis* had contained 683 pounds of bacon, 1,012 eggs, 900 pounds of coffee, and 1,900 rolls of toilet paper (this last fact makes one wonder what explorers of the past, lacking both paper and leaves, used; nothing, presumably; or perhaps lemmings, or snow). Back at the docks of Dartmouth, I personally witnessed stevedores load sixty-four crates of potatoes into a hatch in the *Louis*'s hull. Surely, we hadn't eaten all of them. And so long as we had satellite reception there'd be no need for theatricals. We could instead watch nature documentaries about the Arctic.

In the end Captain Rothwell and the engineers improvised an ingenious remedy, pumping ballast water back and forth from starboard to port, making the ship shimmy and shake, all the while running the props in reverse, until, to a round of cheers, the *Louis* slid free. It had taken forty-five minutes—nothing compared with the *Jeannette*'s twenty months, of course, but then it happened again, and then again. We kept stopping, sticking. No one aboard had experienced anything like it. I asked Erin Clark, the ice pick, if she could explain it. The problem, she said, was that the ice here had begun to thaw but hadn't yet thawed enough. "It's not hard, and it's not soft," she said. "It's sticky." Hard ice is brittle. Soft ice is slushy. This ice was rubbery. Beneath the weight of the ship, it didn't break; it bent. Our average speed fell to just four knots, or less than five miles per hour. We might as well have snowshoed to Cambridge Bay.

Even now, a day later, as Carmack was delivering his lecture in the main lounge, Captain Rothwell was considering giving up. If conditions didn't improve soon, we'd have no choice but to turn around and follow the slushy channel of our wake irresolutely back to Resolute. So much for the record-breaking transit of the Northwest Passage.

Now Carmack clicked the button on his remote control, and onto the screen and across his fleece vest there flashed an incongruous photograph of open water, taken the previous summer not far from the North Pole. "That's a lot of open water that far north," Carmack said. Then onto his screen there flashed a satellite image from 2005. "Everyone was shocked," Carmack said, "when the ice cover collapsed to this position." Then Carmack played the same animation he'd played for me back at the Institute of Ocean Sciences, the stop-motion one depicting the great melt of 2007. Said Carmack: "All the multiyear ice is streaming out, and as the Arctic refreezes, it just refreezes as first-year ice." The meaning for the planet was ominous: the subtropical gyres were expanding, the subpolar fronts were moving north, along with invasive species. The meaning for me was ambiguous.

Ebbesmeyer had based his prediction that the toys would reach New England in the summer of 2003 on a dozen-odd transarctic drifts, intentional and accidental, that he'd found in the historical data record— drifts including those of the *Fram* and the *Jeannette*, but also studies conducted by NOAA in the 1970s. From these precedents, he'd calculated that the icy currents flow at an average pace of around 0.6 miles per day, sometimes much faster, sometimes much slower. That's about how fast—about 0.6 miles per day—that Carmack had expected his bottles to travel when he began his Drift Bottle Project; at most he'd expected them to travel at an average pace of a mile per day. Instead they appeared to be traveling "twice as fast," he'd told me that afternoon in his office—not one mile per day, but two. In short, I was right to question Ebbesmeyer's 2003 prediction. "Any predictions based on old data would no longer be as likely," Carmack said.

There was, however, one other miscalculation that in Carmack's opinion Ebbesmeyer had made. He'd assumed that like the *Fram* and the *Jeannette* the toys, upon passing through Bering Strait, would have caught the current now known as the Transpolar Drift. Catching the Transpolar Drift is like catching the express. Between the coast of Sibe-

ria and the North Pole, it flows directly, or as directly as any Arctic current flows, from the Bering Strait to Fram Strait, the latter of which was
named for Nansen's famous ship. The data from Carmack's Drift Bottle
Project suggested a different route. Carmack's drift bottles tended to
hug the coastline, caught in coastal currents by the Coriolis force. The
coastal currents that flow through the Bering Strait would have carried
the toys around Alaska and into the Northwest Passage, which is why
I'm here. Carmack now put up a slide titled "The Northern Drift Hypothesis," an animated slide that showed animated bottles zipping
around the northern coast of Canada.

Did Carmack's data prove Ebbesmeyer's drift hypothesis wrong?
Not necessarily. But it did prove it to be fallible. Ducks, frogs, beavers,
and turtles could have made it to New England by 2003, Carmack said,
depending on what unforeseen and unforeseeable events transpired during their transarctic voyage. But they also could have made it sooner,
given the acceleration of the currents. Or perhaps they'd never made it
at all. At the time Carmack and I traveled together aboard the *Louis*,
he'd begun collaborating with Ebbesmeyer on a scientific paper analyzing data from the Drift Bottle Project. One conclusion they would eventually draw: "Our data . . . show the boundary between the subarctic
and subtropical waters to be a near-impenetrable wall." Out of 1,184 bottles launched in the eastern Arctic and recovered elsewhere, only one
had headed south, to Puerto Rico. The rest the Gulf Stream and its
branches had swept east, toward Europe. It was, in other words, statistically likely that the duck allegedly glimpsed in Maine was a counterfeit, an impostor, a figment, a will-o'-the-wisp, and that the advert
created by that British cell phone company was as fantastical as any advert, as childlike as *Ten Little Ducks*, more childlike than *Make Way for
Ducklings*, far more childlike than *Paddle to the Sea*, and that my errand,
from the outset, was indeed that of a fool.

MOONWALK

In helicopter 363. Once again in the backseat. Once again in a jumpsuit.
Once again a headset clapped onto my ears. Just because the ducks never
made it to Maine doesn't mean, as per Carmack's Northern Drift hy-

pothesis, that they never made it here, into the Northwest Passage. I've got four days left to search. And I mean to. In fact, I've printed up WANTED posters, illustrated with the blue, green, yellow, and red likenesses of the toys. They're pretty snazzy, I think. I even put the word WANTED in an old-timey, Wild West font, and at the bottom I cut a little fringe of tear-away tags bearing my e-mail address. I've already posted some in Nuuk, on my way home from the Labrador Sea (so far, no replies), and some more in Resolute, and I'm ready to post more on every bulletin board I encounter before flying home for good. Granted, here, in the Canadian Arctic Archipelago, there's a hell of a lot of shoreline—87,000 miles of it—and a hell of a lot of sticky, rubbery ice, and a dearth of bulletin boards.

With me and pilot Chris Swannell in helicopter 363 are a pair of men who, like me, seem to have been arrested by their childish imaginations, men who boarded the *Louis* in Resolute dressed as if for an expedition to the planet Hoth, in white snow pants and white parkas with furry hoods, parkas decorated with a cryptically Hellenistic logo of the sort one might encounter on the facade of a college fraternity. Even in the main mess, at breakfast, these strangers have been striding around in black knee-high jackboots of the sort favored by Napoleonic cavalrymen. This duo belongs to an outfit called the Phaeton Group, named, curiously, for the doomed son of the Greek god, the one who steals his father's solar chariot as if it were a Buick and takes it for a reckless spin, a whimsical crime for which Zeus, like some celestial traffic cop, executes Phaeton with a lightning bolt. The Phaeton Group, according to their literature, is "a science and consulting organization that carries out research and provides communications services to media and educational clients." They offer "public outreach" and "media savvy" and both "enlightenment and excitement." Publicists with the Canadian Coast Guard have hired them to produce educational posters about icebreakers. We're flying out now so that they can collect footage of an icebreaker in action. I'm here because Chris Swannell refuses to be photographed or filmed and they need a human figure to stand in the foreground, on the ice, providing a sense of scale.

In their getups, the duo from Phaeton look—to me, to the crew of the *Louis*, or at least to those members of the crew I overheard gossiping yesterday in the computer lounge—ridiculous, and they don't seem to know it, but who, while looking or being ridiculous, ever does?

Up we go. Helicopter 363 circles the *Louis* at two hundred feet, collecting footage to aft, astern, abeam. The back door is wide open and the Phaeton cameraman, buckled into a harness, the wind ruffling his hood's furry ruff, is leaning out. I hope his glasses don't fall off. I hope he doesn't drop his fancy camera. At his direction, we ascend, Phaetonlike, to a thousand feet. At this height, except for the plume of yellowy exhaust rising from its stack, the *Louis* very much resembles a red toy boat in an icy bathtub. Now we fly miles ahead, over the ice, the hummocks and bummocks flickering below us.[39] The *Louis* disappears, and Swannell looks for a solid floe on which to set us down. If I were Ishmael—Melville's Ishmael, not the Bible's—I'd probably at this point in my narration say something allegorical, about how we are all precariously aloft, about how the door of the helicopter is always thrown open and we are all always leaning out, dressed in ridiculous costumes, imagining ourselves to be something or someone that we aren't: Phaeton, or John Muir or Rachel Carson or Doc from *Cannery Row*, or Ishmael, who, come to think of it, imagines himself to be the biblical Ishmael's second coming; how, buffeted by winds actual and imaginary, held in place by harnesses that we can only hope will hold, we're dangling above a planet too big for one mind to encompass, a planet that in large part thanks to our imaginings and desires and restlessness and ingenuity is changing more quickly than we can comprehend; how we're all flying over hummocks and bummocks and milky-blue pools of slush that make patterns both beautiful and perilous, at once orderly and chaotic; how our cameras, fancy as they are, will never be fancy enough; how we're all searching for colorful objects and meanings that we'll never find and looking for solid ice that maybe just maybe we might.

Swannell picks his floe and sets us down, bouncing three times to test the thickness, and we all unbuckle ourselves and step out, cryonauts, onto the frozen sea, boots crunching on the lunar snow. The hummock Swannell chose, like those he did not choose, is shaped like a sand trap, or an amoeba. Around us ten thousand other hummocks shine and ten thousand pools of melt sparkle. They look delicious, these pools. I'd like to kneel down beside them and drink my fill, but I'm scared I might slip in, and besides I have a role to play, the role of a human in an inhuman landscape. Chocolate-covered cliffs, streaked with snow, rise up to the east and west of us, but we're in the middle of the ocean—the ocean! Most inhuman of all is the silence. There are no sounds besides the crunching of our boots and the rustling of our outerwear. Not even the melt pools lap.

Near the edge of the hummock the Phaeton cinematographer sets up his camera on a tripod and aims it north, the direction from which we expect the *Louis* to approach. "What do we do if a crack does open up?" he asks of our floe.

"I have no idea," says Swannell. "Try to get out. What you really don't want is for the ice to close back in on you. I've seen that happen to seals. They just go *sppppt*"—he pinches two fingers together—"like a pimple."

The *Louis* appears. The red smudge of the ship grows into the red triangle of the looming bow and the faint murmur crescendos. As it breaks the ice, water and snow shoot up around the bow in feathery plumes, and with it comes the sound of ice buckling—the creak as it stretches, the crack as it gives way, a low grumble as big blue-white blocks of ice heave upward, spilling over each other, into ridges that paw along the hull. What the breaking of ice sounds most like out here is fireworks heard from afar.

The cameraman finally gives me my cue, and I stand before the camera, as close to the edge of our hummock as I dare to go, the big red ship passing behind me. When the shoot ends I search my pocket for the yellow duck Ebbesmeyer gave me three years ago. I'd meant to leave it here, Marpol Annex V and environmental impacts and my promise to return it to Ebbesmeyer be damned. I meant to leave it here as a kind of wish. But grapnelling my pocket, I come up empty. Evidently I left the thing back in my cabin.

While the film crew pack up their equipment, I scan the hummocks for bears, having learned from Barry Lopez that the specially adapted fur on a polar bear's paws makes its tread almost silent. Their coats, of course, make them nearly invisible. For all I know there is a bear nearby, but here in Peel Sound, there's no sign of them, nor of any other life. As Swannell and the two cinematographers climb back aboard helicopter 363, I linger a little, making the moment last, trying to imagine what it would feel like to find myself marooned out here alone and wishing again that I'd brought my yellow duck. As consolation I kneel down and scoop up a handful of snow. And eat it. It tastes good, a teensy bit salty, but good. Then I take in one last drink of the scenery. At the edges of some of the melt pools, I notice, a breeze is kicking up capillary waves. And there, on the snow, are our boot prints.

The *Louis* by now has steamed far ahead. The blue-white blocks chopped up by its propellers flow back into its wake, and you can hear them flowing, a slushy river whose currents gradually slow until once again, nothing is moving and nothing besides us makes any sound.

WANTED

We didn't turn back in the end. Captain Rothwell, after a suspenseful conference call with "all the powers that be" at Coast Guard headquarters, had decided to press on, breaking, bubbling, backing and filling, whatever the risks. Expensive, nonrefundable plane tickets hung in the balance, including my own. By the following day the ice conditions had improved—for us if not for the polar bears. Leads opened. The ice parted. Parted because it had melted enough. And would continue to melt. The melt of 2008 would be among the worst on record, coming in second only to that of 2007. It's possible, maybe even probable, that in my lifetime toys from China will be arriving via the Northwest Passage.

My last day aboard the *Louis*, I had one of those unexpectedly memorable moments that justify travel. Marie-Ève Randlett, graduate student, was in the main lounge, playing the piano. And I was out on the bow, playing an air piano, tinkling my cold fingertips on the starboard rail. I could hear Randlett's water music. And in the distance was a water sky, a blackness spreading across the underside of the overhanging clouds. Open water ahead. And I thought, *This rail is cold*. And then I thought, *What a wonderful phrase, "water sky."*

In the Cambridge Bay airport terminal, a small room that scarcely deserved to be called a terminal, itinerant Russian miners in muddy boots were sleeping on the metal benches. A taxidermy musk ox stood inside a knee-high picket fence beside the plate-glass windows. There were muddy bootprints on the floor. As I and other supernumeraries were helicoptering off the *Louis,* the scientists and sailors who would be taking our places for the third leg of the voyage, had assembled here, waiting to get on.

If Resolute was the place of dirt and dust, then Cambridge Bay, the morning we arrived, seemed to be the place of chilly mud and chilly rain. Riding in a taxicab from the airport to town, I noticed, through the scummy windshield, way out on the tundra, white, faceted spheres. These were part of the North Warning System, a chain of radar stations that Canada and America had strung across the Arctic during the Cold War. The North Warning System is the reason Cambridge Bay exists. Until the Cold War, the local Inuit mostly lived, as the Nunavut saying goes, "out on the land."

I'd arranged to spend two days there, beachcombing, posting my WANTED posters, talking to the locals. I had, however, neglected to arrange for lodging, and lodging in Cambridge Bay, during summer months, is hard to come by, it turns out. Lots of gold miners and gold-mining geologists and gold-mining middle managers. Lots of scientists. Erin Freeland-Ballantyne, the elfin Rhodes scholar, rescued me. She had a friend in Cambridge Bay, a young doctoral candidate in archaeology named Brendan Griebel. Griebel happened to be house-sitting for a couple who'd adopted several Inuit children. They were all now on vacation, in Maine. There were plenty of spare rooms. It really was no problem. Which is how I found myself that first afternoon napping on the bunk bed of an adopted Inuit child I'd never met, atop a camouflage bedspread.

The next day Freeland-Ballantyne volunteered to help me in my search, acting as a kind of fixer. We began at the public library, talking to Emily Angulalik ("pronounced like 'Uncle Alec,'" she said), who curated the town's historical archives. I told her the story of the bath toys lost at sea and showed her my poster, which she studied awhile and then said, dryly, "I saw that one," pointing at the frog. "Just kidding." She was curious to know what kind of toys they were, and I told her what I knew about polyethylene and blow-molding and Guangdong. "Are they harmful?" she asked, and I told her about plastic pollution and seabirds, and about the currents that carry plastic and other pollutants to Arctic waters. "Now you've got me thinking," she said.

She suggested that we speak with some of the elders, but then added, "summertime is not the best time to meet. People are out on the land. Or sleeping all day." She contacted one elder to see if he could speak with me, but returned a moment later with bad news: "He's drunk right now." She did have one idea: Try the Ekaluktutiat Hunters and Trappers Organization.

The Ekaluktutiat Hunters and Trappers Organization matched Inuit guides with tourists keen to bag musk ox, or polar bears, or grizzly bears, the last of which had only recently begun appearing this far north. Photos of hunters holding up their dead trophies decorated one wall of the organization's office, maps of Nunavut covered another. Again I told my story and shared my poster. In charge of the office that day was Cathy Aitaok. She had plenty to tell us. First she told us about a rich, white American dude who came up to Cambridge Bay hoping to bag a polar bear and who lost two fingers to frostbite while he was out on the ice, a detail that made her snicker. (Behind her, pinned to a bulletin board, I noticed a postcard featuring a San Francisco cable car.) More pertinently, she suggested that we locate an elder named Annie Anuvaluk. Perhaps, also, Peter Avalak, the DJ at the radio station, would let me send out an on-air announcement. The radio station, a single-story box of a building with brown siding, was straight down on Michuk.

Peter Avalak was spinning classic rock when we poked our heads through the door. He waved us right in. It was an austere studio. A few shelves of discs and records. A table outfitted with microphones and equalizers. Avalak, a small wiry man wearing a black hoodie and a wispy goatee, was excited to learn I'd come all the way from New York. He'd gone to New York once, in high school, for the Model UN. "The first four days I was there I was afraid," he said. "I thought the buildings might fall on me." A song ended. He cued up another. I told him my story and showed him my poster, and he thought it was a pretty good story, and when the next track ended, he put me on the air, and into the homes and shops and boats of greater Cambridge Bay I delivered a nervous, stuttering parole ending with the instructions to give a call to Peter Avalak if you'd seen any of the four plastic animals matching the description I'd given.

Finding Annie Anuvaluk proved far harder than finding the radio station. Peter Avalak had given us directions, and we thought we'd followed them, but there was no Annie Anuvaluk living in the first house we visited. Freeland-Ballantyne, trained in ethnographic etiquette, led the way, knocking on doors at random, asking whoever answered if they knew where Annie Anuvaluk lived. No. No. Sorry, no. Then Freeland-Ballantyne stopped a guy out walking a black Lab. "Just up the road. It's an orange duplex." We found a duplex that looked more yellow than orange, and were told that Annie Anuvaluk did not live there, she lived down that way.

"What color is the house?" Freeland-Ballantyne asked.

"Black."

Freeland-Ballantyne was dressed today in a long sky-blue tuniclike coat that looked influenced by the *amautik* we'd seen back in Resolute. She'd made it herself, she said. The voracious Arctic mosquitoes were biting, and it was cold, and muddy, and overcast, and the ramshackle squalor of the place was making me depressed. Stumbling about the shabby homes and down the broken roads of Cambridge Bay, I couldn't help but think of Nunavut's crippling poverty rate, to say nothing of its notable rates of suicide, both of which have risen steeply since the town was founded a little more than fifty years ago. Here, you can order alcohol by mail, which is expensive, but there are no bars—an ordinance meant to mitigate the epidemic of alcoholism in the area. And there was, I noticed, litter everywhere. I thought of something Charlie Moore had told me. Subtropical islands like Samoa also have a noticeable litter problem, he said. His explanation: the natives had for generations wrapped their food in biodegradable wrappers—leaves and the like. Toss a leaf on the ground, it disappears, as does a bone, as does just about everything. One of the trashiest spots Freeland-Ballantyne and I encountered was the little playground we happened on at an intersection: children scrambling over the jungle gym, the muddy ground below covered in candy wrappers and potato chip bags. By then, I was ready to give up and retreat to my bunk bed in Brendan Griebel's borrowed house. Not Freeland-Ballantyne. The longer we searched, the more determined and cheerful she seemed. Down one road she actually started to skip, the hood of her *amautik* bouncing behind her.

Finally we ran into an old man in coveralls who worked for Qulliq Energy, the local power utility. "See that window up there?" he said, pointing authoritatively up the road. "That house with the Christmas tree?"

The house in question was a red one-story unit up on concrete blocks. The Christmas tree in question was a plastic one intended for a tabletop. Someone had bolted an upside-down crate to the facade and perched the Christmas tree on it, the white cord, unplugged, dangling down. We knocked and a small crinkly-eyed woman answered. Yes, she told us, her name was Annie. Freeland-Ballantyne, as usual, made the introductions, cueing me up to tell my story. Annie offered us coffee. Her "common-law" William joined us in the living room, where, taped all over the walls, were handmade pictures of crosses—the handiwork of a Sunday School class by the look of it. In the midst of the crosses was a big color photo of Inuit in fur parkas sledding on the ice—Annie's grandparents, with whom she still remembered going "hunting and trapping, staying out on the land." Spider-Man curtains hung in the windows. As we spoke Annie and William's three-year-old boy, Sam, played among us, trying to draw the attention of the two strangers who'd come knocking. He landed his Lego airplane on my lap and kept wanting to touch Freeland-Ballantyne's long, rusty hair.

Annie said they'd seen plenty of toys littering the coast but she'd always assumed they were from local kids. Neither she nor William recognized the animals on my poster, though William did have something to offer: "I threw a bottle in the ocean once. Put a note inside. Out on the ocean." We could have left then. Instead we stayed awhile, listening to Annie and William talk about the changes they'd seen in their lifetimes—how there were the grizzly bears now, and the kingfishers. "The ice is really different this year," William said. "There are open spaces where there shouldn't be." Annie added: "Usually you have the cracks that go straight like this." She drew a line with her hand. "This year it's all like zigzags. You have to go around them. We've never seen that before. My parents told me that when I was growing up so many things were changing in their lifetimes." Of the future these changes portended, they were less certain. Annie had heard that the ice age might come back. William suspected that if the opposite happened, if the ice melted away, "all the fish would disappear." This was the sort of information that Freeland-Ballantyne was meant to gather for Carmack, the firsthand observations and ancestral memories of people who have spent their lives up here. A plane was taking off in the distance, and young Sam climbed up onto the back of the couch and watched it through the window. Silently, we all watched the boy watch the plane.

Late in our conversation, as I was checking the spelling of Annie's name, we discovered we'd made a mistake. We'd been speaking to Annie Agligoetok, not Annie Anavaluk. We decided it didn't matter, and after posting a few more posters in town, headed to have dinner with our host, the archaeologist, Brendan Griebel.

That night, my last, we took a walk in the midnight sun. Griebel wanted me to see the land, by which Inuit mean the land beyond the limits of town. Setting out we passed back by the playground, where even at this late hour children were playing: three little girls and, standing a few feet away, looking on, an older kid, who when he saw us greeted Griebel happily. Griebel had been living in Cambridge Bay for months, and he'd done some volunteer work at the local school, and the boy, Puglik, was obviously fond of the archaeologist. He wore his hooded sweatshirt cinched tight into an oval around his face, about which there was something peculiar. He had the face of a child and of an old man. He wore his navy-blue sweatpants tucked into his white sneakers, visibly too big for his feet, and his matching navy-blue sweatshirt was stained here and there. "I'll come with you," he said. "I'm bored."

Although still in the sky, the sun was low and orange. As we walked, boots crunching, side by side, down the dirt road leading out of town, our elongated shadows walked beside us, and after we'd set our pace, Puglik commenced to talk, delivering a rambling monologue, much of which I scribbled in my notebook. It went something like this. "Geese eggs are three times the size of chicken eggs. Loon eggs taste like fish. Loons can't fly so when the ice freezes you throw rocks at them. You know what I'd really like? A Nintendo Wii. My family's scattered all over the place. Lots of adoption."

The road now followed a river flowing north through the tundra toward the sea.

"People sure do catch a lot of fish out here. They're jumping every day. What I do for spare money is I fix up bikes. You know kids can be bullies."

Up along a ridge to our left, backlit by the low sun, there appeared a low cruciform thicket—wooden crosses, tilted this way and that, in silhouette.

"Well, there's the old graveyard. The old cemetery. It's beginning to fall down the hill. Most of the old people who've passed away are out here. There's a great fishing spot not far ahead by the old bridge. You could jump off the old bridge. 'Geronimo!' We call this part of the river 'an Inuit Jacuzzi.'"

We were a ways out now. Dusted across the tundra were little white flowers, and an icy wind blew from the north. I'd dressed warmly in many layers, but even still I could feel the wind sneaking through my collar. Puglik in his navy-blue sweatsuit must have been freezing. He'd pulled his sweatshirt cuffs over his hands and hugged his crossed arms tightly around him. He had a bad case of the sniffles and occasionally wiped his nose on his sleeve.

"Here's the picnic spot. It's nice to get out here, out on the land. It's real quiet, you can just hear the river." He noticed a pair of ATV tracks in the dirt. "Somebody turned real hard right here. This is a good place to practice driving donuts. You have to be careful not to tip over. You don't want to do a Donut of Death. It's pretty, eh? Oh, I saw a rainbow!"

We now reached the archaeological site where Brendan Griebel was helping to excavate a Thule house pit, the buried remains of a seven-hundred-year-old stone cottage. Griebel and his team had stretched strings into a grid across the site. With little shovels and brushes they'd begun sorting through the dirt, square by square, carefully labeling everything they found—charred bone, rock tools.

"Back then they were pretty strong, eh?" Puglik said. "My favorite animal to have would be a musk ox."

"As a friend?" Freeland-Ballantyne asked.

"No, to eat. No, to hunt it down and then eat it. It's starting to melt earlier and earlier, and it's changing and changing. We used to go to Kugluktuk by Ski-Doo and sled."

"Duck!" Freeland-Ballantyne shouted, pointing out at the shallow river across the soggy tundra.

"Where?" Puglik asked. "Might have eggs. Ducks are the last eggs of the year. Best thing about the North is being on the land. It's nice and quiet up here. You can see all the animals. Pretty peaceful."

It was too cold for any of us now, and we were all sleepy, at least we three grown-ups were. On the walk back in we discovered that the Annie we'd interviewed today was Puglik's mom, from an earlier common-law marriage. He didn't always sleep at her house. Sometimes he slept with his other family—wherever there was the most room. "Did they say anything about me?" Puglik asked Freeland-Ballantyne and me, and if he was sad to learn that they hadn't, he didn't show it. "The video game I want to get is *Kung Fu Panda*. Have you guys ever played *Grand Theft Auto: San Andreas*?"

None of us had.

"Good rainbow," Puglik said.

Cambridge Bay, when we reentered its outskirts, seemed deserted. No grown-ups. No police. No one. The place might as well have been a ghost town. Then, in an industrial area near the docks, we passed cones of piled gravel, and from behind them popped two urchin faces, a boy and a girl. Seeing Puglik, they leapt to their dirt bikes and followed us. Her handlebars had pink grips, and a pink plastic tassel on one side, but not on the other. "That's Brenda," Puglik said. "That's Angus." They pedaled around us, circling, ogling us, ogling Freeland-Ballantyne especially. Brenda wanted to know whether Puglik was staying with us in the big house, one of the nicest in town, and he said no, but now I thought I detected a hint of hopefulness in his voice.

When we reached the entrance of our borrowed house, Puglik lingered at the doorstep, waiting, expectantly, still warming his hands in his overlong sleeves. Was he hoping for an invitation to come in? Something more? Griebel offered him a slice of the pizza we'd made for dinner, and Puglik happily accepted it. While the archaeologist and the Rhodes scholar ascended to the kitchen, I stayed with Puglik in the damp warmth of the mudroom, among the many muddy boots.

Sitting on the carpeted steps, I began to unlace my boots. The boy kept his big white sneakers on, but as he was standing there, and I was sitting there, he untied his hood and revealed himself to be totally bald. I suspect he anticipated my surprise, which I tried hard not to show. He wasn't merely bald; he possessed, I noticed now, no eyebrows. No wonder he'd seemed neither young nor old. Puglik wasn't a leukemia patient or anything like that, Griebel later told me; his hairlessness was the symptom of a rare condition, alopecia universalis.

Freeland-Ballantyne appeared, descending the carpeted stairs behind me, bearing a slice of pizza on a plate, which she handed to me, which I handed to Puglik, then she ascended the stairs to help Griebel with the dishes, and once again, Puglik and I were alone. He ate hungrily, and happily. It was only then that I thought to ask him the question that had been bothering me: What were he and Brenda and Angus and the other children of Cambridge Bay all doing up so late, alone?

"In the summer, when there's no school, parents sleep at night, kids stay up till seven"—A.M., he meant—"and sleep in the day so the parents can work. I like to look after the little kids. Kids need looking after, eh?" Then he handed me the now empty plate, sprinkled with crumbs, cinched his hood over his moony skull, and disappeared into the midnight sun. With misgivings, I bolted the door shut.

EPILOGUE

But then he thought that he would just look at the river instead, because it was a peaceful sort of day, so he lay down and looked at it, and it slipped slowly away beneath him . . . and suddenly, there was his fir-cone slipping away too.

— A. A. Milne, "In Which Pooh Invents a New Game and Eeyore Joins In"

"The great shroud of the sea rolled on as it rolled five thousand years ago," Melville famously writes at the catastrophic end of the voyage of the *Pequod*. It's a beautiful, hauntingly apocalyptic sentence, its rhythms and sounds—all those long vowels, the repetition of "rolled"—enacting the eternal rolling of the sea. And five thousand years here is no arbitrary number, but one that harks back to the old Christian estimates of the age of Creation. Poetically satisfying as it is, I now know that the sentence is, scientifically speaking, illusory. The oceans have never been immutable, eternal.

In 1951, Rachel Carson wrote in *The Sea Around Us* that man "cannot control or change the ocean as, in his brief tenancy on earth, he has subdued and plundered the continents." A mere ten years later, she'd revised her opinion. In a preface to the 1961 edition of that same book, Carson wrote, "Although man's record as a steward of the natural resources of the earth has been a discouraging one, there has long been a certain comfort in the belief that the sea, at least, was inviolate, beyond man's ability to change and to despoil. But this belief, unfortunately, has proved to be naïve." What had happened in the interval between those two editions? Nuclear waste, placed in barrels lined with concrete and dumped offshore, was leaking radioactive elements. A year after writing that preface, with the publication of *Silent Spring*, Carson would bring to the attention of the world a much longer list of man-made marine pollutants, forcing her readers to imagine the ocean anew. I know now that it is upon Rachel Carson's ocean, not Melville's, that I've sailed.

No one ever replied to my WANTED posters and I doubt they ever will. Nevertheless, I suspect I know what happened to the toys. Stored for several months in my freezer, the photodegraded duck I'd salvaged on Gore Point, brittle to the touch, was crumbling into pieces. The legend of the ducks that drifted around the world, however, has proved more durable. In the summer of 2007, thousands of yellow ducks swam into the minds of newspaper readers in England. The *Times* of London notified readers that "an armada of rubber ducks" would soon appear over the horizon. "Drake's Other Armada," the *Daily Mail*'s punning headline read. By July 13, the anticipated invasion had begun—or so it seemed. Penny Harris, a retired schoolteacher in Devon, had found a convincingly weather-beaten specimen. "It's covered in brown algae and has got barnacles on it," she reportedly said. The *Times* was convinced. "First of the Plastic Duck Invasion Fleet Makes Landfall on the Devon Coast," it announced on July 14. The tabloid *Sun* appeared to be a bit more cautious— "First of the Duck Armada?" its headline asked—but the accompanying graphic was unequivocal: a speech bubble superimposed onto a photo of Harris's duck read, "I've been at sea for 15 yrs, swum 17,000 miles . . . Drake would be proud of me." That same day, the local Devon paper, the *Western Morning News*, bothered to investigate, determining, correctly, that "it was not the right duck." Neither the *Times* nor the *Sun* took note of the correction, and a day after the *Western Morning News* delivered its disappointing verdict, the *Sunday Mail* published yet another story about Harris's discovery, under the alliterative headline "Found: The First 'Friendly Floatee' Rubber Duck in Britain"—getting the brand name of the Floatees wrong as well as the facts. Five days later the Toronto *Globe and Mail* published a time line of the rubber duck saga. It began in January 1992 with the spill and ended in July 2007 with Harris and her barnacle-encrusted counterfeit. Meanwhile, on the coast of Cornwall, beachcombers are still searching. Maybe it was enough that the ducks had in fact fallen overboard. Enough to know that they had crossed the Pacific. That messages in bottles were traveling the Arctic currents.

Moby-Dick begins as Ishmael's story. The opening chapters suggest a bildungsroman in which a green youth will, in the conventional Victorian manner, undergo an adventurous rite of passage into manhood. But in the middle of the novel, as many readers have noticed, Ishmael recedes into the background, less an actor in the action than an omniscient narrator, and interpreter, and at times even an author of the drama in which Ahab and the Whale now take center stage. Some readers suggest that the grandeur of Ahab's tragic character almost usurps Ishmael's role as protagonist. I'm not so sure.

Not long after becoming a father, I read in Andrew Delbanco's biography that Melville's firstborn son, Malcolm, was born midway through the novel's composition. Rereading the novel with that fact in mind, I couldn't help noticing the emergence of paternity as a major theme. It's almost as if, over the course of the novel, we voyage not only from Manhattan to the South Pacific but from the mercenary and possibly bisexual freedoms of bachelorhood to the sorrows and fidelities—and toward but not quite to the heart-straining joys—of fatherhood.

The theme first appears early in the novel, in Nantucket. When Ishmael shares his misgivings about sailing under the command of a captain named Ahab (in the Bible, Ahab is a wicked king), Peleg, one of the *Pequod*'s owners, offers the young bachelor this reassurance: "my boy, he has a wife—not three voyages wedded—a sweet resigned girl. Think of that: by that sweet girl that old man has a child: hold ye then there can be any utter, hopeless harm in Ahab?" Later, in one curious parenthetical aside, Ishmael uses the phrase "we fathers." We? Has Ishmael become a family man since returning from his look at the watery part of the world? Is Melville speaking autobiographically? The farther the *Pequod* sails, the more numerous the allusions to parenthood become. When Queequeg rescues his fellow harpooneer Tashtego—who in a mishap, while harvesting spermaceti, has tumbled into the cavernous forehead of a dead sperm whale—Ishmael describes the deliverance in satirically obstetric terms: "[Queequeg] averred, that upon first thrusting in for him, a leg was presented." Following the delivery, the mother, "the great head itself" (now sinking into the deep), Ishmael tells us, was "doing as well as could be expected."

If this is not enough, look again. Here, at the heart of a "grand armada" of whales, the oarsman Ishmael is peering down—into what? The inscrutable deep? His own reflection? Not this time. No, he's peering into an underwater, cetological nursery. An umbilical cord—antithesis of the harpoon line—still connects one cow to her newborn calf, a newborn calf whose fins and flukes "still freshly retai[n] the plaited crumpled appearance of a baby's ears newly arrived from foreign parts." Here, other calves, cordless, are nursing, slurping down milk that is, says Ishmael—speaking from experience—so "very sweet and rich . . . it might do well with strawberries"; milk that, though sweet and rich, in the whaling grounds nevertheless sometimes pours forth so gushingly that "milk and blood rivallingly discolor the sea."

After many chapters spent in the murderous, manly battlefield of the whaling grounds, the glimpse into the nursery of whales is both enchanting and jarring, and is made all the more enchanting and jarring by Ishmael's well-informed, fatherly eye. The nursing calves, he compares to "human infants" who "while suckling will calmly and fixedly gaze away from the breast, as if leading two different lives at the time; and while yet drawing mortal nourishment, be still spiritually feasting upon some unearthly reminiscence."

Fatherhood even helps us understand the apostasy that underlies Ahab's monomaniacal quest. "Where is the foundling's father hidden?" Ahab, feeling like a fatherless child, asks both the heavens and himself as he nears his doom. "Our souls are like those orphans whose unwedded mothers die in bearing them: the secret of our paternity lies in their grave, and we must there to learn it." Several chapters later, when St. Elmo's fire lights up the *Pequod*, Ahab thinks he's found the secret of his paternity. "Oh, thou clear spirit," clutching the nautical lightning rod, he shouts at the blue pyrotechnics playing about the sheets, "of thy fire thou madest me, and like a true child of fire I breathe it back to thee."

As *Moby-Dick* nears its end, the *Pequod* first encounters a ship called the *Bachelor*, on the quarterdeck of which "the mates and harpooneers" are "dancing with the olive-hued girls who had eloped with them from the Polynesian Isles." The very next ship the *Pequod* encounters? The *Rachel*, named, allegorically, for the biblical Rachel, figurative mother of that generation of Israelites doomed to exile and wandering. Why does Melville give this ship that name? Because, thanks to the interventions of Moby Dick, the *Rachel*'s captain has lost a boatload of sailors. And among those sailors? "My boy, my own boy is among them," the *Rachel*'s captain tells Ahab, begging him to join in the search.

Now keep in mind that between the appearance of the *Bachelor* and the *Rachel* another fatherly event has occurred: Ahab has adopted the castaway, tambourine-playing cabin boy Pip as a kind of spiritual son. "Thou touchest my inmost centre, boy," Ahab tells his new ward; "thou art tied to me by cords woven of my heart-strings." And Pip, in reply, wishes that his black hand and Ahab's white one might be riveted together. You'd think, given the fatherly warp and woof of his heart-strings, that Ahab would be sympathetic to the appeals of the *Rachel*'s bereft captain. But no. Ahab's reply is terrible, terrible for the bereft father, but also—because of what it implies about his lonely, fatherless, unfatherly fate—terrible for Ahab: "God bless ye, man," the doomed captain says to the bereft one, "and may I forgive myself, but I must go."

One last time. In *Moby-Dick*, Starbuck, first mate, is the consummate father, and as the novel ends his fatherly voice grows loud. On the eve of destruction, Starbuck reminds his captain of his captain's wife and child: "Away with me! let us fly these deadly waters! let us home! Wife and child, too, are Starbuck's—wife and child of his brotherly, sisterly, play-fellow youth; even as thine, sir, are the wife and child of thy loving, longing, paternal old age! Away! let us away!"

For a moment Ahab seems to be persuaded: "By the green land; by the bright hearth stone! this is the magic glass, man; I see my wife and my child in thine eye." For a moment, but only for a moment. Of course, Ahab, the foundling son of a fatherless universe, possessed by his chase, making of his own son a foundling, renounces that homecoming, though he does tell Starbuck not to renounce it—to stay on the *Pequod* rather than chase the white whale, choosing for his second-in-command a happier fate than the one he chooses for himself. Cruelly, Melville drowns Starbuck, too. And Ishmael? Abob on the life buoy of Queequeg's coffin, he's rescued by the *Rachel*, "the devious-cruising Rachel, that in her retracing search after her missing children, only found another orphan."

Two days after my walk on the tundra, after a long series of connecting flights and a ride from JFK on the A train, I turned the key in the lock of my front door and lugged my ergonomic suitcase up three flights of stairs. On this my final return, at least my final one for a good long while, Beth and Bruno had made, with crayons and butcher paper, a big sign that read WELCOME HOME, PAPA. Now almost three, my son had entered an A. A. Milne phase. In the following days I read to him such classics as "In Which Piglet Is Entirely Surrounded by Water" and "In Which Christopher Robin Leads an Expotition to the North Pole."

Of a Sabbath afternoon, not long after my return, I take Bruno down to the park along the Hudson River. We gather pinecones from beneath the pine trees there, Bruno foraging, me watching for dog turds and syringes and fragments of broken glass, of which there are surprisingly few. Parents, like their children, are at the mercy of their nightmares and dreams. Unlike children, they're also at the mercy of memories—and I remember well the pleasure of exploring those secret, shadowy grottoes beneath branches and behind bushes. When Bruno's arms and my hands are full of pinecones, we carry them—dropping a few, among the sunbathers, along the way—to the river's edge. There we divide them into equal piles and take turns throwing them, as if launching a pinecone assault on the condominiums of New Jersey, or on the ferry-boats, behind which the sun has begun to set. To throw his, Bruno has to reach through the railings of the balustrade, a significant handicap. His barely make it into the water. I hurl mine as far as I can, trying to impress him.

Bruno can play this game for hours on end, never tiring of it. So can I. When all the pinecones are adrift, we follow them, hurrying among the joggers and strollers, stopping to peer down. Finally, we let them go, and I tell Bruno where the currents will take them—out of the harbor, onto the Atlantic, into the Gulf Stream, which will sweep them toward northern Europe, odds are, and then, if they chance to remain adrift, perhaps to the Arctic or else into the Sargasso Sea. Perhaps they will end up on some European or Brazilian or African or Asian coast and take sprout—these seeds of Manhattan pines.

This was pure fancy, of course. Pinecones aren't sea beans, after all. They evolved to germinate on a forest floor, not ride the currents. Still, it was fun to imagine.

ACKNOWLEDGMENTS

I've been lucky, before setting out to write this book and since, to have worked with and learned from a number of superlative editors: at *Agni*, Askold Melnyczuk; at *Harper's*, Lewis Lapham, Ben Metcalf, Colin Harrison, John Jeremiah Sullivan, Ellen Rosenbush, Jen Szalai, and Bill Wasik, among others; at *The New York Times Magazine*, Alex Star; at *Outside*, Will Palmer; and at Viking, Joshua Kendall, who faithfully and skillfully piloted this book through calms and storms, even when—especially when—it seemed to have sprung a leak.

But the editor to whom I'm most indebted is Lewis Lapham's successor at the helm of *Harper's*, Roger Hodge. It was Roger who, when I called him from my classroom one afternoon in the spring of 2005 and recounted what little I then knew about the legend of the rubber ducks lost at sea, saw what I saw: the germ of a story, the makings of a quest. At the time I was already under commission to write a piece about the elephants of the Detroit Zoo. "Have you heard about this?" I said to Roger. "In 1992, a bunch of rubber ducks fell off a container ship." Roger's response: the elephants of the Detroit Zoo could wait. Throughout the past five years, he's offered unceasing support, reading many drafts in progress, offering counsel and encouragement.

I'm similarly indebted to many of the people—scientists, beachcombers, naval architects, supernumeraries, toymakers—I met during the course of my travels. Eddy Carmack, Amy Bower, Robie Macdonald, Curtis Ebbesmeyer, Thomas Royer, John Toole, and Willa France were especially generous with their time. I'm also grateful to the Woods Hole Oceanographic Institution, where I underwent a weeklong crash course in oceanography.

I never would have made it home from the Arctic, book in hand, or at least in mind, were it not for the generous support of the Whiting

Foundation and the National Endowment for the Arts. I wouldn't have made it from idea to book without the guidance of my agent, Heather Schroder at ICM. Or from rough draft to final draft without the help of Emily Votruba, copy editor, and my intrepid research assistants, Joseph Bernstein and Justin Stone. Matt Fishbane, Lia Miller, and Ted Ross also helped check many facts. Alice Karekezi, Claire Jeffers, and Matt Flegenheimer spent long, tedious hours transcribing recordings. I've tried hard to get the facts right, but any errors are mine alone.

In Seattle, Pat, John, and Clare O'Connor repeatedly provided me with a safe harbor, fetching me from the airport and from the docks, feeding me, listening to me carry on about seafaring and childhood and currents. In Ann Arbor, Jeremiah Chamberlin and Natalie Bakopoulos let me use their home as a writer's retreat. In Manhattan, John and Angela Chimera provided a heroic amount of grandparenting while their son-in-law was off having a look at the watery part of the world. My own father helped in innumerable ways, not least of all by believing in this possibly foolish errand of mine.

To anyone who thinks that those who can't do, teach, I say, teaching *is* doing; and furthermore, many of those who "do," can't teach; and furthermore, many of those who can teach, also "do." The transition is a keen one, I assure you, from sailor to schoolmaster. I've been waiting decades to acknowledge some of the teachers who taught me: Mrs. Peskin (thanks for *The Little Prince!*); Mr. Rees (for everything); Mr. Tacke; Mr. Wright; Ms. Lyons; David Walker (also for everything); David Young (forgive me); Ralph Lombreglia; Charles Baxter; Eileen Pollack. Then there are the teachers at Friends Seminary with whom I taught— Maria Fahey and Sarah Spieldenner, especially. Then there are the students whom I taught, and from whom I learned—too many to name here, but I will mention one student in particular, the one who, by introducing me to the legend of the rubber ducks lost at sea, changed my life: Evan "Big Poppa" Drellich, no longer pudgy, now a journalist. Evan, may Luck Duck continue to bring you luck.

Finally, to my wife, I say: Beth, it was a long and at times arduous journey, but we made it.

SELECTED BIBLIOGRAPHY

I have swam through libraries and sailed through oceans.

—Herman Melville, *Moby-Dick*

Sitting on a sea-chest, and swaying to and fro because the ship compelled me to a figure of woe, I began to consider whether it was only the books about the sea which I had loved hitherto, and not the sea itself.

—H. M. Tomlinson, *The Sea and the Jungle*

Aristotle, and H. D. P. Lee. *Meteorologica*. Cambridge, MA: Harvard University Press, 2004.

Ariès, Philippe. *Centuries of Childhood: A Social History of Family Life*. New York: Vintage, 1962.

Barr, William. *The Expeditions of the First International Polar Year, 1882–83*. Calgary: Arctic Institute of North America, 2008.

Beebe, William, John Tee-Van, Gloria Hollister, Jocelyn Crane, and Otis Barton. *Half Mile Down*. New York: Harcourt, Brace, and Company, 1934.

Berger, John. *About Looking*. New York: Vintage International, 1991.

Berton, Pierre. *Arctic Grail: The Quest for the Northwest Passage and the North Pole, 1818–1909*. New York: The Lyons Press, 2000.

Browne, Thomas, and Robin Hugh A. Robbins. *Sir Thomas Browne's Pseudodoxia Epidemica*. Oxford: Clarendon, 1981.

Carle, Eric. *10 Little Rubber Ducks*. New York: HarperCollins, 2005.

Carson, Rachel. *The Sea Around Us*. New York: Oxford University Press, 1989.

Carson, Rachel. *Silent Spring*. New York: Mariner Books, 2002.

Clark, Eric. *The Real Toy Story: Inside the Ruthless Battle for America's Youngest Consumers*. New York: Free Press, 2007.

Colling, Angela. *Ocean Circulation*. Boston: Butterworth Heinemann, in Association with the Open University, 2001.

Conrad, Joseph. *Typhoon and Other Tales*. New York: Oxford University Press, 2002.

Conrad, Joseph. *A Personal Record* and *The Mirror of the Sea*. New York: Penguin Books, 1998.

Corfield, Richard. *The Silent Landscape: The Scientific Voyage of HMS* Challenger. Washington, D.C.: The Joseph Henry Press, 2003.

Cross, Gary. *Kids' Stuff: Toys and the Changing World of American Childhood*. Cambridge, MA: Harvard University Press, 1997.

Cudahy, Brian J. *Box Boats: How Container Ships Changed the World*. New York: Fordham University Press, 2006.

Darwin, Charles. *The Voyage of the* Beagle. New York: Harper, 1959.

Davidson, Peter. *The Idea of North*. London: Reaktion, 2004.

Davis, Jodie. *Rubber Duckie: It Floats*. Philadelphia: Running Press Book Publishers, 2004.

De Long, George W. *Our Lost Explorers: The Narrative of the Jeannette Arctic Expedition*. Hartford, CT: American Pub., 1882.

Deacon, Margaret. *Scientists and the Sea: 1650–1900: A Study of Marine Science*. Aldershot, England: Ashgate, 1997.

Delbanco, Andrew. *Melville: His World and Work*. New York: Alfred A. Knopf, 2005.

Ebbesmeyer, Curtis C., W. James Ingraham, Thomas C. Royer, and Chester E. Grosch. "Tub Toys Orbit the Pacific Subarctic Gyre." *Eos* 88.1 (2007): 1–12.

Ebbesmeyer, Curtis C., H. Drost, I. M. Belkin, and E. Carmack. "Bottles and Buoys Define the Atlantic Subarctic Gyre." *Eos* (Forthcoming 2010).

France, W. N. *Incunabulum: A Story of Beginnings in Verse*. West Conshohocken, PA: Infinity, 2007.

France, William N., et al. "An Investigation of Head Sea Parametric Rolling and Its Influence on Container Lashing Systems." *Marine Technology* 40:1 (January 2003).

Fraser, Antonia. *A History of Toys*. New York: Spring Books, 1972.

Glavin, Terry. *The Last Great Sea: A Voyage Through the Human and Natural History of the North Pacific Ocean*. New York: Greystone Books, 2000.

Grahame, Kenneth. *The Wind in the Willows*. New York: Atheneum, 1983.

Hardin, Garrett. "The Tragedy of the Commons." *Science* 162 (1968): 1243–47.

Harvey, Miles. *The Island of Lost Maps: A True Story of Cartographic Crime*. New York: Random House, 2000.

Hayes, Derek. *Historical Atlas of the North Pacific Ocean: Maps of Discovery and Scientific Exploration, 1500–2000*. Seattle: Sasquatch Books, 2001.

Hendrickson, Robert. *The Ocean Almanac: Being a Copious Compendium on Sea Creatures, Nautical Lore & Legend, Master Mariners, Naval Disasters, and Myriad Mysteries of the Deep*. New York: Doubleday, 1984.

Hessler, Peter. *Oracle Bones: A Journey Through Time in China*. New York: Harper Perennial, 2007.

Higonnet, Anne. *Pictures of Innocence: The History and Crisis of Ideal Childhood*. New York: Thames and Hudson, 1998.

Holling, Holling Clancy. *Paddle-to-the-Sea*. New York: Houghton Mifflin, 1941.

Horwitz, Tony. *Blue Latitudes: Boldly Going Where Captain Cook Has Gone Before*. New York: Picador, 2003.

"How Eider Down Is Gathered." *Harper's New Monthly Magazine*, May 1853: 784–86.

Kent, Rockwell. *Wilderness: A Journal of Quiet Adventure in Alaska*. Middletown, CT: 1996.

Kunzig, Robert. *Mapping the Deep: The Extraordinary Story of Ocean Science.* New York: W. W. Norton, 2000.

Levinson, Marc. *The Box: How the Shipping Container Made the World Smaller and the World Economy Bigger.* Princeton, NJ: Princeton University Press, 2006.

Liittschwager, David, and Susan Middleton. *Archipelago: Portraits of Life in the World's Most Remote Island Sanctuary.* Washington, D.C.: National Geographic Society, 2005. Princeton, NJ: Princeton University Press, 2006.

Lopez, Barry. *Arctic Dreams: Imagination and Desire in a Northern Landscape.* New York: Vintage, 2001.

Matthiessen, Peter. *Blue Meridian: The Search for the Great White Shark.* New York: Penguin Books, 1997.

Meikle, Jeffrey L. *American Plastic: A Cultural History.* New Brunswick, NJ: Rutgers University Press, 1997.

Miller, David William. *Exploring Alaska's Kenai Fjords: A Marine Guide to the Kenai Peninsula Outer Coast.* Seward, AK: Wilderness Images Press, 2004.

McCloskey, Robert. *Make Way for Ducklings.* New York: Viking, 1969.

McPhee, John. *Coming into the Country.* New York: Farrar, Straus & Giroux, 1985.

Melville, Herman. *Moby-Dick.* New York: W. W. Norton & Company, 2002.

Melville, Herman. *Typee, Omoo, Mardi.* New York: The Library of America, 1982.

Milne, A. A., and Ernest H. Shepard. *The Complete Tales of Winnie-the-Pooh.* New York: Dutton Children's, 1994.

Mintz, Steven. *Huck's Raft: A History of American Childhood.* Cambridge, MA: The Belknap Press of Harvard University Press, 2004.

Murphy, Dallas. *To Follow the Water: Exploring the Sea to Discover Climate: From the Gulf Stream to the Blue Beyond.* New York: Basic, 2007.

Nansen, Fridtjof. *Farthest North: The Epic Adventure of a Visionary Explorer.* New York: Skyhorse Pub., 2008.

Nash, Roderick Frazier. *Wilderness and the American Mind.* New Haven, CT: Yale University Press, 2001.

Newton, Isaac. *The Principia.* Amherst, NY: Prometheus, 1995.

"North Pole a Hole; Likewise the South." *New York Times,* April 4, 1908.

Pollack, Richard. *The Colombo Bay.* New York: Simon & Schuster, 2004.

Postman, Neil. *The Disappearance of Childhood.* New York: Vintage, 1994.

Pielou, E. C. *A Naturalist's Guide to the Arctic.* Chicago: University of Chicago Press, 1994.

Raban, Jonathan. *The Oxford Book of the Sea.* New York: Oxford University Press, 1992.

Royte, Elizabeth. *Garbage Land: On the Secret Trail of Trash.* New York: Back Bay Books, 2006.

Safina, Carl. *Eye of the Albatross: Visions of Hope and Survival.* New York: Henry Holt, 2002.

Schell, Orville, and David Shambaugh, eds. *The China Reader: The Reform Era.* New York: Vintage, 1999.

Schlee, Susan. *The Edge of an Unfamiliar World: A History of Oceanography.* New York: Dutton, 1973.

Smith, Craig B. *Extreme Waves.* Washington, D.C.: Joseph Henry Press, 2006.

Sontag, Susan. *On Photography.* New York: Farrar, Straus & Giroux, 1977.

Sparke, Penny, ed. *The Plastics Age: From Bakelite to Beanbags*. Woodstock, NY: Overlook Press, 1993.

Spence, Jonathan D. *The Search for Modern China*. New York: W. W. Norton, 1991.

Stearns, Peter N. *Anxious Parents: A History of Modern Childrearing in America*. New York: New York University Press, 2003.

Stevens, Wallace. *The Collected Poems*. New York: Vintage, 1982.

Thoreau, Henry D. *Walden and Resistance to Civil Government*. New York: W. W. Norton, 1992.

Thurman, Harold V., and Alan P. Trujillo. *Essentials of Oceanography*. Upper Saddle River, NJ: Prentice Hall, 2002.

"Told in an Arctic Diary; the Retreat from Fort Conger and Life on Cape Sabine. A Record of Suffering and Endurance. A Hitherto Unpublished Journal of One of the Members of the *Lady Franklin* Bay Expedition—Its Romantic History—Found Three Thousand Miles from Where It Was Written—the Interesting Story of Private Roderick Schneider." *New York Times*, December 3, 1891.

Tomlinson, H. M. *The Sea and the Jungle*. New York: Time Incorporated, 1964.

Traxel, David. *An American Saga: The Life and Times of Rockwell Kent*. New York: Harper & Row, 1980.

Twain, Mark. *Mark Twain's Letters from Hawaii*. Honolulu: University of Hawaii Press, 1975.

Whitman, Walt. *Leaves of Grass, and Other Writings*. New York: W. W. Norton, 2002.

West, Nathanael. *Miss Lonelyhearts* and *The Day of the Locust*. New York: New Directions, 1962.

Wood, Amos L. *Beachcombing the Pacific*. West Chester, PA: Schiffer Publishing, 1987.

Zipes, Jack, et al. *The Norton Anthology of Children's Literature: The Traditions in English*. New York: W. W. Norton, 2005.

NOTES

1   Confusingly, the conventions for indicating the directions of currents contradict those for indicating the direction of winds. An easterly wind blows from east to west; an easterly current flows west to east.

2   Later, I looked up the passage in *The Day of the Locust* in which the phrase "sargasso of the imagination" appears. As he contemplates the jumbled stage sets on that Hollywood backlot, Nathanael West's protagonist is reminded—just as I'd been reminded of *The Day of the Locust* when listening to Ebbesmeyer—of something he'd once read, an adventure novel for boys called *In the Sargasso Sea* by a certain Thomas Allibone Janvier. Curious what West's protagonist had in mind, I procured a copy of *In the Sargasso Sea*. Published in 1898, the novel recounts the fantastical adventures of a twenty-three-year-old mechanical engineer named Roger Stetworth who at the outset of his journey—as he freely admits with the benefit of hindsight—was "very young and very much of a fool." While crossing the Gulf Stream aboard a slave ship called the *Golden Hind*, Stetworth strikes up a conversation with a kindly mate, the only kindly mate aboard, and this kindly mate tells him the legend of the Sargasso Sea, comparing the flotsam stranded there to "a sort of floating island . . . as big as the area of the United States." The second mate's description, Stetworth says, "took a queer deep hold upon me, and especially set me to wondering what strange old waifs and strays of the ocean might not be found in the thick of that tangle if only there were some way of pushing into it and reaching the hidden depths that no man ever yet had seen." Thereafter ensues a series of improbably disastrous and fortuitous plot twists that will—surprise!—deliver Stetworth into those hidden depths, where he discovers what he variously describes as a "strange floating continent," "a hideous sea-labyrinth," "a graveyard" of dead ships, "a hideous wilderness," "a great marine museum."

3   Or, better yet, something like,

> There be many shapes of mystery,
> And many things God makes to be,
> Past hope or fear.
> And the end men looked for
> Cometh not,
> And a path is there where no man sought.
> So hath it fallen here.

(Lines by Euripides, chosen by Evan Connell as the epigraph for *Notes from a Bottle Found on the Beach at Carmel*.) The message inside the Orbisons' bottle reads as follows: "SOS. I'm on a small Pacific island in the South Pacific. Paul." Saint Paul, perhaps?

4   Sitting in the Orbisons' living room, I remember something I'd read in Miles Harvey's *The Island of Lost Maps: A True Story of Cartographic Crime*. Harvey asks the psychologist Werner Muensterberger, author of *Collecting: An Unruly Passion*, to explain the mind of the obsessive collector, and in reply Muensterberger tells him about a man who stopped hunting big game and started buying African art. In the psychologist's opinion, both trophy hunting and art collecting could be traced to the primal animism of hunters and gatherers. "There is reason to believe that the true source of the habit is the emotional state leading to a more or less perpetual attempt to surround oneself with magically potent objects," Muensterberger writes in his book. I can't help wondering if the buying habits of Americans don't derive from that emotional state that Muensterberger describes. Perhaps we are hunter-gatherers of the mall. Perhaps our perpetual dissatisfaction derives from the ease of the hunt and the never-ending supply of magically potent objects that beckon us, only to lose their potency once possessed, because enchantment has an evanescent half-life. Perhaps this explains the ten thousand varieties of rubber duck.

5   In fact the name of the town is a contraction of *shee atika*, Tlingit for "the ocean side of Shee Island." What exactly *shee* means, no one seems to know.

6   On the first floor of his condo in a kind of rec room was a great ziggurat of cardboard boxes that he hadn't bothered to unpack, and upstairs in the carpeted living/dining room were a few Christmas decorations he hadn't bothered to put away—two stuffed snowmen on a windowsill, a bowl of sparkly balls on a counter, a string of unlit Christmas lights entangled in an ivy plant, the leafy vines of which tumbled from ceiling to floor down a macramé hanger. Neither had Pallister bothered to remove Jane's message from the answering machine. He still hoped to win her back.

As it happened, Jane was there in the kitchen that night, but only briefly, to bake a batch of homemade cookies. They were for her three sons, who'd been out at Gore Point for the past two weeks. In their absence, I felt a bit like a surrogate, like an exchange student adopted by a host family of strangers, a dysfunctional host family of strangers. It would be the next day before Pallister confided in me about Jane's desertion. But like many people brought up in unhappy families, I felt I possessed a kind of sixth sense for marital discord. It didn't take long for me to detect in that kitchen, beneath the homey smell of melting marshmallows and morsels, the old familiar signals—the unrequited glance, the small talk freighted with simmering subtextual resentments that threatened to boil over any second into a quarrel. Pallister plucked an oatmeal cookie from a baking sheet. Jane flashed him a glare, which he ignored.

7   "We came to this new land, a boy and a man, entirely on a dreamer's search," Kent writes; "having had a vision of a Northern Paradise, we came to find it. With less faith it might have seemed to us a hopeless thing exploring the un-

known for what you've only dreamed was there. Doubt never crossed our minds. To sail uncharted waters and follow virgin shores—what a life for men!" Kent wasn't the first to dream that dream of a Northern Paradise, nor was he the last. What distinguishes him from other dreamers is the way in which he tried, and in some ways succeeded, to reconcile painting and adventuring with fatherhood.

Rowing across Resurrection Bay on a "calm, blue summer's day" just like this one, the Rockwell Kents had chanced upon an old man in a dory, to whom they explained their wish of finding some forgotten cabin in which to spend a year. "'Come with me,' the old man cried heartily, 'come and I show you the place to live.'" He led them to Fox Island, named for the fox ranch that the old man had set up there, and offered them an old cabin rent-free.

It's hard to imagine a father repeating Kent's experiment today—pulling his son out of school and dragging him off to spend a sub-zero winter, a continent away from his mother, on a remote island without medical facilities or telephone lines or playmates other than magpies and porcupines. The paintings and drawings and woodcuts Kent completed there made his reputation and rescued his family from financial ruin. As for the son, fifty years later, now a balding, six-foot-four biologist, Rockwell the Younger would tell his father that the "year we spent together on Fox Island was the happiest of all my life," or so his father reports in his preface to the 1970 edition of *Wilderness*.

In their one-room cabin, they shared everything. They slept in the same bed. They cleared trees together, cut firewood together, cooked together, ice-skated together, "holding hands like sweethearts." They even drew and painted together. For entertainment, Kent brought along a small library of books, and at bedtime, by lamplight, he would read to his son from *Robinson Crusoe* or *Treasure Island* or the fairy tales of Hans Christian Andersen. *A Journal of Quiet Adventure*, Kent's book is aptly subtitled. It contains moments of genuine danger, but little of the testosterone addled man-versus-wild drama one usually encounters in louder adventure narratives, and when those moments of danger do arise, young Rockwell responds to them in poignantly childish ways. One day, rowing back from Seward, where they'd gone for supplies, father and son are ambushed by a storm. "Father," young Rockwell pipes up as the elder Kent is laboring furiously at the oars, "when I wake up in the morning sometimes I pretend my toes are asleep, and I make my big toe sit up first because he's the father toe."

8 A landscape similar to the one Kent encountered upon arriving at Fox Island. Fox Island: "Twin lofty mountain masses flanked the entrance and from the back of these the land dipped downwards like a hammock swung between them, its lowest point behind the center of the crescent. A clean and smooth, dark-pebbled beach went all around the bay, the tide line marked with driftwood. . . . Above the beach a band of brilliant green and then the deep, black spaces of the forest."

9 The woodcuts and drawings and paintings that Rockwell Kent produced during his Alaskan retreat are far more moving to me than the landscapes that inspired them. What makes them moving isn't the landscapes, per se, but the human figures whose inner lives the landscapes serve to dramatize. Some of

Kent's images realistically depict the daily life of father and son—father and son sawing a log; father and son sharing a meal; father and son surveying the world from the summit of a mountain; father greeting the sunrise with outspread arms; son riding a driftwood stick down a pebble beach, pretending to be Sir Lancelot.

But then nature, bearer of truth, begins to intrude: "Is it to be believed that we are here alone, this boy and I, far north out on an island wilderness, seagirt on a terrific coast!" Kent exclaims as fall is turning to winter. "It's as we pictured it and wanted it a year and more ago—yes, dreams come true." A month later, as the sunless Alaskan winter sets in, so does disenchantment. The dream shades into nightmare. Kent looks "to the sun's going with a kind of dread." Having come to Fox Island seeking artistic inspiration, what he often feels, contemplating a day's just-finished work, is "repugnance." Even when the work goes well, the triumph is fleeting. "Over to-day's painting I'm filled with pride," he wrote on October 14; "it will be equaled by to-morrow's despair over the very same pictures." The capricious muse isn't his only demon. With time he grows desperately lonely: "I have terrible moments, hours, days of homesick despondency . . . for my family. There are times when if I could I'd have fled from here in any raging storm."

This homesickness might come across as a touchingly sentimental bit of familial devotion; might, but only if one knows nothing of Kent's biography. Out on Fox Island, he may have been a model—or even a magical—father to Rockwell III, if not to the three children he'd left behind in upstate New York. But he was no family man. It seems that Kent's wife, Kathleen, whom he would later divorce, only permitted her nine-year-old to tag along with her husband on his latest extravagant saunter because she feared that if she didn't, he'd instead bring his German mistress, Gretchen, who had, says one biographer, helped comfort Kent through previous "suicidal depressions"— depressions he'd fled to Fox Island hoping to escape. He swore to Kathleen that he'd broken off the affair, but on Fox Island he and Gretchen continued to correspond. At the same time, Kent had nightmares that Kathleen was cheating on him. From these nightmares, he'd wake in a jealous rage. Kathleen didn't accompany him to Fox Island herself because of a previous misadventure in Newfoundland, during which all their children had caught a nearly fatal case of whooping cough.

One gets the impression that without the company of his son, Kent's lonely thoughts that winter on Fox Island might once again have turned suicidal. And one also gets the impression that without animal playmates, the younger Kent might have gone bonkers too.

Just as little Rockwell delights in the local flora and fauna, the elder Rockwell delights in his son, who increasingly becomes the journal's main subject. "He is beautiful after his bath," Kent writes at one point. At another: "We have a good time washing dishes, racing—the washer, myself—to beat the dryer. Rockwell falls down onto the floor in the midst of the race in a fit of laughter." What Kent discovers on Fox Island isn't mainly the transporting beauty of Alaska but the poignant beauty of his son.

10   Compared with Kent's moody, intimate account of his long dark winter of the soul, John Muir's *Travels in Alaska*, the genre's seminal work, is almost un-

readably monotonous. Wandering among glaciers and mountains, Muir proceeds from one ecstatic rhapsody to another, scattering abstract adjectives like wildflower petals as he goes. Notes the historian Roderick Frazier Nash in the "Alaska" chapter of *Wilderness and the American Mind*, for Muir, "everything was 'glorious' or 'sublime' or 'grand' or 'glowing.'" The human figures, Muir himself included, recede into the background, outshone by the blinding glory of the landscapes.

In his defense, it helps to remember that at the time of his Alaskan travels, before the Klondike gold rush of 1898, most Americans still considered Secretary Seward's territorial purchase a folly. Alaska was a wilderness in the old, pejorative sense of the word—the sort of wilderness that writers used to describe as "howling"; an inhospitable wasteland of worthless ice the furry resources of which the Russians had already depleted; in short, a damn'd unhappy part of the world. Muir helped the American public reimagine Alaska, and with it the meanings of wilderness, meanings with which the word "Alaska" has since become synonymous, at least in the minds of outdoorsmen and ecotourists. In Muir's time and ours, *wilderness* was *civilization*'s antonym, but thanks to Muir and other Romantic pantheists, the connotations of the antonyms have blurred or even, in some minds—Raynor's and Pallister's, for instance—reversed. Now civilization is often the wasteland, and wilderness the source of all that is beautiful and redemptive and good.

In certain misanthropic moods, I feel the same. Mostly, though, I feel an irresolvable ambivalence, torn between my fondness for hominids, who are splendid and hilarious as well as idiotic, and my wish that we could populate the world without ruining it.

11   Allow me to attempt to ignite similar intimations in yours. Exterior:
From the wooded bank of a sparkling stream, the camera frames an Indian paddling a canoe, a canoe almost identical to the miniature one that, in *Paddle-to-the-Sea*, goes on that long journey to the Great Salt Water. Its bow and stern curl into matching crescents like the toes of elfin shoes. Its hull appears to be made of moose skin, lashed to the wooden frame perhaps with venison thongs, and the Indian paddling it appears to be made of Indian. He isn't.

Now the camera leaps into the canoe's bow. On the sound track, a timpanist plays a martial tattoo meant to evoke a war dance. Close-up of Indian looking stoical in his buckskin getup as he paddles once to starboard, once to port, fringe dangling Daniel Boone–like from his sleeves. A big feather protrudes from his wig, the long black braids of which make him look almost schoolgirlish. The timpani give way to snares. Cellists play an ersatz Cherokee theme.

The forest stream broadens, and so does the symbolism. The Indian's canoe, once again viewed from the wooded shore, is a bicorn silhouette amid dazzling sparkles. As in *Paddle-to-the-Sea*, stream has led to harbor—a Californian harbor, judging from the branches of the trees in the foreground, which if I'm not mistaken are those of a Monterey cypress. For just a moment, we enter the Indian's point of view. Indian-cam. In the water, he spies a foreboding sign: a floating scrap of newspaper, which with an oar-stroke to starboard Espera "Iron Eyes Cody" De Corti sends swirling. To make the meaning of this portent clear the sound track turns chromatic and modernist. An ugly crescendo of brass as the camera pans back, revealing—*the horror, the horror*—a port: cranes, freighter, tugboat, and on a hillside in the background, those squat silos resembling giant tin cans that one often sees in the vicinity of ports.

Now the theme, orchestra in full swell, sounds like something from a Hollywood western, a bit like the theme to the TV show *Big Valley*, as if the Indian were paddling into battle. Shot of refineries, smoke billowing sinisterly around them. At last the Indian hauls out on a trashy little beach. The music goes quiet and elegiac, and, as the Indian strides in a kind of stupor of stoicism to the shoulder of a highway, a somber voice-over begins: "Some people have a deep, abiding respect for the natural beauty that was once this country"—paper bag of what appears to be orange peels and french fries, flung from window of passing car, explodes across Indian's moccasins—"and some people don't." Iron Eyes Cody looks into the camera, which zooms in on the big shiny tear rolling down his cosmetically tanned Sicilian cheek. Tagline: "People start pollution. People can stop it."

12   In hindsight, looking back from the summer of 2010, those 200,000 gallons, dwarfed by the millions gushing into the Gulf of Mexico, would seem like a portent, and the Deepwater Horizon disaster less an aberration than the culmination of one oil company's negligence.

13   By way of illustration, an anecdote: If you've ever seen a photograph of a bald eagle, the odds are good it was taken in Homer. The odds are good, in fact, that it was taken just a few hundred yards from the end of the Homer Spit, across the road from Brad Faulkner's house. There, between a muddy little campground and a row of three-story waterfront condominiums erected not long ago by a real estate developer, is a little shack decorated with deer antlers. A fence encircles the shack's tiny yard, and protruding above the fence is a dead tree. If you look closely you will see talon scrapings on the branches of the dead tree. You will also see a great deal of eagle guano.

Until she died not long ago, the woman who lived in the shack, Jean Keene, was locally known as the Eagle Lady. Keene moved to the Homer Spit in 1977 and found a job at a nearby fish-processing plant. Soon thereafter she began bringing home scraps of fish to feed to the local population of bald eagles. Her flock steadily grew, and so did the flock of photographers. The eagles came for the free fish. The photographers came for the free eagles. By the time Keene died, between two hundred and three hundred bald eagles had become regular customers. The number of photographers is harder to determine, but by one estimate, 80 percent of the photographs of bald eagles sold every year—to illustrate newspapers and magazines and advertisements and the *Colbert Report*, but also souvenir place mats and calendars and postcards—were taken beside that campground overlooking Kachemak Bay, whose silvery waters, snowcapped mountains rising up behind them, make for a spectacular backdrop.

In such photographs, you'll see bald eagles in close-up, yellow beak turned in profile, so that the famous eagle eye gazes out at you. You'll see eagles in flight, their black wings outspread angelically, yellow talons outthrust. You'll see action shots of an eagle swooping down onto the silver water in an explosion of spray, a silver fish visible in its yellow claws. What you won't see, except in news stories about her, is the Eagle Lady or her shack. You might see the bespattered branches of her dead tree, but if so the photo will be cropped, leading the viewer to imagine an old-growth forest. You won't see the campers and their colorful tents, or the hand-painted school bus inhabited by latter-day merry pranksters, or the aluminum fishing boat up on cinder blocks in Brad Faulkner's yard, or the muddy cars and trucks in the campground parking lot, or the waterfront condominiums. Or the flock of bald eagles scrambling after fish scraps like pigeons after crumbs.

14 By then I'd received from Pallister the following e-mail:

> Hi Donovan,
>
> You won't believe this . . . *Opus* was involved in a motorcycle accident. Yesterday, a lady across the street lost control of a new Harley Sportster her husband was teaching her to ride. She slammed into the back of *Opus*. She nearly severed her leg in a couple of places when it got caught between the bike and the end of the boat trailer frame. She also crushed a wrist and bruised her face. Her husband came over last night and said she would recover, that the docs had bolted and plated her back together. She will suffer a long recovery. *Opus* took a pretty good lick, tore a large hole in the transom, broke off the hydraulic cylinder for the trim tab, bent the bottom of the outdrive, tore off a trailer light and wrapped the trailer fender around the tire. I can fix most of it pretty quickly, but it will still take a couple of days.
>
> We can't get a helicopter to Gore Point until August 19–20. The guys decided unanimously to stay out there and continue cleaning adjacent bays until the helicopter comes whether they get paid for that or not. They've all caught Ted's mission fever. They found another beach on the southwest corner of Port Dick they refer to as Mini Gore Point. It was blasted with debris, but the area isn't nearly as big as the isthmus. They have finished cleaning all of Port Dick and now are moving into Tonsina Bay.
>
> You made a good decision on Montague . . . the weather socked in and will remain that way for awhile.
>
> Take care.
>
>
> Chris

In the following days, other unforeseen events would occur. Way out on Resurrection Bay, the *Johnita I*, John Cowdery's other yacht, burst into flames. Fishermen came to Cowdery's rescue. Cowdery attributed the fire to faulty wiring, but now a pesky reporter from the *Anchorage Daily News* was sniffing around for evidence of arson and insurance fraud. She wouldn't find it, but she would, eventually, find incriminating evidence of graft—not enough to get Cowdery indicted. Later, though, on unrelated corruption charges, he'd serve three years in prison. During the course of her investigations, the reporter called Pallister with questions about the *Johnita II* and his partnership with

Cowdery. The *Anchorage Daily News* would run a story that in passing mentioned GoAK's attempt to win funding from the state legislature—how
Cowdery had helped secure the allocation, how Pallister had hired two lobbyists. "I told [the lobbyists] right out of the get-go, 'Listen, John Cowdery
and I are partners in a boat. Keep him the hell out of this,'" the article would
quote Pallister as saying. "'I don't want any linkage with John Cowdery or
Veco or anything else.'"

To me—without a shred of evidence, let it be clear—Pallister would insin-
uate his own suspicions. He was convinced he knew who'd put this *Daily
News* muckraker on his trail: Bob Shavelson. "Next year I might fold up shop,"
Pallister would tell me. "I didn't get into this to go through this crap—have
people digging through my personal business, telling lies about me. I got into
this to clean up beaches, you know?"

15 The toxicity of PCBs is well established, and their story has become a famous
chapter in the annals of environmentalism.

PCBs were first synthesized in 1881 and first brought to market as a non-
flammable insulator in electrical transformers, a safer alternative to mineral
oils, in 1929. In the U.S., until they were banned in 1979, their primary producer was the Monsanto Corporation, whose sorcerers of the lab, by playing
variations on the theme of chlorine and carbon, eventually found myriad other
commercial applications for this family of compounds. They polymerized it
and sold it as an insulator for copper wiring. They added it to varnishes and
paints. By the 1970s, PCBs were ubiquitous, despite mounting empirical evidence of their toxicity. In response to that evidence, Monsanto and its competitors deployed the tactics pioneered to great success by the tobacco companies:
they produced studies that minimized the risks and confused the public. In the
industrialized nations of the West, commercial production of the substances
had all but ceased by 1989, but to this day in the environment PCBs remain
ubiquitous, seeping out of landfills and superfund sites and electrical transformers, persisting in sediments and drifting on the surface of the sea.

What's so bad about PCBs? Until the mid-eighties, according to the industry, nothing. Fears flamed by academic and government scientists were
overblown, Monsanto and its competitors claimed. Finally, in 1987, they surrendered. In a study funded by the EPA, independent scientists evaluated the
literature and deemed the evidence for the "carcinogenicity of PCBs" to be
"overwhelming," and a peer-reviewed, industry-funded study found that
"every commercial PCB mixture tested caused cancer." For oceanographers
the data were more troubling still: "It is very important to note," the EPA
notes, "that the composition of PCB mixtures changes following their release into the environment. The types of PCBs that tend to bioaccumulate in
fish and other animals and bind to sediments happen to be the most
carcinogenic."

And PCBs aren't only carcinogenic. They impair the immune system, increasing vulnerability to opportunistic viruses. They reduce fertility. In laboratory tests conducted on monkeys, they've been shown to stunt "neurological
development," specifically "short term memory and learning." They concentrate in breast milk. And according to the laws of biomagnification, the predators at the top of the food chain carry the heaviest contaminant burden. The

killer whales of the North Pacific are among the most intoxicated mammals on earth. So are human populations that eat a fishy diet. So are the albatrosses that nest on Laysan Island. Look again at the plastic duck Bryan Leiser found just north of Gore Point; you won't see any of this. You'll see a faded, sea-battered bath toy, an icon of childhood. "Everything tells a story," Ebbesmeyer likes to say. Perhaps, but not all stories are visible, no matter how illimitably long you study the evidence. Some stories only a mass spectrometer can tell.

16 Additionally you will pass a sign reading MARK TWAIN MONKEY POD TREE. In his Hawaiian travelogue, Twain makes mention of riding a mule to the summit of Kilauea, of visiting the scene of Captain Cook's death, of trying in vain to procure coconuts by hurling rocks at them, of enjoying himself lecherously at a hula dance, of investigating the burial cave of a dead Hawaiian king (where he and his traveling companion blundered upon a skeletal hand in a burial canoe). He even makes mention of a cistern tree that collects freshwater and a mango tree that collects exceptionally delicious mangoes. But nowhere does he mention planting a monkey pod tree, or of planting any other sort of tree. Perhaps he did and neglected to mention it. The tree that he supposedly planted blew down in a hurricane in 1957. This impostor grew from a salvaged shoot.

17 Whether or not it was now, Moore's Spanish hadn't always been perfect: he'd intended to name his research foundation after an endangered species of Mexican seaweed that he thought was called *algalita*. In fact, the Spanish for the weed is *alguita*—little algae. Hence the name of his catamaran. Until his fateful detour into the North Pacific in 1998, protecting and restoring kelp forests had been one of the foundation's main missions.

18 In the spring of 2010, cetologists in Washington State would use similar forensic methods, investigating the stomach of a thirty-seven-foot-long gray whale that had washed up, dead, near Seattle. Among the items they found: duct tape, electrical tape, fabric (miscellaneous), sock, sweatpant leg, towel, fishing line, golf ball, green rope, nylon braided rope, red plastic cylinder, black fragments, CapriSun juice pack, miscellaneous bag material (times twenty-six), red plastic stake, sandwich bag, ziplock bag, rubbery string, surgical glove, "unknown shell-like material, possibly natural."

19 It didn't help that in a book on oceanography, I'd recently learned why NOAA research vessels no longer permit swim calls—not long ago, a scientist swimming in the Caribbean lost a leg to a tiger shark. Nor did it help that Amy Young had told me about a surfer friend of hers who'd lost a ham-size hunk of thigh to the jaws of a shark. Nor did it help that, aboard the *Alguita*, I'd been reading and admiring Peter Matthiessen's *Blue Meridian*, about a search for the great white. Matthiessen includes many impressively detailed accounts of attacks by what South Africans call the "white death."

20 Like those facing the Laysan albatross, the threats to monk seals, a biologist named Bud Antonelis explained to me, are legion. There's the increase in shark predation caused in part by the dredging of lagoons. There's the loss of breeding grounds to rising, warming seas. There's the toxic waste dumped by the U.S. military, the toxoplasmosis contracted from cat pee, the spread of

West Nile virus. Of all the perils monk seals face, the most memorable one to my mind is this: changing phocine demographics have led to a shortage of females, and in response to this shortage horny bull seals have grown murderously, pedophilically aggressive, drowning and suffocating female pups while attempting to mate with them, thereby reducing the female population further still. And although the number of entanglements has fallen since the cleanup efforts began, the rate of entanglement in 2004 was actually seven times higher than in 2000. What accounts for the spike? During El Niño years such as 2004, the boundaries of the North Pacific Subtropical Convergence Zone shift south, engulfing the monk seal's habitat. Even in more typical years, the ocean deposits an estimated fifty-two tons of debris on the Northwestern Hawaiian Islands.

21 If I don't make it to Kamilo Beach then at least I'd like to make it to Green Sand Beach, which I read about in *The Rough Guide to Hawaii*: "It is greenish in a rusty-olive sort of way, but if you're expecting a dazzling stretch of green sand backed by a coconut grove you'll be disappointed. The only reason to venture here is if you feel like braving a four-mile hike along the oceanfront, with a mild natural curiosity at the end." The green sand is pulverized olivine. Olivine sounds like a cosmetic product or else a butter substitute but is in fact a mineral forged in volcanoes.

22 Later, traveling alone in Guangzhou, I would hire a twenty-two-year-old freelance translator who called herself Amy, a name she preferred to the Chinese one her parents gave her. Her ideas of America were strongly influenced by her favorite television show, *Sex and the City*, the appeal of which was obvious: the lives of its characters were like fantasy versions of Amy's own. She too was a single, independent woman who'd come to the big city seeking excitement and glamour, and although she'd so far attained more of the former than the latter, she was by China's standards a success, an entrepreneurial escapee from Guangdong's formerly rural countryside, lifted up by her good English from the life of drudgery to which some of her friends and family are still condemned. When she was growing up, everyone in her town worked in the fields. Now they work in a factory making electrical cables, she said.

She was the first woman in her family to have gone to college, the first to have moved away, the first to have traveled abroad on business (to Hong Kong, Vietnam, and once to Stuttgart). Even though she struggled to pay her rent, she was helping to put her younger brother through school. Evenings she taught English to salespeople. By day she worked as a translator and sourcing agent for foreign businessmen who came to inspect factories or find suppliers. "My mother's Buddhist," she told me, "but I'm not anything," except, that is, "a workaholic." She almost never returned home to visit her family—no trains go there, you have to take three different buses, and furthermore, the mosquitoes there are really big. "It's crazy!" she said of the mosquitoes. "Crazy" was her favorite English word. The characters on *Desperate Housewives* were "really crazy." Her friends told her she was crazy for working so much. Someday, she'd like to live abroad, preferably in Australia, though she'd settle for Germany. In the meantime, she spent most of her free time on her cell phone or at her computer, surfing the Web, keeping up with her clients via Skype, the Internet phone service. Afraid that she wasn't skinny or pretty enough, yet unafraid to speak her mind ("Put me in your book!" she said when I told her I was a writer), she reminded me a lot of the young Americans I used to teach.

23   Vendors sell black-market goods openly on the streets of Guangzhou. One night outside my hotel there, a man would approach me with what looked like a deck of cards, fanning them with his thumb and saying, "Would you like some girls?" Each card showed a different photograph.

24   MGA Entertainment, the maker of Bratz dolls, insisted at the time that the National Labor Committee had mischaracterized the conditions in this factory, though similar practices have been documented at other Guangdong toy factories.

25   As the ferry made its way up the Pearl River, there would materialize out of the smog, like ghost ships, idle barges and little junks and great container ships awaiting repair. At one point the span of a bridge suddenly loomed overhead, an unfinished bridge; the span terminated in midair. Zooming toward China, I paged through Jonathan D. Spence's *In Search of Modern China,* and stumbling upon the following passage, written in the early 1800s by a scholar named Gong Zizhen, I thought of the Pearl River's smog: "When the wealthy vie with each other in splendor and display while the poor squeeze each other to death; when the poor do not enjoy a moment's rest while the rich are comfortable; when the poor lose more and more while the rich keep piling up treasures; when in some ever more extravagant desires awaken, and in others an ever more burning hatred; when some become more and more arrogant and overbearing in their conduct, and others ever more miserable and pitiful until gradually the most perverse and curious customs arise, bursting forth as though from a hundred springs and impossible to stop, all of this will finally congeal in an ominous vapor which will fill the space between heaven and earth with its darkness."

26   About that breakdown and vanishing act, it's the old, familiar story: the parent, sometimes a sad-sack father, sometimes a clinically depressive mother, takes flight from his or her life of quiet desperation and runs off in search of some Northwest Passage of the mind or heart. My mother searched for hers every few years and finally, failing to find it, tried unsuccessfully to overdose on tricyclic antidepressants. There was institutionalization. She recovered, partially. Today she lives alone, unemployed, in a condominium her ex-husband, my father, bought for her so that she would stop showing up, needy and homeless, on his sons' doorsteps. On the rare visit I pay her, she will occasionally try to resurrect that old alliterative sobriquet, Donovan Duck, speaking in a baby voice, as if I were still two, as if time could be turned back. Every so often, my mother will dig out a snapshot of me as a child and mail it to me. Her reasons for choosing a particular photograph are always a bit mysterious. I study them for significance. Not long ago, she sent a photograph in which, naked, ten months old, sitting in the bath across from my brother, I appear to be attempting to gnaw through my rubber duck's skull. The picture is dated January 1973.

27   Writes the art historian Anne Higonnet, "The modern child is always the sign
     of a bygone era, of a past which is necessarily the past of adults, yet which,
     being so distinct, so sheltered, so innocent, is also inevitably a lost past, and
     therefore understood through the kind of memory we call nostalgia."

     Take a look, for instance, at one of the most famous representations of
     childhood, Gainsborough's *The Blue Boy*, for which the painter dressed up a
     neighbor's son in a luxurious costume so dazzling it seems spun from sky. To
     a contemporary eye, the lad's blue satin suit, beribboned shoes, and lace collar
     look old-fashioned, but no more so than the painting itself. It was painted in
     1770, after all, when people favored such fancy getups, we assume. In fact, the
     outfit was already old-fashioned in Gainsborough's time, as was the pristine
     natural landscape over which the boy so aristocratically presides. What to a
     contemporary eye appears to be the portrait of a noble scion, commissioned
     no doubt by his proud parents, is actually a bit of *mise-en-scène*, an eighteenth-
     century precursor to photographs of children dressed like shepherdesses or
     cowboys and posed before painted scrims at Coney Island or Sears.

     A hundred years after Gainsborough painted *The Blue Boy*, the Pre-
     Raphaelite painter John Everett Millais gave the world another famous portrait
     of childhood, *Bubbles*. Here again is a child fancifully dressed in a lace collar
     and an antiquated suit, this time of green velvet rather than blue satin. Millais's
     preschooler is younger than Gainsborough's tween, and he is also cuter—
     porcelain-faced, pink-lipped, with a curly mop of strawberry blond hair.
     Whereas the Blue Boy gazes directly at the viewer, assuming the conventional,
     almost cocky pose of an adult, Millais's moppet has been transfixed by the bub-
     ble he has just blown from a pipe. He is, it seems, assuming no pose at all. The
     original title of the painting was *A Child's World*—that realm so unlike our own
     into which the painting offers a voyeuristic glimpse. Rather than presiding over
     a landscape, the boy sits on a block of weathered stone in a garden of potted
     plants, the sort of place one might encounter in the pages of Beatrix Potter. The
     scenery has grown more domestic, more suburban, but it still evokes a bygone,
     preindustrial past. Thanks to Pears Soap, which purchased the copyright to
     Bubbles and turned it into a print advertisement, Millais's mass-produced image
     eventually eclipsed Gainsborough's museum piece in the iconography of child-
     hood, adorning collectible plates and ephemeral toiletries as well as the pages of
     magazines. Soap bubbles have been a symbol of innocence ever since.

     In her amply illustrated study *Pictures of Innocence*, Anne Higonnet identifies
     several subgenres of child portraiture. Along with children in costumes (the
     genre to which *Bubbles* and *The Blue Boy* belong), there is, for instance, the genre
     of children with pets. Antique costumes, Higonnet argues, make the child seem
     timeless; pets make them seem like animals—less conscious than us, less human,
     more natural: "Usually the pets are small and cuddly—kittens, puppies and bun-
     nies were favorite choices—cueing the viewer's interpretation of the child."

28   The reefers must be plugged into special electrified bays or their contents will
     spoil. Hazardous chemicals must be stored in special compartments be-
     lowdecks to minimize the danger of a conflagration or spill. The lightest con-
     tainers must go on top, or the forces impinging on the bottom containers
     when the ship begins to pitch and roll may be greater than they can bear, and
     the entire stack will topple, sending the top containers tumbling overboard.

The weight to starboard must balance the weight to port, or the ship may dangerously list.

29   A surprisingly good poet, considering that her autobiographical, book-length poem, called *Incunabulum*, is self-published. A passage:

> I've woven dreams from streams and rivers
> and seas of water. See what has become
> of the vision that guided my hand: a strand
> of fresh Lake Wisota water from the cove
> sheltering sailboats of a seven year old's
> imagination. Threads from the swift Chippewa,
> that swept my message in a bottle down
> the Mississippi to the sea. Whole bolts
> of wild, salty spray, rough-nubbed twill
> from Greenland's southern coast. Grief-
> stitched, musty, smelling of the unrescued,
> of the drowned, and the faint, sickly-sweet
> stain of fuel slicking the spot where
> their ship had been.

30   While the customs agents inspected my passport, I thought of Melville, who, in 1866, once again living in the insular city of the Manhattoes, stopped trying to earn a living from his decreasingly profitable writing and took a day job as a customs agent. For two decades he clung, Hawthorne's son-in-law later put it, "like a weary but tenacious barnacle to the N.Y. Custom House."

31   One afternoon at Woods Hole I listened to a behavioral biologist explain the surprising discovery that killer whales exhibit the rudiments of culture; hunting methods, taught to the young, can vary from pod to pod. In the Woods Hole necropsy lab, I visited a sub-zero meat locker where dolphins in yellow body bags hung like dry cleaning from a motorized rack. "Know how you anesthetize alligators?" the lab technician quipped. "Stick 'em in the fridge."

32   What is the ocean?

Tyler (17, totally blind since birth)

> The ocean is a vast area of salt water. The salt from the water creates a unique scent in the atmosphere. The fast moving waves create a sound that is pleasing to the ear. The swiftly moving water is pleasing to the touch when one is standing or swimming in it. The ocean and its many effects provide many benefits to one's physical and emotional well-being.

Jon (16, partially blind)

> THE OCEAN
>
> Vast treacherous waves
> Ships travel to continents
> Gone for months on end.

Minh (17, totally blind since birth)

The sea reminds me of romance, and meditation. It's a rough and dangerous world. It's like our world except in an animal way. The sea also has moods, for example being stormy, or it can be as calm as a river. Or it can flow smoothly like a stream. The sea can be known as a bubbling pot of soup with all the waves. Some seas have different temperatures. Sometimes there is more seasoning than most.

Igor (16, totally blind since birth)

The sea is a vast body of water. It is filled with seaweed. The sea is filled with waves. You would body surf. As you swim in the sea, you would absorb the salt. The salt helps you float. As you walk along the shore, you step on the sand. The sand sticks to your toes. Shells wash up and your feet are tangled up in seaweed.

Michelle (17, partially blind)

It is very big and full of fish. My impression of the ocean: Whoosh, Whoosh, blub, blub . . . .

33   Northern fulmars breed on Arctic cliffs, and when nesting have a memorable weapon with which to defend their single, precious egg: if any predator approaches, a nesting fulmar will vomit onto it a jet of stinky and potentially lethal stomach oil. They can also desalinate seawater, expressing the salt through the tubes on their beaks. (Thus the name "tubenoses.") Like albatrosses, they forage at the surface, and like albatrosses they end up swallowing a lot of plastic, Fifield says, even up here in Arctic waters, far from the Garbage Patch.

34   By the estimate of the naturalist E. C. Pielou, in the seas east and west of Greenland, there are some ten thousand icebergs afloat at any time and their numbers are greatest here, in Baffin Bay, a body of water rimmed by calving glaciers. The biggest icebergs, Pielou reports, weigh ten million tons and can rise to heights of two hundred feet or more—550 feet being the record. And the part you can see, above the waterline, represents only the uppermost fraction of the thing. An iceberg that rises 200 feet above the waterline might extend 1,600 feet beneath it. Because of their deep keels, the wind has little effect on their motion. It's the currents that determine their fate, and the currents of Baffin Bay, like those of the North Pacific Subtropical Gyre, converge. Few icebergs escape into southern waters. The one that sank the *Titanic* was an aberration. Most stay up here, circling around, diminishing with every lap.

35   In 1884, anyone who read the newspaper, even a Danish newspaper, would have heard of the USS *Jeannette*. In 1878, James Gordon Bennett, owner of the *New York Herald*, had approached the U.S. Navy with an extraordinary offer. If the navy would organize a sea voyage to the North Pole, Bennett would finance it. All previous attempts to reach the pole by sea had approached it from the east, by way of the North Atlantic, and for this reason, Bennett believed, those attempts had failed. This expedition, unlike the others, would approach the pole from the west, by way of the Bering Strait. In theory, Bennett's plan was a good one. The icy currents of the Arctic, we now know, do tend to flow

in an easterly direction, and ships entering from the North Atlantic have to sail or steam against them.

The voyage wouldn't cost the federal government a dime, Bennett promised. Not surprisingly, both Congress and the navy were happy to accept the gift, and on July 8, 1879, while crowds in San Francisco waved their hats and handkerchiefs from shore, the *Jeannette*, a three-masted steam yacht that Bennett had purchased from the Royal Navy, sailed through the Golden Gate (not yet spanned by the famous red bridge) under the command of George De Long. In the last surviving portrait taken of him, De Long, thirty-five, is wearing an ascot and a pair of pince-nez spectacles, from the right corner of which drips a silver chain. With his walrus mustache and puffy cheeks, he looks a bit like a cross between Teddy Roosevelt and Marcel Proust.

After three uneventful weeks at sea, the *Jeannette* stopped in Unalaska to stock up on fur and coal. Ten days later, at the island of St. Michael's in St. Lawrence Bay, it stopped again to take on forty sled dogs, three dogsleds, and two dogsledding Inuit, bringing the ship's hominid company to thirty-three. Along with the officers and enlisted men and the dogsledding Inuit, the company included two civilian scientists, the naturalist Raymond Newcomb and the meteorologist Jerome Collins, the latter having agreed to serve as the *New York Herald*'s correspondent. From St. Lawrence Bay Collins sent, aboard the *Jeannette*'s southbound supply ship, the *Herald,* a prescient dispatch. "All before us now is uncertainty," he wrote, "because our movements will be governed by circumstances over which we can have no control."

On August 28, the *Jeannette* passed through the Bering Strait and turned west, coasting along Siberia, where, with the help of Chukchi natives, three members of the ship's company located a camp abandoned the year before by a Swedish expedition. The Swedes had left behind a stash of tinned food and what Lieutenant John Danenhower, the *Jeannette*'s navigator, would later delicately call "some interesting pictures of professional Stockholm beauties." From the natives, the foresighted Americans purchased both the Swedish food and the Swedish porn. Then the *Jeannette* turned north.

On September 6, southeast of Wrangell Island, having already ventured farther into the ice pack than the seasoned captains of whaling ships dared to, Captain De Long selected a lead—that is, a long, navigable, canal-like crack in the ice—and piloted the *Jeannette* into it. The lead tapered. Then terminated. "The ground which we are going to traverse is an entirely new one," De Long had said in a speech to the California Academy of Sciences a few days before the *Jeannette*'s departure. "After reaching the seventy-first parallel of latitude we go out into a great blank space, which we are going to endeavor to delineate and to determine whether it is water or land or ice." Already, De Long had his answer: from the lofty vantage of the crow's nest, he trained his telescope north. All there was to see, stretching to the far horizon, was ice. What to do? Turn back? Sail home in defeat?

Perhaps De Long considered his reputation. Perhaps he considered the money James Bennett had invested in the expedition. Perhaps, like so many other Arctic explorers, he was enthralled by the object of his quest. Or perhaps, looking aft, he made a purely tactical calculation. In the wake of the *Jeannette*, the lead had closed. There was ice ahead, ice behind.

De Long had prepared for this eventuality. The *Jeannette* had three years' worth of rations in its stores, and back in California shipwrights had fortified the hull, bracing it with oak beams. Persuaded that his ship could survive the winter frozen in, De Long gave the command to charge on, under full steam and full sail. In the memoir he would survive to write, Lieutenant Danenhower describes the moment of impact: "We met with the young ice, and forced our way through it by ramming. This shook the ship very badly, but did not do her any damage; indeed the ship stood the concussions handsomely."

By late on that same afternoon, however, the floes had become impassable. The *Jeannette* was beset, and its thirty-three men and its forty dogs were now at the mercy of the currents—currents that had never been traveled, let alone charted; currents the very existence of which had previously remained in dispute.

"We banked fires, secured the vessel with ice-anchors, and remained," Danenhower writes. "Our position was not an enviable one. At any moment the vessel was liable to be crushed like an egg-shell among this enormous mass of ice, the general thickness of which was from five to six feet, though some was over twenty where the floe pieces had overrun and cemented together and turned topsy-turvy. Pressures were constantly felt. We heard distant thundering of the heavy masses, which threw up high ridges of young ice that looked like immense pieces of crushed sugar."

You might imagine that, faced with either imminent shipwreck or prolonged imprisonment in the icy labyrinth, De Long and his men would have despaired. Were I among them, I think I might well have helped myself to a few extra rations of rum, jumped overboard, and been done with it. But the *Jeannette* withstood the pressures of the ice, and its crew the pressures of light deprivation and claustrophobia and fear, passing that first winter in a paradoxical sort of perilous ease.

They went ice-skating on the floes. They hunted seal, walrus, and polar bear. They made meteorological and astronomical observations. The ship's doctor conducted monthly medical examinations, checking for signs of scurvy. On New Year's Day, the ship's cook served a multicourse feast the menu of which included spiced salmon, roast seal ("Arctic turkey," they called it), green peas, succotash, canned plum pudding, mince pie, muscat dates, sherry, stout, French chocolate, French coffee, and cigars. When the feast was done, enlisted men put on a "minstrel show."

One would like to imagine this scene—the virile, unwashed explorers, after months in close quarters, frozen into the ice pack, tipsy on rum and stout, entertaining each other with "magic lantern" shows and minstrel songs called "The Spanish Cavalier" and "What Should Make You Sad" and with an orientalist drama described on the playbill as follows: "The great 'Ah Sam' and 'Tong Sing' in their wonderful tragic performances."

In January, when temperatures had dropped to minus 42 degrees Fahrenheit, the ship sprang a leak. The carpenter managed to repair it, and Engineer George Melville managed to pump the water out, but the incident, in hindsight, would prove to be portentous. By February, the *Jeannette* had drifted, circuitously, fifty miles, on a northwesterly bearing. Some days it made three nautical miles, others, nine. On the windiest days, it made twelve. Summer came, temperatures rose, and with them, so did hopes. De Long had assumed that the summer melt would set the *Jeannette* free. The anticipated emancipation never came. "The surface of the floe-pieces was now of a hard, greenish blue, and flinty, being covered in many places with thaw-water," Danenhower would recall. "There were numerous cracks near the ship, but no leads that went in any definite direction, and there was no chance to move, for the ship was imbedded in the ice so firmly that a whole cargo of explosives would have been useless."

If the *Jeannette* survived, De Long and Danenhower reasoned—correctly—the northwesterly currents would carry them over the pole and out into the North Atlantic. But the *Jeannette*, of course, did not survive. On June 12, 1881, after twenty-one months adrift, the hull was sundered by ice. Water poured in. The ship heeled over 23 degrees to starboard. De Long gave the order to abandon ship. The following night, while most of the company was asleep in tents pitched on a floe, the *Jeannette*, suddenly released from what Danenhower called "the monster's grip," sank into seas thirty-eight fathoms deep, the adjacent floes snapping the spars of its masts like twigs.

Pulling sledges over the ice, the shipwrecked explorers now beat a desperate, southward retreat, toward Siberia. At the edge of the ice pack, they abandoned their sledges and took to their three boats, sailing for the mouth of the Lena River. One of the boats was lost at sea. Another, piloted by Lieutenant Danenhower, now suffering from snow blindness, eventually delivered its crew to safety. The third, captained by De Long, made it to the Siberian shallows, where it kept running aground. Carrying whatever provisions they could, De Long and his men had no choice but to wade ashore through the hypothermically cold waves. Frostbitten, already starving, they now found themselves on a wild Arctic coast bereft of inhabitants, human or otherwise. There, on the tundra, five of the seven members of De Long's party, including De Long, died in the usual way. The two survivors? Louis Noros and F. C. Nindermann. Somewhere during the course of their retreat, the former had lost a pair of oilskin pants, the latter a woolen cap.

36 Me, I'm more inclined to attribute Nansen's success to Nansen than to miracles, or fortune, or providence. Read his account of the voyage, *Farthest North: The Epic Adventure of a Visionary Explorer*, and you'll repeatedly come upon moments of great peril and great self-doubt, moments in which I would have lost my wits and probably my life, like this one, recorded in his logbook in October 1894, after a year adrift in the ice:

> Personally, I must say that things are going well with me; much better than I could have expected. Time is a good teacher; that devouring longing does not gnaw so hard as it did. Is it apathy beginning? . . . Oh! sometimes it comes on with all its old strength—as if it would tear me to pieces! But this is a splendid school of patience. Much good it does to sit wondering whether they are alive or dead at home; it only almost drives one mad.

Only almost, here, are the decisive, characterological words. Nansen continues:

All the same, I never grow quite reconciled to this life. It is really neither life nor death but a state between the two. It means never being at rest about anything or in any place—a constant waiting for what is coming; a waiting in which, perhaps, the best years of one's manhood will pass. It is like what a young boy sometimes feels when he goes on his first voyage. The life on board is hateful to him; he suffers cruelly from all the torments of sea-sickness; and being shut in within the narrow walls of the ship is worse than prison; but it is something that has to be gone through. Beyond it all lies the south, the land of his youthful dreams, tempting with its sunny smile. In time he arises, half dead. Does he find his south? How often it is but a barren desert he is cast ashore on!

37   The 1957 undertaking was, in fact, called the IGY, for International Geophysical Year, since the field of study was more global in its scope. Nevertheless, it is still counted as the third such event, and the most recent IPY, therefore, as the fourth.

38   That "polar bears are rather retiring and unaggressive, especially in comparison with grizzly bears"—to which, genetically, they're closely related, so closely related that polar bears and grizzlies can breed. Big polars standing up are twelve feet t... to seals and sled... walk about 2.5 m... males. The fema... bear's longest h... hollow. And cle... its owner's foots... in the snow. The... pair of back mus... In fact, it contai... discovered the h...

39   A hummock is ... bummocks, pool... create an orderly... derly and uninter...